MACHINE DESIGN
Second Edition

Machine Design
Second Edition

Anthony Esposito
Miami University

J. Robert Thrower
Massasoit Community College

 Delmar Publishers Inc.®

NOTICE TO THE READER

Cover Photo Courtesy of: GE Aircraft Engines
Cover Design by: Juanita Brown

Delmar Staff:
Senior Administrative Editor: Michael McDermott
Project Editor: Carol Micheli
Production Supervisor: Larry Main
Design Supervisor: Susan C. Mathews
Art Supervisor: John Lent

For information, address Delmar Publishers Inc.
Two Computer Drive West, Box 15–015
Albany, New York 12212

Copyright © 1991
By Delmar Publishers Inc.

Printed in the United States of America
published simultaneously in Canada
by Nelson Canada,
a division of The Thomson Corporation

10 9 8 7 6 5 4 3

Library of Congress Cataloging-in-Publication Data

Esposito, Anthony.
   Machine design / Anthony Esposito, James R. Thrower.—2nd ed.
      p. cm.
   Includes index.
   ISBN 0-8273-4081-8
   1. Machinery—Design. I. Thrower, James R. II. Title.
TJ230.E7   1991
621.8'15—dc20   90–14060
                                                     CIP

CONTENTS ▬▬▬▬▬▬▬

Chapter 13   Dynamic Loads on Machine Members and Dynamic Balancing of Shafts   445

Appendices   461

Index   469

TABLE OF TABLES

PREFACE

The purpose of the second edition of *Machine Design* is to provide the student with a basic background of the vast field of machine design. It is written for 2-yr technicians and 4-yr engineering technologists in the mechanical, manufacturing, and industrial disciplines.

It is assumed the student has a basic understanding of mechanical drawing, manufacturing processes, mechanics and strength of materials. Calculus is not a prerequisite for this text.

The derivation of a number of design equations are presented in those instances appropriate to the student's present knowledge. This should increase his, or her, understanding of the equations. Also fundamental equations, rather than specific equations, are used in a number of instances to deepen student understanding of concepts. However, the emphasis is placed on applying the principles of machine design to the analysis and design of machine elements.

A discussion of the SI metric system and a number of metric examples and student assignment problems are included in each chapter to increase the students' understanding of the metric system. However we do not recommend substituting metric data into equations without careful thought. The reason is that the proper conversion of any empirical factors or other constants may not be readily apparent or available. Conversion tables are presented in the appendix.

This second edition takes into account the changes that have occurred in industry and in machine design instruction since the first edition was published. Each chapter now has an initial section devoted to objectives and a final section summarizing the chapter contents. Procedures for solving problems with numerous steps are included because many students who are capable of handling the mathematics involved have trouble deciding where to start and in what order to proceed. This should reduce the heavy reliance many students place on trying to match exactly the given problem with a sample problem.

Computer programs (in BASIC) have been added to the instructor's manual. Students interested in programming can try improving the programs. The number of design problems (as compared to analysis problems) has been increased. This will allow the student to develop a stronger expertise in the design process.

Anthony Esposito
Oxford, Ohio

J. Robert Thrower
Danvers, Massachusetts

August 1990

1

Introduction to Machine Design

Objectives

After completing this chapter, you will

- Have an understanding of the broad field of activities identified by the term *machine design*.
- Be familiar with the problems and concerns of a machine designer.
- Know the objectives of this text.

1.1   WHAT IS MACHINE DESIGN?

Machine design is the application of science and technology to devise new or improved products for the purpose of satisfying human needs. It is a vast field of engineering technology that not only concerns itself with the original conception of a product in terms of its size, shape, and construction details, but also considers the various factors involved in the manufacture, marketing, and use of the product.

A *product* can be defined as any manufactured item, including machines, structures, tools, and instruments. Figure 1.1 shows a very familiar product, commonly called a *lawn chair*. As shown by the enlarged section, acorn-type Palnut lock nuts are used to fasten the tubular seat assembly to the tubular leg and arm section. Figure 1.2 depicts a robotic hand in action. It should be apparent that the lawn chair and the robot are products that serve human needs.

People who perform the various functions of machine design are typically called *designers* or *design engineers*. Machine design is basically a creative activity. However, in addition to being innovative, a design engineer also must have a solid background in the fundamentals of engineering technology.

1.2   FUNDAMENTAL BACKGROUND FOR MACHINE DESIGN

A design engineer must have working knowledge in the areas of mechanical drawing, kinematics, mechanics, materials engineering, strength of materials, and manufacturing processes. The following statements will indicate how each of these basic background subjects relates to machine design.

Mechanical Drawing.   The lawn chair in Figure 1.1 and the robotic hand in Figure 1.2 are clearly represented for visual purposes by photographs. However, these products

FIGURE 1.1   Aluminum lawn chair using Palnut fasteners.
(Courtesy of The Palnut Company)

FIGURE 1.2   This robotic hand can be programmed to apply the paint as required. *(Courtesy of Hitachi, Ltd.)*

could not be manufactured solely from photographs. Detailed drawings must be prepared that note the exact shape, size, and material composition of each component; assembly drawings that show how the total product is put together by fastening each part in proper sequence are also needed.

Kinematics.   *Kinematics* is the study of motion. For example, suppose the robotic hand in Figure 1.2 is required to spray paint on a vertical surface. The spray nozzle must be held 6 in. from and perpendicular to the surface. Knowledge of kinematics would permit analysis of the motion of the hand, forearm, and upper arm of the robot, and a designer must know the type of motion of each part in order to create a satisfactory design.

Mechanics.   *Mechanics* is that division of physics that deals with the action of forces on objects. For example, forces act on the lawn chair in Figure 1.1 when a person sits in the chair. Obviously, a person can damage the lawn chair by carelessly jumping on the seat. This motion, in effect, applies dynamic loading instead of the gradually applied loading taken into consideration when the lawn chair was designed. The end result of this misuse is excessively large forces that can cause permanent damage. Therefore, using the laws of mechanics, a reasonable amount of dynamic loading should be taken into account during the early design phase of the chair.

Materials of Engineering.   Because a lawn chair is commonly used in an outdoor environment, the tubing is made of aluminum to resist corrosion. The webbing is made of plastic material that will not readily deteriorate with sustained exposure to sunlight and moisture. Obviously, the proper selection of materials is a vital area of machine design.

Strength of Materials.   Strength of materials is concerned with whether or not a part is strong enough to sustain the forces it will experience as evaluated in terms of mechanics. For example, the size and shape of the aluminum tubular sections of the lawn chair are specified in such a way that failure will not occur (under normal use) as a result of, excessive stresses and deflections. The magnitude of stresses and deflections depends on the size and shape of a given part as well as on its material, composition, and actual loads.

Manufacturing Processes.   As can be seen in Figure 1.2, the robotic hand is a rather complicated machine. How each component is produced and how the entire robot is assembled are established by using methods learned in manufacturing technology. It is here that the designer comes to grips with the reality of costs.

1.3   PHILOSOPHY OF MACHINE DESIGN

The following poem, entitled "The Designer," was written by an unknown author. It relates how a design engineer may enjoy making a design so complex that manufacture of the product is virtually impossible.

THE DESIGNER

The designer bent across his board
Wonderful things in head were stored.
Said he as he rubbed his throbbing bean,
"How can I make this tough machine?
Now if I make this part just straight
I know that it will work first rate
But that's too easy to shape and bore
It never would make the machinist sore.
So I better put an angle there—
Then watch those babies tear their hair.
And there are the holes that hold the cap
I'll put them down where they're hard to tap.
Now this won't work, I'll bet a buck,
It can't be held in a shoe or chuck.
It can't be drilled and it can't be ground, In
fact, the design is exceedingly sound."
He looked again and cried: "At last!
Success is mine—it can't even be cast."

Obviously, this poem is a satire. However, it clearly emphasizes the importance of a design engineer in establishing the manufacturability of a product. The design engineer must constantly communicate with the manufacturing engineer during the design process. In fact, Lee Iacocca, in his autobiography, made the following comments concerning his rescue of Chrysler Corporation: "Interaction among the different functions in a company is absolutely critical. People in [design] engineering and manufacturing almost have to be sleeping together."

Inventions, discoveries, and scientific knowledge by themselves do not necessarily benefit people; only if they are incorporated into a designed product will a benefit be derived. It should be recognized, therefore, that a human need must be identified before a particular product is designed. To illustrate, Figure 1.3 shows a product used as a vital aid in heart surgery. It is a delicate tool that extends the surgeon's reach and provides the rotary motion to lock a prosthetic heart valve securely into living tissue. A flexible shaft, 7½ in. long, made of Monel material, fits the curved shank of the instrument and gently locks the new valve into place. The tool is then detached and removed.

Machine design requires knowledge of the fundamentals of engineering rather than mere memorization of facts and equations. There are no facts or equations that alone can be used to provide all the correct decisions required to produce a good design. On

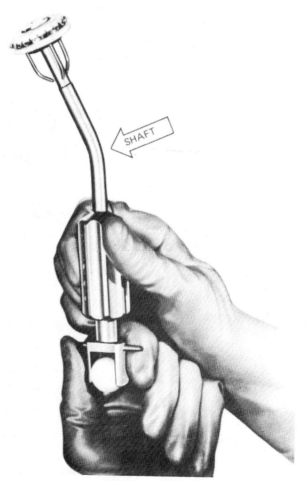

FIGURE 1.3   Flexible shafts used in heart surgery instrument. *(Courtesy of S.S. White Division, Pennwalt Corp.)*

the other hand, any calculations made must be done with the utmost care and precision. For example, a misplaced decimal point can ruin an otherwise acceptable design.

Good designers must try new ideas and be willing to take a certain amount of risk. A designer must have patience, since there is no assurance of success for the time and effort expended. Creating a completely new design generally requires that many old and well-established methods be put aside. This is not easy, since many people cling to familiar ideas, techniques, and attitudes. A design engineer should constantly search for ways to improve an existing product and must decide what old, proven concepts should be used and what new, untried ideas should be incorporated. It should be emphasized, however, that if a design does not warrant radical new methods, such methods should not be applied merely for the sake of change. The designer must constantly keep in mind the following questions:

1. Does the design really serve a human need?
2. Will it be competitive with existing products of rival companies?
3. Is it economical to produce?
4. Can it be readily maintained?
5. Will it sell and make a profit?

The designer also must keep in mind many human operator features, such as sizes and locations of handwheels, knobs, switches, and so forth; working space; ventilation; colors and lighting; strength of the operator; safety features; monotonous operator motions; and operator acceptance. These considerations comprise the *human engineering* aspect of design.

1.4   MAJOR AREAS OF MACHINE DESIGN

Machine design is a vast field of engineering technology. As such, it begins with the conception of an idea and continues through the various phases of design analysis, manufacturing, and marketing. The major areas of consideration in the general field of machine design are as follows:

— Initial design conception
— Strength analysis
— Materials selection
— Appearance
— Manufacturability
— Economy
— Safety
— Environmental effects
— Reliability and life
— Legal considerations

Designing a machine does not necessarily mean that the designer has to invent something new for the machine. Most commonly, it means applying known methods to a new situation, new requirements, or new materials. Here is an example of applying a known method to a new situation: In the 1950s, a grocery chain wanted to patent their idea of using a movable frame at the checkout counter so the operator could slide the groceries along the countertop until they were within his or her reach. However, the courts refused the patent on the grounds this was simply a variation of the triangular billiard ball rack used on a pool table.

Most manufacturers are redesigning their products to take advantage of automated machinery. Airplane and automobile manufacturers are redesigning their products to take advantage of the new lightweight composite materials, such as carbon filaments combined with plastic.

1.5 OBJECTIVES OF THIS BOOK

As you can see from the preceding explanations, machine design can be a rather complex undertaking. It is also too broad a field to be covered by one text alone. The first objective of this text is to develop your understanding of how many of the most current machine components operate. The second objective is to show you how to apply the information learned in previous mechanics and strengths of materials courses to solve the design problems encountered when using these components. This text will cover such components as bearings, shafts, gears, belts, and so on, as listed in the Contents. The third objective is to provide the experienced designer with information to refresh his or her memory.

1.6 PROBLEMS

The following problems are designed to permit you to use your creativity in applying yourself to practical design situations. You most likely will have to refer to handbooks or appropriate texts in your library. Do not perform stress or deflection analysis. Sketch in part sizes where appropriate. Compare and defend your ideas with others in the class.

1. A pulley can be connected to a shaft in many different ways. Make freehand sketches of three different methods, all of which permit disassembly of the pulley from the shaft. What are the advantages and disadvantages of each method?
2. There are many different ways to prevent a nut from loosening on a bolt. Sketch three possible methods, and explain the advantages and disadvantages of each.
3. What are some of the advantages and disadvantages of two-cycle gasoline engines and four-cycle engines. Explain why two-cycle engines are used almost exclusively for outboard motors and four-cycle engines are most common in automobiles.
4. Compare the advantages and disadvantages of (a) an in-line 8-cylinder engine versus a V-8 engine and (b) aluminum versus cast iron engine blocks.
5. A concession stand handles large quantities of pennies, nickels, dimes, and quarters. Make a freehand sketch of a device that will sort these four types of coins into four separate chutes.

6. Compare the advantages and disadvantages of flat-bottom boats versus V-bottom boats.

7. Design a product that accomplishes something useful and either competes with an existing product or, as far as you know, does not exist. The product must satisfy a human need that you, in fact, have experienced.

    (a) Make a freehand sketch of your product.

    (b) Explain how it works.

2
Failure Analysis and Dimensional Determination

Objectives

After completing this chapter, you will

- know why mechanical parts fail.
- Have an understanding of how the dimensional characteristics of a part are determined.

2.1 INTRODUCTION

It is absolutely essential that a design engineer know how and why parts fail so that reliable machines that require minimum maintenance can be designed. Sometimes a failure can be serious, such as when a tire blows out on an automobile traveling at high speed. On the other hand, a failure may be no more than a nuisance. An example is the loosening of the radiator hose in an automobile cooling system. The consequence of this latter failure is usually the loss of some radiator coolant, a condition that is readily detected and corrected.

The type of load a part absorbs is just as significant as the magnitude. Generally speaking, dynamic loads with direction reversals cause greater difficulty than static loads, and therefore, fatigue strength must be considered. Another concern is whether the material is ductile or brittle. For example, brittle materials are considered to be unacceptable where fatigue is involved.

Many people mistakingly interpret the word *failure* to mean the actual breakage of a part. However, a design engineer must consider a broader understanding of what constitutes failure. For example, a brittle material will fail under tensile load before any appreciable deformation occurs. A ductile material, however will deform a large amount prior to rupture. Excessive deformation, without fracture, may cause a machine to fail because the deformed part interferes with a moving second part. Therefore, a part fails (even if it has not physically broken) whenever it no longer fulfills its required function. Sometimes failure may be due to wear that can change the correct position of mating parts. The wear may be due to abnormal friction or vibration between two mating parts. Failure also may be due to a phenomenon called *creep*, which is the plastic flow of a material under load at elevated temperatures. In addition, the actual shape of a part may be responsible for failure. For example, stress concentrations due to sudden changes in contour must be taken into account. Evaluation of stress considerations is especially important when there are dynamic loads with direction reversals and the material is not very ductile.

In general, the design engineer must consider all possible modes of failure, which include the following:

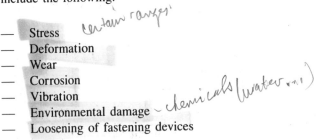

— Stress
— Deformation
— Wear
— Corrosion
— Vibration
— Environmental damage
— Loosening of fastening devices

The part sizes and shapes selected also must take into account many dimensional factors that produce external load effects, such as geometric discontinuities, residual stresses due to forming of desired contours, and the application of interference fit joints.

2.2 TENSILE STATIC STRENGTH

The discussion in this section will use several descriptive terms for steel that must be defined. Steels are generally classified as either *plain carbon steels* or *alloy steels*. Besides iron and carbon, alloy steels contain at least one other element, which gives the steel its distinctive properties. For example, the principal alloying ingredient for stainless steel is chromium. Plain carbon steels are usually separated into three groups, depending on the amount of carbon present. *Low carbon steel* (carbon content less than 0.30%) is commonly referred to as *mild steel*. Mild steel cannot be effectively hardened by heat treating. *Medium carbon steel* (carbon content in the range between 0.30% and 0.70%) is also known as *medium steel* or *machinery steel*. Medium steel can be easily hardened by heat treatment. *High carbon steel* (carbon content in the range between 0.70% and 1.4%) also may be referred to as *hard steel* or *tool steel*.

Lack of understanding of the behavior of materials under actual service conditions has been the cause of many serious failures. A great deal of information can be revealed by performing laboratory tests. One such experimental evaluation of static strength is the tensile test, in which a standard test specimen with threaded ends to fit the screw grips of a testing machine is used. Figure 2.1 shows the specimen, which has a circular cross section. The shank diameter of 0.505 in. corresponds to a cross-sectional area A of 0.2 in.^{2} The testing machine gradually applies a tensile force F that is measured by a dial in units of pounds. A device called an *extensometer* is attached to the specimen at the two gauge marks that are separated initially by a distance l of 2 in. The extensometer measures the elongation Δl of distance l between the gauge marks.

The stress S is calculated by Equation 2.1. Since *stress* is defined as the force per unit area, the units for stress are pounds per square inch (lb/in.^{2}).

$$S = \frac{F}{A} \qquad\qquad\qquad\qquad (2.1)$$

Strain is defined as the change in length divided by the original length. Hence strain ϵ can be found by using

$$\epsilon = \frac{\Delta l}{l} \qquad\qquad\qquad\qquad (2.2)$$

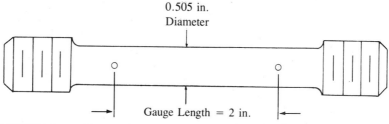

FIGURE 2.1   Standard Tensile Test Specimen with Threaded Ends.

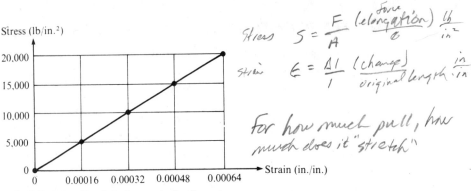

FIGURE 2.2   Partial stress-strain diagram

As the test is performed, values of loads and elongation are recorded. The results shown in Table 2.1 represent the initial portion of the tensile test. Figure 2.2 shows a plot of the data from Table 2.1.

A complete tensile test is obtained by continuing to gradually increase the load until the specimen breaks. A plot of the entire test is called a *stress-strain diagram*. A typical stress-strain diagram for a ductile material such as mild steel is shown in Figure 2.3.

The following information should be clearly understood relative to a tensile test.

Modulus of Elasticity. The stress-strain diagram is linear up to point B, which is the proportional limit. The slope of the straight line of this relationship (called the *modulus line*) is defined as the modulus of elasticity E and is mathematically represented by:

$$E = \frac{\text{Stress}}{\text{Strain}} = \frac{S}{\epsilon} \qquad (2.3)$$

Substituting data from Table 2.1 into Equation 2.3 provides the value of the modulus of elasticity for mild steel:

TABLE 2.1   Partial Data for Tensile Test of Mild Steel

Load, F (pounds)	Elongation, $\Delta\ell$ (inches)	Calculated Stress, $S = \dfrac{F}{A} = \dfrac{F}{0.2}$ (lb/in.²)	Calculated Strain, $\epsilon = \dfrac{\Delta l}{l} = \dfrac{\Delta l}{2}$ (in./in.)
1000	0.00032	5,000	0.00016
2000	0.00064	10,000	0.00032
3000	0.00096	15,000	0.00048
4000	0.00128	20,000	0.00064

$$E = \frac{5000}{.00016} = 31 \times 10^6 \text{ lb/in.}^2$$

Proportional Limit. If the load is released at any value below point B, the test specimen will contract exactly along the modulus line until point A is reached, where stress and strain both equal zero. Thus the *proportional limit* is defined as the highest stress a part can sustain and yet return to its original shape and size when the load is removed. The modulus line from A to B represents the elastic portion of the stress-strain diagram.

Yield Strength. If the proportional limit is exceeded, the material begins to yield plastically. This means that there is a small increase in stress while the strain increases appreciably. Assume that the load is increased until point D is reached, at which time the load is removed. The remarkable result is that the specimen returns along the dashed line (which is parallel to the modulus line) until point G is reached. Obviously, the part has yielded plastically, because there is a permanent strain even though the load is zero. Since the permanent strain at point G is 0.002 in./in., point D is called the *0.2% yield strength*. Other percents of yield strength, such as 0.1%, could just as well have been chosen. These values simply indicate the stress levels required to produce corresponding amounts of permanent strain. Sometimes the term *yield strength* is used without a percent value. The yield strength (commonly called the *yield point*) is the stress level where the stress-strain diagram temporarily flattens out, as shown in Figure 2.3 at point C.

Ultimate Strength. As loading progresses beyond point C, the specimen continues to elongate plastically. The load dial on the test machine and the extensometer output show that the load is increasing very slowly compared to the elongation. Maximum

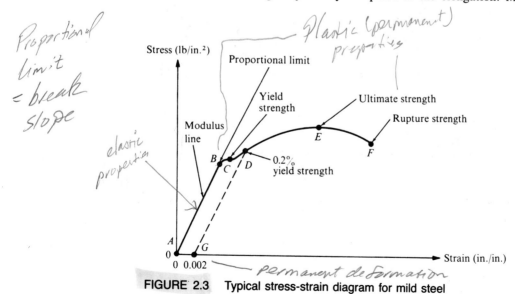

FIGURE 2.3 Typical stress-strain diagram for mild steel

load and, thus, maximum calculated stress are reached at point E. The stress level corresponding to point E is called the *ultimate strength*.

Rupture Strength.   When point E is reached, the specimen loses its ability to resist load. The load dial indicates an automatic drop in load until point F is reached and the specimen ruptures. The stress at point F is called the *rupture stress* or *rupture strength*.

Necking.   The two broken sections of the test specimen are removed from the machine grips and positioned as shown in Figure 2.4. Inspection reveals a thinning of the shank in the area of the rupture. This phenomenon is called *necking,* and it occurs only with ductile materials. Notice that the shape of the necked-down section reveals a cup and cone configuration.

Actual Stress-Strain Curve.   It should be recognized that all the stress calculations have been based on the original cross-sectional area of 0.2 in.[2] This one value has been used for ease of testing, since it would be very time-consuming to continuously monitor the actual values of area. Also, note that the area changes by a negligibly small amount up to the 0.2% yield strength. This fact is demonstrated in Figure 2.5 where the actual (using actual areas) and apparent (using original area) stress-strain diagrams are superimposed. Notice that in reality the stress continues to increase throughout the test even though the dial on the test machine shows a decreasing load beyond the ultimate strength. The actual stress-strain diagram does not really provide any additional valuable information because most machine parts are designed to operate at stress levels below the 0.2% yield strength.

Elongation and Reduction in Area.   The percent elongation and percent reduction in area, two very important parameters, can be obtained from examination of the specimen after rupture.

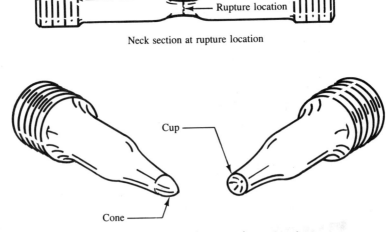

Neck section at rupture location

FIGURE 2.4   Necking with typical cup-and-cone rupture

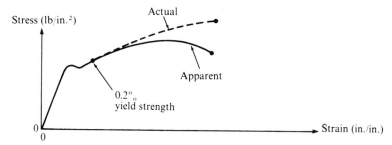

FIGURE 2.5 Comparison of actual and apparent stress-strain diagrams

Modulus of Elasticity and Proportional Limit for Steel. Figure 2.6 shows that the modulus of elasticity (at a given temperature) is the same for different types of steels. In reality, what differs is the value of the proportional limit. Specifically, all types of steel possess a modulus of elasticity of approximately 30×10^6 lb/in.^{2} at room temperature.

Brittle Material Deformation. A brittle material does not exhibit large plastic deformations prior to rupture. This is shown in Figure 2.7 for cast iron and concrete. Also, there is no necking down of a brittle material loaded to failure, because there is no significant amount of plastic deformation. Recall that for a ductile steel part loaded to failure, over 95% of the deformation is plastic as compared to elastic.

2.3 DESIGN PROPERTIES OF MATERIALS ▬▬▬

The following design properties of materials are defined as they relate to the tensile test.

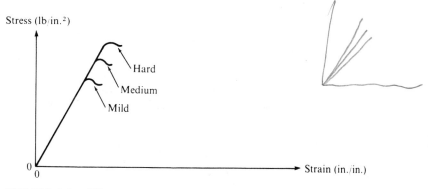

FIGURE 2.6 Effect on proportional limit of carbon content in steel

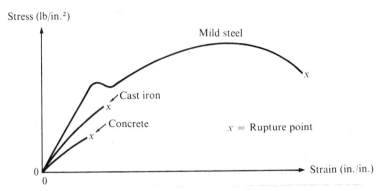

FIGURE 2.7   Stress-strain diagram comparison between ductile and brittle materials

Static Strength.   The *strength* of a part is the maximum stress that the part can sustain without losing its ability to perform its required function. Thus the static strength may be considered to be approximately equal to the proportional limit, since no plastic deformation takes place and no damage theoretically is done to the material.

Stiffness.   *Stiffness* is the deformation-resisting property of a material. The slope of the modulus line and, hence, the modulus of elasticity are measures of the stiffness of a material.

Resilience.   *Resilience* is the property of a material that permits it to absorb energy without permanent deformation. The amount of energy absorbed is represented by the area underneath the stress-strain diagram within the elastic region. (See Figure 2.8a.)

Toughness.   Resilience and toughness are similar properties. However, *toughness* is the ability to absorb energy without rupture. Thus toughness is represented by the total area underneath the stress-strain diagram, as depicted in Figure 2.8b. Obviously, the toughness and resilience of brittle materials are very low and are approximately equal.

Brittleness.   A *brittle* material is one that ruptures before any appreciable plastic deformation takes place. Brittle materials are generally considered undesirable for machine components because they are unable to yield locally at locations of high stress because of geometric stress raisers such as shoulders, holes, notches, or keyways.

Ductility.   A *ductile* material exhibits a large amount of plastic deformation prior to rupture. Ductility is measured by the percent reduction of area and percent elongation of a part loaded to rupture. A 5% elongation at rupture is considered to be the dividing line between ductile and brittle materials.

Malleability.   *Malleability* is essentially a measure of the compressive ductility of a material and, as such, is an important characteristic of metals that are to be rolled into sheets.

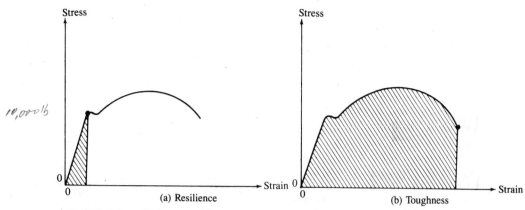

FIGURE 2.8   Diagram comparison of design properties resilience and toughness

Hardness.   The *hardness* of a material is its ability to resist indentation or scratching. Generally speaking, the harder a material, the more brittle it is and, hence, the less resilient. Also, the ultimate strength of a material is roughly proportional to its hardness.

Machinability.   *Machinability* is a measure of the relative ease with which a material can be machined. In general, the harder the material, the more difficult it is to machine.

2.4   COMPRESSION AND SHEAR STATIC STRENGTH ▬▬▬

In addition to the tensile tests, there are other types of static load testing that provide valuable information.

Compression Testing.   Most ductile materials have approximately the same properties in compression as in tension. The ultimate strength, however, cannot be evaluated for compression. As a ductile specimen flows plastically in compression, the material bulges out, but there is no physical rupture as is the case in tension. Therefore, a ductile material fails in compression as a result of deformation, not stress.

Shear Testing.   Shafts, bolts, rivets, and welds are located in such a way that shear stresses are produced. A plot of the shear stress-shear strain diagram reveals a similar pattern as compared to the tensile test. The ultimate *shearing strength* is defined as the stress at which failure occurs. The ultimate strength in shear, however, does not equal the ultimate strength in tension. For example, in the case of steel, the ultimate shear strength is approximately 75% of the ultimate strength in tension. This difference must be taken into account when shear stresses are encountered in machine components.

2.5 DYNAMIC LOADS

An applied force that does not vary in any manner is called a *static* or *steady load*. It is also common practice to consider applied forces that seldom vary to be static loads. The force that is gradually applied during a tensile test is therefore a static load.

On the other hand, forces that vary frequently in magnitude and direction are called *dynamic loads*. Dynamic loads can be subdivided to the following three categories.

Varying Load.   With varying loads, the magnitude changes, but the direction does not. For example, the load may produce high and low tensile stresses but no compressive stresses.

Reversing Load.   In this case, both the magnitude and direction change. These load reversals produce alternately varying tensile and compressive stresses that are commonly referred to as *stress reversals*.

Shock Load.   This type of load is due to impact. One example is an elevator dropping on a nest of springs at the bottom of a chute. The resulting maximum spring force can be many times greater than the weight of the elevator. The same type of shock load occurs in automobile springs when a tire hits a bump or hole in the road.

2.6 DYNAMIC STRENGTH

Parts designed on the basis of the proportional limit are acceptable for static loads. However, these same parts will not function properly if they are subjected to a large number of load reversals. A different material parameter called *fatigue strength* must be considered in machines with moving parts. For example, it is possible for a part to fail even though it has never been stressed to its proportional limit. This is due to a phenomenon called *fatigue*. To understand the mechanism of fatigue, refer to Figure 2.9, which shows a fatigue testing machine. The test specimen, which has a polished surface, is mounted in bearings as shown. A motor rotates the test specimen, which is actually a circular shaft of known size and material composition. A second set of bearings supports a trunnion that permits the application of specified amounts of weight W at the midpoint position of the length of the test specimen. By adding or removing weights, the desired magnitude of load can be selected. We have, in effect, a rotating shaft that is experiencing a bending load.

Figure 2.10*a* shows the applied loads when the rotating shaft is being treated simply as a supported beam. From strength of materials considerations, the shear and bending moment diagrams are developed and depicted in Figure 2.10*b* and *c*, respectively. As shown, the central portion of the test specimen experiences pure bending (no shear) and the constant bending moment is the maximum value $Wm/2$. Notice that l is the distance between the outboard shaft support bearings and m is the distance between each outboard shaft support bearing and its adjacent trunnion support bearing.

The maximum bending stress S_{max} can be calculated by the use of

$$S_{max} = \frac{M_{max}C}{I} \tag{2.4}$$

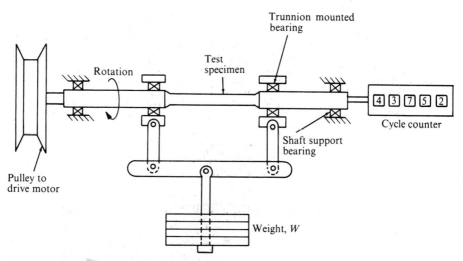

FIGURE 2.9   Fatigue testing machine

where M_{max} = maximum bending moment (lb·in)

             C = radius of test specimen (in.)

             I = moment of inertia of test specimen cross-section (in.⁴)

Substituting into Equation 2.4, we obtain the following result, where d is the test specimen diameter.

$$S_{max} = \frac{\left(W\frac{m}{2}\right)\left(\frac{d}{2}\right)}{\frac{\pi d^4}{64}} = \frac{16}{\pi}\frac{Wm}{d^3} \qquad (2.5)$$

From Equation 2.5, we see that the maximum bending stress in the test specimen is a function of dimension m, diameter d, and the applied load W. Thus, for a given load, the stress is known.

Under load, the shaft bends slightly. This is shown in Figure 2.11 in exaggerated fashion for illustration purposes.

Notice that the shaft's outermost fibers experience compression at point A and tension at point B. However, after 180° of shaft rotation, the stresses are reversed. Thus we have stress reversals as shown in Figure 2.12, where T is the period of time for one complete cycle or stress reversal. The period T in units of minutes can be obtained from

$$T = \frac{1}{n} \qquad (2.6)$$

where n equals the shaft revolutions per minute.

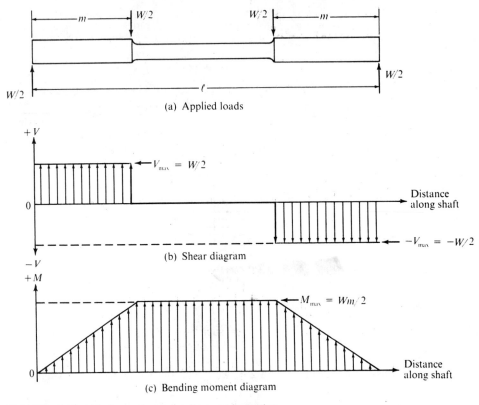

(a) Applied loads

(b) Shear diagram

(c) Bending moment diagram

FIGURE 2.10   Test specimen loading configuration

The number of cycles can be found from

$$N = \frac{t(\text{min})}{T(\text{min})}$$

(2.7)

where t is the time over which the load is applied.

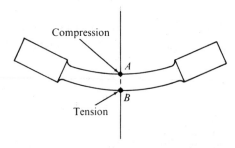

FIGURE 2.11   Shaft bending deformation

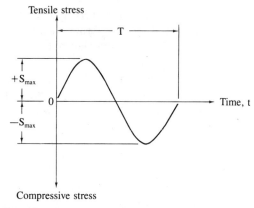

FIGURE 2.12   Single stress reversal of test specimen

2.7 FATIGUE FAILURE—THE ENDURANCE LIMIT DIAGRAM ━━━

The test specimen in Figure 2.10a., after a given number of stress reversals will experience a crack at the outer surface where the stress is greatest. The initial crack starts where the stress exceeds the strength of the grain on which it acts. This is usually where there is a small surface defect, such as a material flaw or a tiny scratch. As the number of cycles increases, the initial crack begins to propagate into a continuous series of cracks all around the periphery of the shaft. The conception of the initial crack is itself a stress concentration that accelerates the crack propagation phenomenon. Once the entire periphery becomes cracked, the cracks start to move toward the center of the shaft. Finally, when the remaining solid inner area becomes small enough, the stress exceeds the ultimate strength and the shaft suddenly breaks. Inspection of the break reveals a very interesting pattern, as shown in Figure 2.13. The outer annular area is relatively smooth because mating cracked surfaces had rubbed against each other. However, the center portion is rough, indicating a sudden rupture similar to that experienced with the fracture of brittle materials.

This brings out an interesting fact. When actual machine parts fail as a result of static loads, they normally deform appreciably because of the ductility of the material.

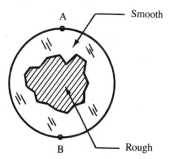

FIGURE 2.13   Fatigue failure of shaft test specimen

Thus many static failures can be avoided by making frequent visual observations and replacing all deformed parts. However, fatigue failures give no warning. Fatigue failures occur suddenly without deformation and, thus, are more serious. It has been estimated that over 90% of broken automobile parts have failed through fatigue.

The *fatigue strength* of a material is its ability to resist the propagation of cracks under stress reversals. Endurance limit is a parameter used to measure the fatigue strength of a material. By definition, the *endurance limit* is the stress value below which an infinite number of cycles will not cause failure.

Let us return our attention to the fatigue testing machine in Figure 2.9. The test is run as follows: A small weight is inserted and the motor is turned on. At failure of the test specimen, the counter registers the number of cycles N, and the corresponding maximum bending stress is calculated from Equation 2.5. The broken specimen is then replaced by an identical one, and an additional weight is inserted to increase the load. A new value of stress is calculated, and the procedure is repeated until failure requires only one complete cycle. A plot is then made of stress versus number of cycles to failure. Figure 2.14a shows the plot, which is called the *endurance limit* or *S-N curve*. Since it would take forever to achieve an infinite number of cycles, 1 million cycles is used as a reference. Hence the endurance limit can be found from Figure 2.14a by noting that it is the stress level below which the material can sustain 1 million cycles without failure.

The relationship depicted in Figure 2.14 is typical for steel, because the curve becomes horizontal as N approaches a very large number. Thus the endurance limit equals the stress level where the curve approaches a horizontal tangent. Owing to the large number of cycles involved, N is usually plotted on a logarithmic scale, as shown in Figure 2.14b. When this is done, the endurance limit value can be readily detected by the horizontal straight line. For steel, the endurance limit equals approximately 50% of the ultimate strength. However, if the surface finish is not of polished quality, the value of the endurance limit will be lower. For example, for steel parts with a machined surface finish of 63 microinches (μin.), the percentage drops to about 40%. For rough surfaces (300 μin. or greater), the percentage may be as low as 25%.

The most common type of fatigue failure is that due to bending. The next most frequent is torsion failure, whereas fatigue failure due to axial loads occurs very seldom. Spring materials are usually tested by applying variable shear stresses that alternate from zero to a maximum value, simulating the actual stress patterns.

In the case of some nonferrous metals, the fatigue curve does not level off as the number of cycles becomes very large. This continuing toward zero stress means that a large number of stress reversals will cause failure regardless of how small the value of stress is. Such a material is said to have no endurance limit. For most nonferrous metals having an endurance limit, the value is about 25% of the ultimate strength.

2.8 STRESS CONCENTRATION

Basic formulas for stress, such as Equations 2.1 and 2.4, assume that there are no irregularities in the shape of the part undergoing load. However, it is impossible to

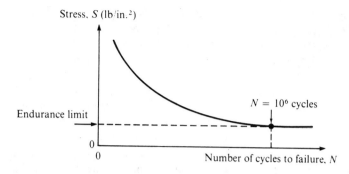

(a) Rectangular plot of S-N diagram

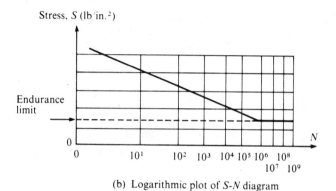

(b) Logarithmic plot of S-N diagram

FIGURE 2.14 Typical endurance limit diagram

design a machine without allowing discontinuities in the contour of parts. The following are a few examples where changes in contour are necessary:

— Shafts with shoulders to accommodate the seating of bearings
— Keyways in shafts that use keys to secure pulleys, cams, and gears
— Threads on one end of a bolt and a head on the other end
— Fillets at the base of gear teeth

Geometric irregularities cause the stress in the area of the discontinuity to exceed that predicted by equations. The physical discontinuity is called a *stress raiser* or a region of *stress concentration*. A stress concentration factor K is defined as the maximum actual stress divided by the average calculated stress based on the minimum cross-sectional area. This definition is represented mathematically by

$$K = \frac{S_{\text{max actual}}}{S_{\text{avg calc}}} \tag{2.8}$$

Stress concentration factors have been experimentally determined for various geometric and load configurations; several of the popular experimental methods used are photoelastic study, strain gauge application, and the Moire technique. You will find that graphs provide detailed stress concentration factors in reference books, such as *Stress Concentration Design Factors,* by R. E. Peterson.

It is important for design engineer to have an intuitive grasp of the relative effects of various geometric configurations on stress concentration. This can be done by assuming that forces and, hence, stresses flow through loaded parts. For example, Figure 2.15a shows a uniform bar with no hole, and Figure 2.15b shows an identical bar with a circular hole. Each bar is loaded by a tensile force *F.* In Figure 2.15a, the stress flow lines are equally spaced because the bar is uniform and, thus, the stress is a constant. Also, notice that the stress lines are actually straight lines with no bends. However, in Figure 2.15b, as the stress flow lines approach the hole, the lines closest to the hole have to bend to get around the hole, and the stress concentration factor is proportional to the amount of this bending from the normal path of the stress flow lines. Therefore, the maximum stress occurs at the edges of the hole.

Figures 2.16a and b shows the corresponding stress distribution for the same two bars. Each bar is cut in half, and the left half is represented as a free-body diagram.

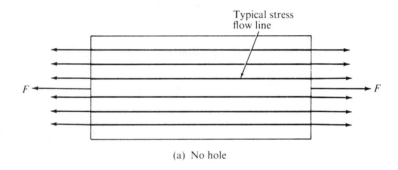

(a) No hole

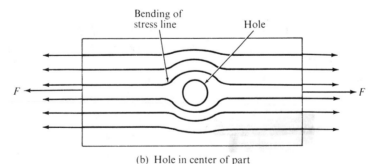

(b) Hole in center of part

FIGURE 2.15   Stress flow lines in uniform tensile bar

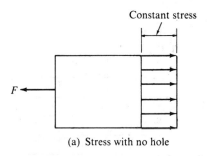

(a) Stress with no hole

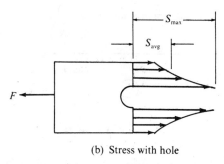

(b) Stress with hole

FIGURE 2.16   Stress distribution in uniform bar

The applied force F is shown on the left end, while the actual resulting stress distribution is depicted on the right end. As can be seen, the hole produces a stress raiser such that S_{max}, occurring at the edge of the hole, is appreciably greater than S_{avg}. This condition is represented by

$$S_{av} = \frac{F}{A_B - A_H} \tag{2.9}$$

$$S_{max} = S_{avg} \times K \tag{2.10}$$

where K = stress concentration factor
A_B = cross-sectional area of the bar
A_H = cross-sectional area of the hole

It should be noted, of course, that the value of the average stress at the hole section is greater than the value of the constant stress occurring at a section with no hole. This is due to the reduced cross-sectional area.

Figure 2.17 shows that for a uniform bar loaded in tension, the stress concentration factor is a function of the diameter d of the hole and width, w of the bar. As the hole diameter becomes smaller, the stress concentration becomes larger. This relationship can be reasoned intuitively by visualizing that a stress flow line must bend more

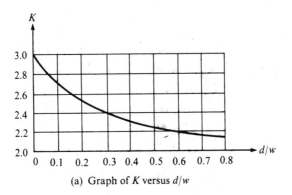

(a) Graph of K versus d/w

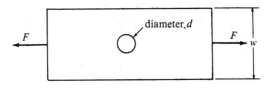

(b) Bar configuration

FIGURE 2.17   Stress concentration factors for uniform
tensile bar with hole stress raiser

abruptly to get around a small hole than around a large hole. From Figure 2.17 we conclude that the stress concentration factor reaches a value of 3 as the hole diameter approaches zero.

SAMPLE PROBLEM 2.1

Stress Concentration

PROBLEM: Figure 2.18 shows a rectangular bar containing a hole and undergoing a tensile force of 10,000 pounds. In this case, $t = 2$ in., $d = 0.5$ in., and $w = 1.5$ in.

(a) Find the tensile stress at a no-hole section.

(b) Find the average stress at the hole section, assuming no stress concentration.

(c) Find the maximum stress at the hole section taking into consideration the effect of the stress raiser.

Solution

(a) Equation 2.1 is applied to the gross cross sectional area opposing the tensile forces.

$$S_{no-hole} = \frac{F}{A_{no-hole}} = \frac{F}{tw} = \frac{10,000}{2(1.5)}$$

$$= 333 \text{ lb/in.}^2$$

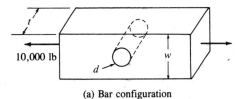

(a) Bar configuration

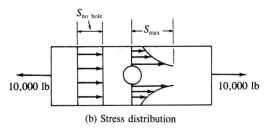

(b) Stress distribution

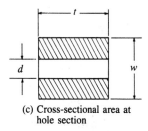

(c) Cross-sectional area at hole section

FIGURE 2.18   Rectangular bar under tensile load

(b) Equation 2.9 uses the net area to find the average stress.

$$S_{\text{avat hole}} = \frac{F}{A_B - A_H} = \frac{10,000}{2(1.5) - 2(0.5)}$$

$$= 5000 \text{ lb/in.}^2$$

(c) Equation 2.10 and Figure 2.17a are used to find the maximum stress.

$$\frac{d}{w} = \frac{0.5}{1.5} = 0.333.$$

From Figure 2.17a, $K = 2.3$.

$$S_{\max} = K\, S_{\text{avg}} = (2.3)(5000) = 11,500 \text{ lb/in.}^2$$

2.9 METHODS TO REDUCE STRESS CONCENTRATION

Stress concentration factors can be reduced by carefully examining the geometry and making minor changes in the shapes of parts. A general rule can be made as follows: *Make all transitions in contour as gradual as possible.*

Figure 20.19a shows a cross-sectional view of a shaft with a keyway. The stress concentration can be reduced by drilling a hole on both sides of the keyway as shown in Figure 2.19b. Notice that the stress flow line does not bend as abruptly for the shaft with the holes.

A second example in which the stress concentration is reduced by removing material is shown in Figure 2.20. The reduction can be made by initially machining a fillet at the junction between the shaft and shoulder (see Figure 2.20b). A further reduction

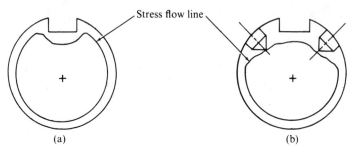

Stress flow line

(a)　　　　　　　　　　　(b)

FIGURE 2.19　Stress reduction by drilling holes on both sides of keyway in shaft

in stress is then achieved by machining a groove as shown in Figure 2.20c. Observe the amount of bending of the stress flow line: the no-fillet case has the greatest amount of bending, while the combination of fillet and groove produces the least. The machining of the groove is again a case where material is removed to reduce stress concentration.

Figure 2.21 shows how to reduce the stress concentration factor for a threaded cylindrical part. Threaded parts are machined so that the thread root diameter equals its adjacent shank diameter. Notice that the stress flow line bends a small amount in Figure 2.22b and none at all in Figure 2.21c.

Sometimes a gear, pulley, cam, or bearing is press-fitted to a shaft. A stress concentration occurs because of the abrupt contour, as shown in Figure 2.22a. The addition of a shoulder and fillet provides a reduction in stress concentration, as depicted in Figure 2.22b.

The study of stress concentration is a highly complex subject. In reality, not only does a stress concentration factor depend on the geometry of the part, it also depends on whether the material is ductile or brittle. Stress concentration is primarily a localized effect, which means that the maximum stress occurs only in a small region in the vicinity of the stress raiser. Hence, for a ductile material experiencing a static load, the material will yield locally. Therefore, the effect of the stress concentration is

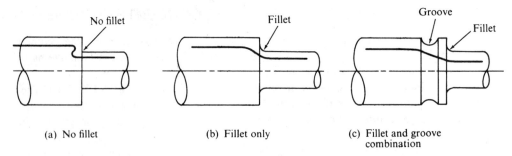

No fillet　　　　　　　Fillet　　　　　　　Groove　　　Fillet

(a) No fillet　　　　　　(b) Fillet only　　　　　(c) Fillet and groove combination

FIGURE 2.20　Stress reduction by use of fillet and groove

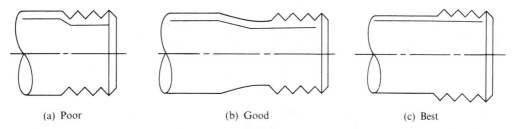

(a) Poor   (b) Good   (c) Best

FIGURE 2.21   Stress concentration reduction for threaded cylindrical part

minimized. However, for a ductile material undergoing stress reversals, the full value of the stress concentration factor should be applied when performing a stress analysis of a given machine.

2.10 ALLOWABLE STRESS AND FACTOR OF SAFETY ▬▬▬

The stress at which a part is designed to operate based on calculation is called the *allowable*, or *design, stress*. It would be risky to design machine parts to operate at the yield strength, where some permanent deformation takes place. In deciding on an allowable stress value, the design engineer must take into account the consequences of failure, such a danger to humans and high costs of repair. The types of loads and the number of unknowns that affect the life and reliability of a machine also must be considered.

As a result, a factor of safety (FS) is applied to the yield strength for ductile materials. A different factor of safety is applied to the ultimate strength when brittle materials are used. The reason for using the ultimate strength in the latter case is that brittle materials do not have a pronounced yield strength.

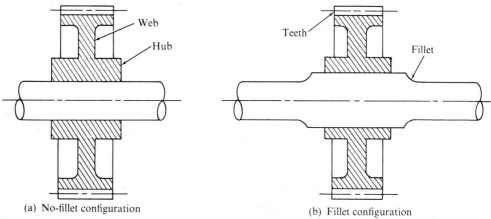

(a) No-fillet configuration   (b) Fillet configuration

FIGURE 2.22   Reduction of stress concentration in gear press-fitted to shaft assembly

We can therefore define the allowable stress mathematically by Equations 2.11 and 2.12 for ductile and brittle materials, respectively:

$$S_{allowable} = \frac{S_{yield\ strength}}{FS_{ductile\ material}} \qquad (2.11)$$

$$S_{allowable} = \frac{S_{ultimate\ strength}}{FS_{brittle\ material}} \qquad (2.12)$$

Factor of safety is, in a large measure, an ignorance factor that compensates for various possible unknowns, including the following:

— Exact type and magnitude of all loads
— Material property variations
— Precise stress concentration effects
— Extremes of environmental conditions, such as heat and moisture
— Approximate stress analysis formulas
— Residual stresses produced during manufacturing

Consequently, determination of a realistic factor of safety is a difficult task and is usually handled by a senior design engineer. If the factor of safety is excessive, the cost and sizes of parts increase. On the other hand, too small a factor of safety results in premature failure, which can be very costly in terms of human safety as well as in dollars.

Equation 2.13 defines a factor of safety as a function of four main parameters:

$$FS = a \times b \times c \times d \qquad (2.13)$$

The following examples show the significance of each parameter. Note that the values are approximate and can vary somewhat depending on the judgment of the design engineer.

1. *Factor a* considers the variability of the applied loads:

    $a = 1$ for constant magnitude and direction loads
    $a = 2$ for complete load reversals in order to take fatigue into account

2. *Factor b* takes into account the abruptness with which the loads are applied:

    $b = 1$ for gradually applied loads
    $b = 2$ for suddenly applied loads
    $b = 3$ or more (depending on the severity) for impact loads.

3. *Factor c* considers the consequences of failure as they relate to human safety, cost, and so on:

> *c* normally varies between 1.2 and 2, depending on the seriousness of a failure

4. *Factor d* differentiates FS ductile material from FS brittle material:

$$d = \frac{\text{ultimate strength}}{\text{yield strength}}$$

Factor *d* is used only when basing the factor of safety on the ultimate strength, which would normally be done for brittle materials. Otherwise, *d* equals unity.

The value of allowable stress obtained from Equations 2.11 or 2.12 is the value that the design engineer attempts to match during the design and stress analysis process. It should be noted that all stress concentrations are included as part of the stress analysis procedure. Therefore, the design should be such that a thorough stress analysis predicts that the maximum actual operating stress will equal the allowable stress. Also note that the strengths of some parts such as pressure vessels, are dictated by government codes that include the use of specific equations. Caution must be exercised in the use of these equations, since the desired factor of safety is usually built in.

The following sample problem shows that the factor of safety based on the yield strength can typically vary from a low of 1.2 to a high of 12.0 depending on the total design requirements.

SAMPLE PROBLEM 2.2 _____

Factor of Safety

PROBLEM: For the rectangular bar of Sample Problem 2.1 (Fig. 2.18), the material has a yield strength of 50,000 lb/in.^{2} Determine the factor safety (FS), the allowable stress ($S_{allowable}$), and the allowable load F for the following two cases:

(a) Load does not vary and is gradually applied. The consequence of failure is minimal.

(b) Load is applied with mild impact and with complete reversals of direction. The consequence of failure is very serious.

Solution:

From Sample Problem 2.1, the stress concentration factor K is 2.3 and the minimum cross-sectional area A_{min} is 2 in.^{2}

(a) $FS = a \times b \times c \times d$

$\qquad = 1 \times 1 \times 1.2 \times 1 = 1.2$

$$S_{\text{allowable}} = \frac{S_{\text{yield strength}}}{\text{FS}} = \frac{50{,}000}{1.2} = 41{,}700 \text{ lb/in.}^2$$

$$S_{\text{max}} = S_{\text{allowable}} = 41{,}700 = \frac{F}{A_{\text{min}}} \times K = \frac{F}{2} \times 2.3$$

$$F = 36{,}300 \text{ lb}$$

(b) $\quad$ FS $= a \times b \times c \times d = 2 \times 3 \times 2 \times 1 = 12$

$$S_{\text{allowable}} = \frac{50{,}000}{12} = 4170 \text{ lb/in.}^2$$

$$4170 = \frac{F}{A_{\text{min}}} \times K = \frac{F}{2} \times 2.3$$

$$F = 3630 \text{ lb}$$

2.11 EFFECTS OF TEMPERATURE ON YIELD STRENGTH AND MODULUS OF ELASTICITY

Generally speaking, when stating that a material possesses specified values of properties such as modulus of elasticity and yield strength, it is implied that these values exist at room temperature. At low or elevated temperatures, the properties of materials may be drastically different. For example, many metals are more brittle at low temperatures. In addition, the modulus of elasticity and yield strength deteriorate as the temperature increases. Figure 2.23 shows that the yield strength for mild steel is reduced by about 70% in going from room temperature to 1000°F.

Figure 2.24 shows the reduction in the modulus of elasticity E for mild steel as the temperature increases. As can be seen from the graph, a 30% reduction in modulus of elasticity occurs in going from room temperature to 1000°F. In this figure, we also can see that a part loaded below the proportional limit at room temperature can be permanently deformed under the same load at elevated temperatures.

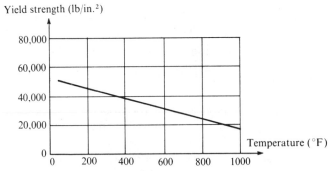

FIGURE 2.23 Reduction of yield strength with temperature increase for mild steel

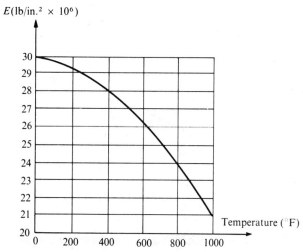

$E(\text{lb/in.}^2 \times 10^6)$

FIGURE 2.24   Reduction of modulus of elasticity with temperature increase for mild steel

2.12 CREEP: A PLASTIC PHENOMENON

Temperature effects bring us to a phenomenon called *creep,* which is the increasing plastic deformation of a part under constant load as a function of time. Creep also occurs at room temperature, but the process is so slow that it rarely becomes significant during the expected life of the part involved. If the temperature is raised to 300°F or more, the increasing plastic deformation can become significant within a relatively short period of time. The *creep strength* of a material is its ability to resist creep, and creep strength data can be obtained by conducting long-time creep tests simulating actual part operating conditions. During the test, the plastic strain is monitored for given materials at specified temperatures.

Since creep is a plastic deformation phenomenon, the dimensions of a part experiencing creep are permanently altered. Thus, if a part operates with tight clearances, the design engineer must accurately predict the amount of creep that will occur during the life of the machine. Otherwise, problems such binding or interference can occur.

Creep also can be a problem in the case where bolts are used to clamp two parts together at elevated temperatures. The bolts, under tension, will creep as a function of time. Since the deformation is plastic, loss of clamping force will result in an undesirable loosening of the bolted joint. The extent of this particular phenomenon, called *relaxation,* can be determined by running appropriate creep strength tests.

Figure 2.25 shows typical creep curves for three samples of a mild steel part under a constant tensile load. Notice that for the high-temperature case the creep tends to accelerate until the part fails. The time line in the graph (the *x*-axis) may represent a period of 10 years, the anticipated life of the product.

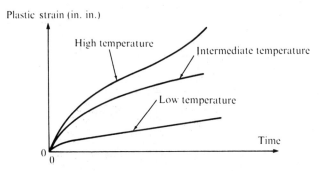

FIGURE 2.25   Typical creep curves for mild steel

2.13   THERMAL STRESSES

expand + contract

Another temperature effect of concern in machine design is that of thermal expansion and contraction. Most materials expand when heated and contract when cooled. This phenomenon is defined mathematically by

$$\delta = L\alpha(\Delta T) \tag{2.14}$$

where δ = thermal deformation in inches

L = part length in inches

α = coefficient of thermal expansion in inches per degree F

ΔT = change in temperature in degrees F

The coefficient of thermal expansion α is a property of a material. It is the change in length of a part per unit length per degree Fahrenheit. For example, the coefficient of thermal expansion for steel is 6.5×10^{-6} in./in.°F. This means that for every degree Fahrenheit change in temperature, the length of a steel part will change by 0.000,006,5 in. per inch of length of the part. Thermal deformation, if allowed to occur freely, is completely elastic. Thus a part will return to its original size if the temperature is returned to its original value. However, if free thermal deformation (expansion or contraction) is prevented from occurring, extremely large stresses can build up in machine parts undergoing temperature changes. The following sample problems show how thermal deformation can be of great concern in machine design.

SAMPLE PROBLEM 2.3

Deformation due to Temperature Change

PROBLEM: How much separation should exist between successive 50 ft steel railroad rails to allow for thermal expansion as shown in Figure 2.26? The rails are laid when the temperature is 20°F, and the maximum expected temperature is 120°F.

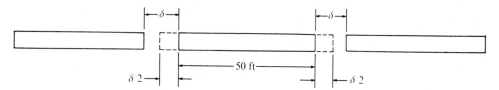

FIGURE 2.26   Successive 50-ft. rails with thermal gaps

Solution   Use Equation 2.14 for thermal deformation:

$\delta = L\alpha(\Delta T)$

$\quad = 50 \text{ ft} \times 12 \text{ in./ft.} \times 6.5 \times 10^{-6} \text{ in./in.°F} \times 100°F$

$\quad = 0.39 \text{ in.}$

SAMPLE PROBLEM 2.4

Stress due to Temperature Change

PROBLEM: Figure 2.27 shows a steel bar welded to two nonyielding supports. What tensile stress will the bar experience if the temperature drops 200°F? Assume that the bar is initially unstressed.

Solution   Two values must be obtained. The coefficient of thermal expansion value is given in Section 2.13 and the modulus of elasticity is obtained from the table of properties in the appendix.

$\alpha = 6.5 \times 10^{-6} \text{ in./in.°F} \quad \text{and} \quad E = 30 \times 10^{6} \text{ lb/in.}^2$

The bar attempts to contract thermally, but the nonyielding supports prevent this from happening. The net effect is that a tensile force is applied by the nonyielding supports to the bar. The magnitude of the tensile force is such that the mechanical strain equals the thermal strain to keep the length of the bar constant.

From Equation 2.14 we have the relationship for thermal deformation:

$\delta = L\alpha(\Delta T)$

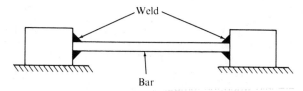

FIGURE 2.27   Bar undergoing thermal stresses

Strain equals deformation per unit length:

$$\text{Thermal strain} = \frac{\delta}{L} = \alpha(\Delta T) = 6.5 \times 10^{-6}(200)$$

$$= 0.0013 \text{ in./in.}$$

From Equation 2.3, we obtain the mechanical strain:

$$\text{Mechanical strain} = \epsilon = \frac{S}{E} = 0.0013 \text{ in./in.}$$

The final result for the stress becomes

$$S = 0.0013E = 0.0013 \times 30 \times 10^6 = 39,000 \text{ lb/in.}^2$$

Notice that the length of the bar does not affect the thermal stress. The thermal stress is directly proportional to the coefficient of thermal expansion, the modulus of elasticity, and the temperature change. The thermal force produced can be found by multiplying the stress by the cross-sectional area of the bar.

2.14 INTERFERENCE FITS   *by Temp-*

The concept of thermal stress can be applied utilizing interference fits as a means of fastening one part to another. Figure 2.28 shows a hub that has an inside diameter d and length l. Next to the hub is a shaft having a diameter $d + i$ and a length l. Thus the diameter of the shaft is larger than the diameter of the bore of the hub. The difference between the two diameters is called the *interference i*.

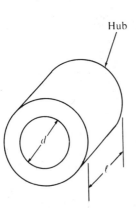
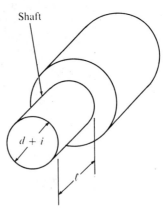

Hub

Shaft

$d + i$

FIGURE 2.28   Interference fit of hub and shaft

The hub can be heated to a sufficient temperature so that the bore will expand by an amount equal to or exceeding the interference. The shaft can then be readily inserted into the bore of the hub. As the hub cools, it contracts and forms a strong joint via a shrink interference fit. A *light shrink fit* is one with an interference of 0.00025 in. per inch of hole diameter, a *medium shrink fit* has twice as much interference, and a *heavy shrink* fit has four times as much interference as a light shrink fit. The contracting hub creates normal forces on the shaft and hole surfaces, and these forces permit torque to be transmitted across the joint.

A useful formula permitting the calculation of the transmitting torque capacity will now be developed under the following two assumptions:

1. The shaft diameter does not decrease (shaft is perfectly rigid).
2. The hub wall thickness is not too large and thus permits use of the following thin-walled pressure vessel equation:

$$S = \frac{pd}{2t} \qquad (2.15)$$

where S = tensile stress in the hub in pounds per inch^2

d = shaft diameter in inches

t = hub thickness in inches

p = normal pressure between hub and shaft in pounds per inch^2

From Equation 2.3, we have the following relationship between stress, strain, and modulus of elasticity:

$$S = E\epsilon$$

But

$$\epsilon = \frac{\text{diametral deformation}}{\text{diameter}} = \frac{i}{d}$$

so

$$S = E\frac{i}{d} \qquad (2.16)$$

Equating S's from Equations 2.15 and 2.16, we have

$$p = \frac{2Eit}{d^2} \qquad (2.17)$$

We also know that torque T equals $F \times d/2$, where F is the friction force between the hub and shaft. But $F = \mu N$, where μ is the coefficient of friction between the hub and shaft and N is the total normal force. Also, $N = pA$, where A is the cylindrical area of contact between the hub and shaft. $N = p\,(\pi dl)$. Thus we have

$$T = F \times \frac{d}{2} = \mu(N) \times \frac{d}{2} = \mu(pA) \times \frac{d}{2} = \mu(p)\,(\pi dl) \times \frac{d}{2}$$

Substituting for p from Equation 2.17 yields

$$T = \mu\left(\frac{2Eit}{d^2}\right)(\pi dl)\left(\frac{d}{2}\right) \qquad \text{or} \qquad T = \pi\mu Eitl \qquad\qquad \textbf{(2.18)}$$

From Equation 2.18, we conclude that the torque-transmission capability is directly proportional to the coefficient of friction, the modulus of elasticity of the hub, the amount of interference, the wall thickness of the hub, and the length of the hub. Since the shaft does, however, contract somewhat, Equation 2–18 predicts a somewhat higher value than would actually be attained.

In order to determine the temperature to which the hub must be raised, we can use Equation 2.14:

$$\delta = L\alpha(\Delta T)$$

Substituting corresponding terms, we have

$$i = d\alpha(\Delta T) \qquad \text{or} \qquad \Delta T = \frac{i}{d\alpha} \qquad\qquad \textbf{(2.19)}$$

An alternate method of producing a shrink fit is to cool the shaft and then insert it into the hub. Quite often bushings are shrunk by cooling and are then inserted into housings. This is usually done when the housing is difficult to heat because of its large size. One other practical application is the insertion of valve seats into huge blocks. The following sample problems shows the use of equations in solving a shrink-fit problem.

SAMPLE PROBLEM 2.5

Interference Fit

PROBLEM: A steel ring, $\frac{1}{4}$ in. thick and 1 in. long is to be joined around a 1 $\frac{1}{2}$ in. diameter shaft. The stress in the ring is not to exceed 30000 lb./in.^{2}

(a) Find the maximum interference permitted.

(b) Find the torque transmission capacity.

(c) Find the temperature increase needed for the ring.

Solution

(a) From Equation 2.16, we have

$$i = \frac{Sd}{E} = \frac{30{,}000 \times 1.5}{30 \times 10^6} = 0.0015 \text{ in.}$$

Thus we have a heavy shrink fit with an interference of 0.001 in. per inch of shaft diameter.

(b) Using Equation 2.18 we have

$$T = \pi \mu E i t l$$
$$= (3.14)\,(0.3)\,(30 \times 10^6)\,(0.0015)\,(0.25)\,(1)$$
$$= 10{,}600 \text{ lb·in.}$$

Obviously, a very large amount of torque can be transmitted.

(c) Equation 2.19 is then used:

$$\Delta T = \frac{i}{d\alpha} = \frac{0.0015}{(1.5)\,(6.5 \times 10^{-6})} = 154°F$$

2.15 METRIC (SI) UNITS

Conversion to the metric or SI system of measurement is underway in the United States. You will no doubt be called on by industry to make conversions to or from the U.S. Customary System (USCS) and the SI system or be required to use data or charts in the SI system. The conversion to and from kilograms, newtons, and pounds can be confusing. Through long usage in the United States, the *pound* (a force unit) has been used to compare masses of two objects. In metric countries, the *kilogram* (a mass unit) is used to compare masses. The force unit in the SI system is the *newton*. When solving problems in the SI system, it is common practice to identify loads on machine members in kilograms. This must be changed to newtons for use in equations requiring force units. Also, in most equations, force units must be in newtons, length units must be in meters (not millimeters or kilometers), and stress units must be in pascals.

A thermal stress sample problem will demonstrate the use of SI units.

SAMPLE PROBLEM 2.6

Thermal Stresses in SI Units

PROBLEM: A 30 mm diameter steel bar is 1.5 m long. It connects two rigid machine parts. At 20°C, it is unstressed. Determine the compressive stress and force it exerts on the machine parts when its temperature is raised to 55°C. The coefficient of thermal expansion of steel is 12×10^{-6} m/m °C (or mm/mm °C). $E = 207$ GPa.

Solution   From Equation 2.14, we get thermal strain:

$$\epsilon = \alpha \times \Delta T$$
$$\epsilon = 12 \times 10^{-6} \times 35°C = 420 \times 10^{-6}$$

(a) Since mechanical strain equals thermal strain, Equation 2.3 applies:

$$S = \epsilon \times E = (420 \times 10^{-6}) \times (207 \times 10^9)$$
$$= 86.9 \text{ MPa}   (Round \text{ } to \text{ } 87 \text{ } MPa.)$$

(b) Apply Equation 2.1 to find the force developed:

$$F = S \times A = (86.9 \times 10^6) \times \frac{\pi(0.030m)^2}{4}$$

$$F = 61,430 \text{ N} \quad \text{(Round to 61 kN.)}$$

2.16 SUMMARY

The machine designer must understand the purpose of the static tensile strength test. This test determines a number of mechanical properties of metals that are used in design equations. Such terms as *modulus of elasticity, proportional limit, yield strength, ultimate strength, resilience,* and *ductility* define properties that can be determined from the tensile test.

Dynamic loads are those which vary in magnitude and direction and may require an investigation of the machine part's resistance to fatigue failure. Stress reversals may require that the allowable design stress be based on the endurance limit of the material rather than on the yield strength or ultimate strength.

Stress concentration occurs at locations where a machine part changes size, such as a hole in a flat plate or a sudden change in width of a flat plate or a groove or fillet on a circular shaft. Note that for the case of a hole in a flat plate or bar, the value of the maximum stress becomes much larger in relation to the average stress as the size of the hole decreases. Methods of reducing the effect of stress concentration usually involve making the shape change more gradual.

Machine parts are designed to operate at some allowable stress below the yield strength or ultimate strength. This approach is used to take care of such unknown factors as material property variations and residual stresses produced during manufacture and the fact that the equations used may be approximate rather than exact. The *factor of safety* is applied to the yield strength or the ultimate strength to determine the allowable stress.

Temperature can affect the mechanical properties of metals. Increases in temperature may cause a metal to expand and creep and may reduce its yield strength and its modulus of elasticity. If most metals are not allowed to expand or contract with a change in temperature, then stresses are set up that may be added to the stresses from the load. This phenomenon is useful in assembling parts by means of interference fits. A hub or ring has an inside diameter slightly smaller than the mating shaft or post. The hub is then heated so that it expands enough to slip over the shaft. When it cools, it exerts a pressure on the shaft resulting in a strong frictional force that prevents loosening.

2.17 QUESTIONS AND PROBLEMS

Questions

1. Why is machine design considered to be an inexact science?

2. Why is it necessary for a design engineer to have a good knowledge of strength of materials and the properties of engineering materials?

3. Why are brittle materials undesirable for use as machine parts?

4. Explain why it is possible for a part to have failed even though it has not fractured.

5. What is a stress concentration, and what are its causes?

6. Explain what is meant by the terms *endurance limit*.

7. Why are surface roughness and geometric discontinuities so important when considering the fatigue life of a machine part?

8. What property of materials determines its suitability for use in a part requiring great stiffness?

9. What is the difference between a shock load and a suddenly applied load?

10. What is the difference between a varying load and a reversing load?

11. Discuss the difference between ductile and brittle material behavior at a stress concentration when the load is static.

12. What is creep? Name two instances in which creep must be considered in machine design.

13. Explain briefly how an endurance test is run. Sketch a typical endurance limit diagram.

14. What is meant by the term *allowable stress?*

15. Name three factors upon which factor of safety depends.

16. What is the cause of thermal stresses?

17. How do thermal stresses relate to the application of joints using shrink fits?

18. Discuss whether or not the factor of safety used in Problem 3 is adequate. The load is suddenly applied, and human safety is a very strong concern if failure occurs.

Problems

You must choose the appropriate equation or equations in the section of the chapter that relates to the problem. If you cannot locate the correct equation, look for the equation where all the factors are known except one.

1. A steel tensile test specimen has a cross-sectional area of 0.2 in.^{2} The proportional limit was reached at a load of 10,000 lbs. If the modulus of elasticity is 30 x 10^6 lb/in.^{2}, find the resilience of the specimen material.

2. Figure 2.29 shows a shrink-fit assembly in which a bronze bushing is cooled and then inserted into its large steel housing. Initially, at room temperature the housing hole diameter is 2.000 in. and the bushing outside mating diameter is 2.002 in.

   (a) Find the minimum decrease in bushing temperature to permit the bushing to be freely inserted. $\alpha_{bronze} = 11.2$ x 10^{-6} in./in.°F.

   (b) What will be the resulting compressive stress in the bronze bushing? $E_{bronze} = 15$ x 10^6.

   (c) What would be the answers to parts (a) and (b) if the bushing were made of steel instead of bronze?

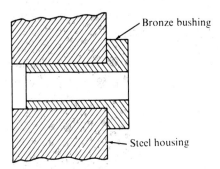

FIGURE 2.29   Shrink fit of bushing inside housing

3. The load of the bolt of Figure 2.30 varies from 5000 lb tension to 5000 lb compression. The thread has a root diameter of 1.5 in. and a stress concentration of 2.5. If the yield strength is 50,000 lb/in.^{2}, find the factor of safety.

4. The hub of a crank for a milling machine vise is to be shrunk onto the ⅞ in. diameter shaft at the end of the vise screw. The crank handle is located 3.5 in. radially from the axis of the screw. The hub length is 2 in., and its thickness is ¼ in. The modulus of elasticity of the hub is 30×10^6 lb/in.^{2}, and the coefficient of friction between the hub and shaft is 0.25. If the allowable tensile stress in the hub is 25,000 lb/in.^{2}, what force can be applied to the handle? Assume that the shaft is perfectly rigid.

5. Prove mathematically whether or not the following statement is true: A shrink interference fit of 0.001 in. per inch of hole diameter will produce a tensile stress of 30,000 lb/in.^{2} for a steel hub mounted on a rigid shaft.

6. A large gear is made in two pieces. The gear teeth are machined on a steel ring that will be fitted onto a steel wheel. The ring has an inside diameter machined to 36.000 in. The wheel has an outside diameter of 36.018 in.
    (a) Is this considered a light, medium or heavy shrink fit?
    (b) What tensile stress will be developed in the ring once it is placed on the wheel?
    (c) What temperature increase must be applied to the ring so that it can be slipped over the wheel?

SI Unit Problems

7. Refer to Figures 2.9 and 2.10. Construct shear and moment diagrams for a test specimen 200 mm long; the *m* distance is 30 mm, and the weight is 20 kg. The

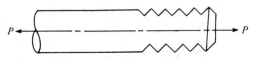

FIGURE 2.30   Sketch for Problem 3

shear diagram should be in units of *newtons,* and the moment diagram should be in units of *newton-meters.*

8. Convert the maximum stress in Sample Problem 2.1 to pascals.

9. How much separation should exist between successive 15 m steel rails to allow for thermal expansion as shown in Figure 2.26? The rails are laid when the temperature is -6°C, and the maximum expected temperature is 50°C. Give the answer in millimeters.

10. A steel hub 10 mm thick and 30 mm long is to be joined around a 50 mm diameter shaft. The stress in the hub is not to exceed 200 MPa.

    (a) Find the maximum interference permitted (in millimeters).

    (b) Find the pressure developed.

3
Journal Bearings, Thrust Bearings, and Lubrication

Objectives

After completing this chapter, you will be able to

- Analyze the frictional forces acting on journal bearings.
- Calculate the pressures acting on bearings.
- Describe the lubricant characteristics and the action of lubricants in journal and thrust bearings.
- Calculate the required length of a journal bearings.
- Calculate the operating temperature and the required cooling oil flow of bearings.

3.1 INTRODUCTION

A *bearing* can be defined as a member specifically designed to support moving machine components. The most common bearing application is the support of a rotating shaft that is transmitting power from one location to another. One example is the crankshaft bearings of the automotive engine; another example is the shaft bearings used in all types of electric motors. Since there is always relative motion between a bearing and its mating surface, friction is involved. In many instances, such as the design of pulleys, brakes, and clutches, friction is desirable. However, in the case of bearings, the reduction of friction is one of the prime considerations: Friction results in loss of power, the generation of heat, and increased wear of mating surfaces.

Journal and antifriction bearings are the two general types of bearings in existence. Journal bearings operate with sliding contact, whereas antifriction bearings operate under predominantly rolling contact. The amount of sliding friction in journal bearings depends on the surface finishes, materials, sliding velocities, and type of lubricant used. The principal motion-retarding effect in antifriction bearings is called *rolling resistance* rather than rolling friction. This is so because the resistance to motion is essentially due to the deformation of the rolling elements and, hence, is not a sliding phenomenon. Antifriction bearings will be discussed in Chapter 4.

To reduce the problems associated with sliding friction in journal bearings, a lubricant is used in conjunction with compatible mating materials. When selecting the lubricant and mating materials, one must take into account bearing pressures, temperatures, and rubbing velocities. The principal function of the lubricant in sliding contact bearings is to prevent physical contact between the rubbing surfaces. Thus the maintenance of an oil film under varying loads, speeds, and temperatures is the prime consideration in sliding contact bearings.

3.2 JOURNAL BEARINGS

A *journal bearing,* in its simplest form, is a cylinderical bushing made of a suitable material and containing properly machined inside and outside diameters. The *journal* is usually the part of a shaft or pin that rotates inside the bearing. Figure 3.1 shows such

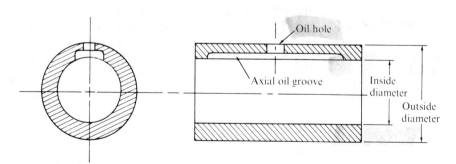

FIGURE 3.1 Sleeve bearing with provisions for lubrication

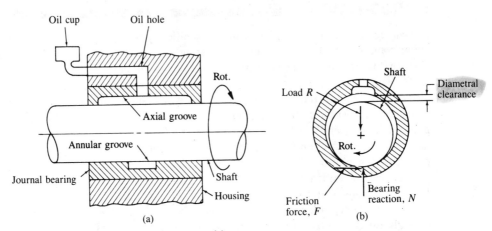

FIGURE 3.2   Journal bearing assembly

a bearing, which is also commonly called a *sleeve bearing.* Notice the oil hole that leads to an axial oil groove. Such a bearing is designed to provide lubrication to reduce friction.

The bore (inside diameter) of a journal bearing is machined to accept a rotating shaft, as shown in Figure 3.2a. The bearing outside diameter is supported in a properly machined bore of a fixed housing. An oil cup is shown attached to one side of the housing, where a passageway leads to the oil hole of the bushing. Notice that the axial oil groove in the bushing does not extend all the way to the ends of the bearing. This prevents undue loss of oil from the bearing. In Figure 3.2b, the diametral clearance C between the bushing and the shaft is greatly exaggerated for illustration purposes. Generally, the diametral clearance is between 0.001 and 0.002 in. per inch of shaft diameter. Also, the thickness of the bushing is often made approximately equal to 1/8 in. per inch of shaft diameter. Moreover, notice the annular groove in the bushing in Figure 3.2a; such grooves ensure more complete distribution of the lubricant. Since there will normally be a load R coming into the bearing, a normal force N and friction force F will exist between the shaft and bushing, as shown in Figure 3.2b.

Even though bearings are usually lubricated, there is friction and some wear (see Section 3.6). With time, the friction will result in wearing down of the bushing bore. Proper selection of bearing materials is an important design consideration. The following material properties are considered desirable for journal bearings.

Compatibility.   Certain combinations of materials do not rub well together. These materials may have a tendency to fuse where there is metal-to-metal contact. This results in damaged surfaces, and the materials are said to be *incompatible.* In general, dissimilar materials are more compatible than similar materials.

Embedability.   Dirt and other foreign particles will always work their way inside a bearing. A bearing material with outstanding embedability allows these foreign particles to become embedded into the bushing, which prevents scratching and other damage from occurring on the surface of the journal or shaft.

TABLE 3.1 Relative Quality of Bearing Materials

Bearing Material	Compati-bility	Embeda-bility	Conforma-bility	Corrosion Resistance	High Thermal Conductivity	High Fatigue Strength
Lead bronze	3	2	2	4	3	3
Tin-base babbitt	5	5	4	5	2	2
Lead-base babbitt	5	4	5	4	2	2
Aluminum	3	1	2	5	3	4
Silver	3	2	1	5	5	5

5 = highest quality.

1 = lowest quality.

 Conformability. A good bearing material should have a tendency to compensate for small amounts of misalignment and shaft deflection. A material with a low modulus of elasticity has this property of conformability. *for low load*

Corrosion Resistance. Lubricants generally have lubricity improvement additives that may corrode some metals. The bearing material must not corrode when the recommended lubricant is used.

High Thermal Conductivity. A high thermal conductivity is desirable to remove heat rapidly from the bearing.

High Fatigue Strength. Since bearing loads continuously vary in magnitude, a high fatigue strength is desirable.

Table 3.1 provides the relative quality of typical bearing materials for each of the preceding desirable properties. A rating system of 1 through 5 is used, where 1 is the lowest and 5 the highest quality. As shown in the table, no single bearing material possesses all the desirable properties at the highest level. Bronze is an alloy of copper and tin, but lead is normally added for bearing applications. The addition of the lead results in a lower coefficient of friction and better conformability. Babbitt consists of approximately 90% tin or lead. The remaining 10% is a combination of antimony and copper. A lead-base babbitt has a lower modulus of elasticity than a tin-base babbitt and thus has greater conformability. Lead also produces a lower coefficient of friction. Aluminum bearings are becoming more popular because they have high fatigue strength and corrosion resistance and are also inexpensive. Silver, although it has many desirable properties, is very expensive and has practically no conformability.

It should be noted that many nonmetallic materials, including nylon, Teflon, and rubber, are often used for bearings. Bearings in water pumps are sometimes made of rubber because the rubber, owing to its embedability, will allow small particles of sand to pass through the bearing without scratching the shaft surface.

FIGURE 3.3   Automotive connecting rod showing crank-pin and wrist pin bearings *(Courtesy of Central Foundry, Division of General Motors Corporation)*

Before selecting a bearing material, the designer must evaluate all the significant application parameters, which include the following:

— Misalignment between the shaft centerline and the centerline of the bushing bore
— Shaft deflection due to shaft bending loads
— Type of shaft material
— Bearing loads and temperatures
— Shaft speed
— Characteristics of the lubricant

3.3   JOURNAL BEARING FRICTION ANALYSIS

Many machines, such as the piston engine, consist of a series of links interconnected by pin joints, which, in effect, are journal bearings. For example, Figure 3.3 is a photograph of an automotive connecting rod showing the crankpin bearing at one end and the wrist pin bearing at the other end.

Figure 3.4 is a photograph of a six-cylinder automotive crankshaft. Observe that there is space for a main bearing on each side of each crank where the connecting rod is attached.

FIGURE 3.4   Six-cylinder automotive crankshaft *(Courtesy of Central Foundry, Division of General Motors Corporation)*

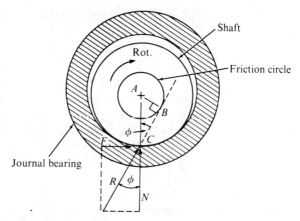

FIGURE 3.5   Bearing friction circle

It is important to analyze the resistance to rotary motion produced by journal bearings. One method that uses the concept of a friction circle will now be developed in reference to Figure 3.5. The figure shows the end view of a shaft rotating in a journal bearing. The normal force N and the friction force F are shown. The friction force opposes the shaft rotation. The total resultant bearing reaction R is inclined by the friction angle ϕ from the normal force N. The line of action of R is extended as a dashed line. A friction circle whose center is the center of the shaft is drawn tangent to the extended line of action of R. The size of the friction circle is derived as follows (the symbol for the coefficient of friction is μ (Greek letter mu):

$$\mu = \frac{F}{N} = \tan \phi \quad \text{By definition.}$$

Notice that triangle ABC is a right triangle and that $\tan \phi$ equals approximately $\sin \phi$ because ϕ is usually less that $15°$ (at $15°$ there is less than a 4% difference, and at $10°$ there is less than a 2% difference). Let r_f equal the radius of the friction circle and r_S equal the radius of the shaft. Therefore, we can say

$$\mu = \sin \phi = \frac{AB}{AC} = \frac{r_f}{r_S}$$

Rewriting the equation, we have

$$r_f = \mu r_S \tag{3.1}$$

Since μ is usually known, the radius of the friction circle, which is an imaginary circle, can be readily found.

Two observations should be noted at this time:

1. The total bearing reaction R, if extended, does not go through the center of the shaft.

2.   The line of action of R is tangent to the side of the friction circle where motion is opposed.

Friction circles are commonly used to determine the forces required to operate rocker arms and eccentric linkages. Eccentric linkages are sometimes referred to as "engine eccentrics" or "Scotch yokes" (see Chapter 6 for a discussion of the Scotch yoke cam). Look at the figures used in Sample Problems 3.1 and 3.2. The objective of Sample Problem 3.1 is to determine the force required to overcome friction and start motion of the rocker arm. The objective of Sample Problem 3.2 is to determine the force that can be delivered by the torque of a rotating shaft.

PROCEDURE

Analyzing Journal Friction by the Friction Circle Method

Step 1 Draw a free-body-diagram (FBD) of the rocker arm or other linkage being investigated.

Step 2 Solve for the required force or forces disregarding friction (this is a check on your final solution). Use the summation of moments method or vector analysis, whichever applies.

Step 3 Calculate the size of the friction circle(s) and sketch them in your FBD.

Step 4 Position the force(s) tangent to the friction circle(s) and then solve for the required force. (This step requires judgment and analysis and will be explained in detail below.)

Remember that friction always opposes motion (or impending motion), thus making it harder to cause the motion. Basically, you have to position the forces on the friction circles to either increase or decrease the moment arms (or torque arms). This calls for good judgment on your part. Generally, if the force *causing* motion acts through a bearing, then it will be placed on the side of the friction circle that *decreases* its moment arm. If a force is resisting the motion, then it is placed on the side of the friction circle that *increases* its moment arm. For the force acting at the fulcrum point (pivot point), move the force (and therefore the pivot point) in a direction to *decrease* the moment arm of the force causing motion.

Sample Problems 3.1 and 3.2 illustrate the use of friction circles to solve linkage problems. The summation of moments method ($\Sigma M = 0$) is used.

SAMPLE PROBLEM 3.1

Determination of Forces on a Rocker Arm

PROBLEM: Refer to the sketch of the setup in Figure 3.6, and determine the force P necessary to start motion of the rocker arm in a counterclockwise

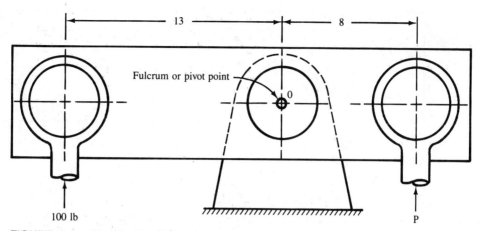

FIGURE 3.6   Sketch of rocker arm setup

(CCW) direction. The coefficient of friction is 0.3, and all bearings are 1.5 in. in diameter. Assume that all information is accurate to three significant figures.

Solution

The summation of moments about the fulcrum point (pivot point) will be used.

Step 1   See Figure 3.7.

Step 2 $\Sigma M_0 = 0 = +100 \times 13 - P \times 8$

$P = 162.5$ lb   (Round to $P = 163$ lb.)

Step 3   Radius of friction circle $(r_f) = \mu \times r_S$. Thus

$$r_f = 0.3 \times \frac{1.5}{2} = 0.225 \text{ in.}$$

Step 4   The FBD is redrawn with friction circles and the forces relocated, as shown in Figure 3.8.

$\Sigma M_0 = 0 = +100 \times 13.45 - P \times 7.55.$
$P = 178$ lb

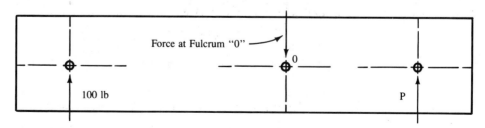

FIGURE 3.7   FBD of rocker arm—no friction circles

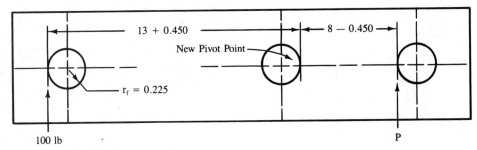

FIGURE 3.8   FBD of rocker arm—with friction circles

SAMPLE PROBLEM 3.2

Determination of Forces Acting on an Engine Eccentric

Refer to the engine eccentric in Figure 3.9. The 7.0 in. diameter circular cam is mounted eccentrically on a rotating shaft. The distance from the center of the shaft to the center of the cam is 2.5 in. The coefficient of friction is 0.35. The rotating shaft can supply a maximum torque of 500 lb-in. What is the maximum resisting force *P* that the shaft torque can overcome when the assembly is in the position shown?

Solution

Step 1   See Figure 3.9.

Step 2   Torque = force × torque arm

$$500 = P \times 2.5$$

$$P = 200 \text{ lb}   (\text{No friction})$$

Step 3 $r_f = \mu \times r_s$ (*Note*: r_s is the radius of the cam.)

$$= 0.35 \times 3.5$$

$$= 1.225 \text{ in.}$$

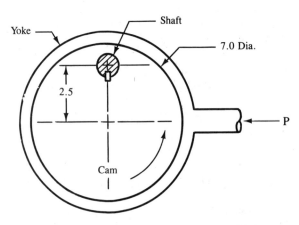

FIGURE 3.9   FBD of engine eccentric—no friction

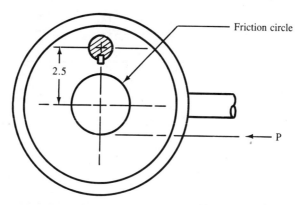

FIGURE 3.10   FBD of engine eccentric—with friction

Step 4 See Figure 3.10.

$$T = F \times D$$
$$500 = F \times (2.5 + 1.225)$$
$$F = 134 \text{ lb}   \text{(With friction.)}$$

3.4 JOURNAL BEARING PRESSURE ▬▬▬▬▬

A rotating shaft usually transmits power by means of devices such as pulleys, cams, clutches, and gears (these devices will be discussed in subsequent chapters). During the power transmission cycle, the shaft absorbs bending loads, which, in turn, are sustained by the supporting bearings, as shown in Figure 3.11. The symbols F_1 and F_2 represent shaft loads that, for example, can result from the tensile forces in the belts of pulleys or from the mating-tooth forces of gears. The bearing reactions R_1 and R_2 are shown as concentrated forces. The effect that these loads have on shafts will be covered in Chapter 5.

Journal bearings have length. They do not, in reality, support a shaft at a single point. Hence a bearing pressure ρ exists rather than a concentrated force R. This is depicted in Figure 3.12, where ρ_1 and ρ_2 are shown as constant-magnitude bearing pressures. L_1 and L_2 represent the lengths of the two journal bearings.

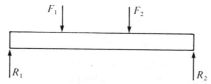

FIGURE 3.11   Shaft loaded as a simply supported beam

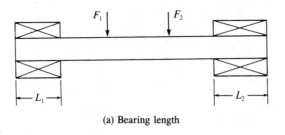

(a) Bearing length

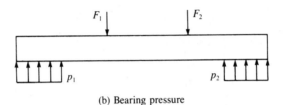

(b) Bearing pressure

FIGURE 3.12 Bearing pressure assumed constant in magnitude

Since the shaft experiences a bending deflection (exaggerated in Figure 3.13a), the bearing pressure will not be uniform but may have a distribution something like that shown in Figure 3.13b. Thus the maximum bearing pressure can be substantially greater than the uniform value. This maximum pressure is the result of not only shaft deflection, but also misalignment between the shaft and supporting housing. A bearing material possessing excellent conformability can reduce the peak pressures and subsequent rapid bearing wear. It should be noted that a very long bearing is not desirable when considering the reality of shaft deflection and shaft-to-housing misalignment.

Because journal bearing pressure distribution is so complex, only the average value (constant magnitude) is normally calculated. From this base, allowable values of pressure are developed for specific applications on the basis of successful operating experience. Figure 3.14 shows the lower half of a bearing with a bearing load R. The

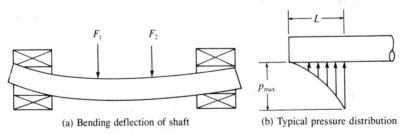

(a) Bending deflection of shaft   (b) Typical pressure distribution

FIGURE 3.13 Effects of shaft deflection: (a) Bending deflection of shaft; (b) Typical pressure distribution

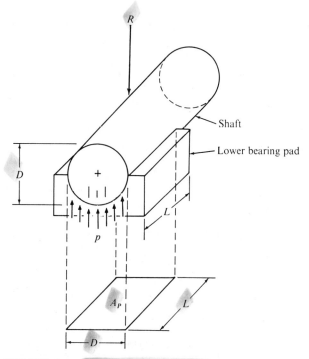

FIGURE 3.14 Bearing projected area

resulting pressure p is assumed uniform not only in the axial direction, but also in the circumferential direction, as shown. Actually, the pressure not only varies with the length of the bearing, as mentioned earlier, but also from side to side, as shown in Figure 3.15. The average pressure is determined on the basis of the bearing projected area A_p, which equals the bearing length L multiplied by the shaft diameter D. Thus the average pressure equals the bearing load divided by the projected bearing area:

$$p = \frac{R}{A_P} = \frac{R}{LD} \tag{3.2}$$

It is common practice to design journal bearings to allowable recommended values. Table 3.2 provides values of allowable pressure and recommended L/D ratios for typical bearing application and also gives the recommended oil viscosities at 100°F.

3.5 LUBRICANT CHARACTERISTICS

A *lubricant* is a special substance whose principal function is to form a film between mating bearing surfaces, preventing metal-to-metal contact. Such lubrication reduces the abrasive action that would cause wear and surface damage. Figure 3.16 shows (in exaggerated form) a lubricant film separating the surface irregularities of a moving

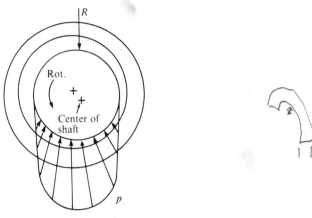

FIGURE 3.15   Typical pressure distribution across the shaft

$\frac{L}{D} = 1.7 =$

$1 = \frac{L}{5}$

$L = 5$

TABLE 3.2   Recommended Design Data for Typical Bearing Applications

Bearing Type	Allowable Pressure (lb/in.²)	Recommended L/D Range	Recommended Viscosity (Z) at 100°F (cP)*
Electric motors	100–200	1.5–2.0	35
Pumps	80–100	1.5–2.0	35
Automobiles:			
Main bearing	500–600	0.5–1.0	130[†]
Crankpin	1000–1200	0.5–1.0	130
Wrist pin	1200–1800	0.8–1.2	130
Air compressors:			
Main bearing	125–250	1.0–2.0	100
Crankpin	300–600	1.0–1.7	100
Wrist pin	500–1000	1.5–2.0	100

*This unit of viscosity, centipoise, will be explained in the next section.
[†]130 cP is close to oil specified as SAE 30.

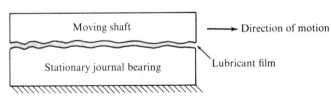

FIGURE 3.16   Separating action of lubricant film

shaft from those of the stationary journal bearing. This complete separation, of course, assumes a stable and adequately thick lubricant film.

The most common type of lubricant is oil, although other types of fluids are used, such as gasoline, water, and air. Grease is also considered a lubricant and is normally used where adequate sealing of a liquid is difficult to achieve and on parts that are not readily accessible.

Also in common use are solid lubricants such as graphite, which comes in either powdered or flake form. When applied to a part, graphite tends to strongly adhere to the surface and thus becomes a very stable lubricant. Quite often graphite is used in applications where high temperatures are encountered because it is stable up to 600°F, a temperature at which most oils and greases burn.

Since oil is the most common lubricant, let us now discuss some of its important characteristics relative to journal bearing design:

— *Pour point:* lowest temperature at which oil will pour
— *Flash point:* temperature at which the vapor above the oil surface will ignite if exposed to a flame
— *Fire point:* temperature at which oil releases enough vapor to support combustion
— *Oiliness:* ability to adhere to a surface
— *Detergency:* ability to clean surfaces of deposits
— *Stability:* ability to resist oxidation, which can produce acids and sludge
— *Foaming:* formation of bubbles in the oil that speed up oxidation
— *Viscosity:* measure of the sluggishness with which a fluid flows
— *Viscosity index:* measure of the sensitivity of an oil's change in viscosity with a change in temperature

Viscosity Measurements. *Viscosity* is probably the least understood (and the most important) of the foregoing fluid properties. There are several measures of viscosity that the technical person must be aware of.

1. *Absolute viscosity (Z)* is a measure of the internal frictional resistance of a fluid to motion. It employs the SI unit centipoise (cP). One poise is equal to one dyne-second per square centimeter, and a centipoise is 1/100 of a poise. The centipoise can be converted to the U.S. Customary System, which employs the unit pound-second per square inch, called the reyn. You may run across the term *reyn* in reference books, but we will not convert to it in this text. Such a value relates the internal shearing stress developed in a fluid to the velocity of the moving machine member and the thickness of the fluid (usually the distance between the machine parts). Absolute viscosity can be obtained under laboratory conditions using the appropriate equations, and it is useful in mathematical relationships (note its use in the graph of Figure 3.20). However, absolute viscosity cannot be readily measured in the shop.

2. *Saybolt viscosity* (SSV) The viscosity of a fluid is commonly measured by a device called a *Saybolt viscosimeter,* which is shown schematically in Figure 3.17. Basically, the viscosimeter consists of an inner chamber containing a sample of the oil to be tested. A separate outer compartment, which completely surrounds the inner chamber, contains a quantity of water whose temperature is controlled by an electric thermostat and heater (not shown). A standard orifice is located at the bottom of the oil chamber.

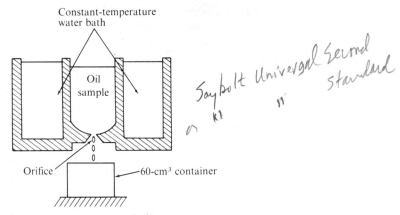

Saybolt Universal Second standard

FIGURE 3.17   Saybolt viscosimeter

When the oil sample is at the desired temperature, the orifice plug is removed. The time it takes to fill a 60 ml container through the metering orifice is then recorded. This time (*t*), measured in seconds, is the viscosity of the oil in official units called *Saybolt seconds universal* (SSU or SUS). In order to convert from Saybolt viscosity to absolute viscosity, another term, *kinematic viscosity,* must be introduced.

Centistoke
cS

3. *Kinematic viscosity* is related to the flow time of a fluid through a viscosimeter capillary. Its symbol is the Greek letter gamma (γ), and it is measured in centistoke (cS). Saybolt viscosity is first converted to kinematic viscosity and then to absolute viscosity. The conversion from Saybolt seconds universal to kinematic viscosity is provided by the following empirical equations:

Time Adjuster

$$\gamma = 0.226t - \frac{195}{t} \quad \text{where } t < 100 \text{ s} \tag{3.3}$$

$$\gamma = 0.220t - \frac{135}{t} \quad \text{where } t > 100 \text{ s} \tag{3.4}$$

where $\quad \gamma$ = the kinematic viscosity in centistoke (cS)

$\quad\quad\quad t$ = the time in Saybolt seconds universal (SSU or SUS)

Kinematic viscosity is defined as absolute viscosity divided by the specific gravity (SG) of the oil. In equation form,

$$\gamma = \frac{Z}{SG} \tag{3.5}$$

where $\quad Z$ = the absolute viscosity in centipoise (cP)

$\quad\quad\quad$ SG = the specific gravity   (unitless)

SSU → y → Z
SSV   cS   cP

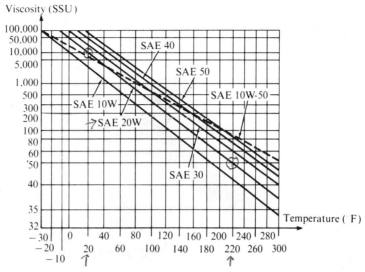

FIGURE 3.18   Viscosity-temperature relationships of motor oils

Usually we know the SSU viscosity and can convert to centistokes with the preceding equations. Also, most oils have a specific gravity of about 0.9. Therefore, we can rearrange Equation 3.5 to find the absolute viscosity:

$$Z = 0.9\gamma \tag{3.6}$$

Note that the viscosity of water at 68°F is 1 cS and also 1 cP (since the specific gravity of water = 1). Also, as general information, the Saybolt viscosity (SSU) of water at 68°F is about 31 s.

Virtually all types of oils become more viscous when the temperature drops, and vice versa. Figure 3.18 shows actual graphs of viscosity versus temperature for some oils graded by the Society of Automotive Engineers (SAE). Oils with the suffix *W* (such as SAE 10W oil) are based on viscosities measured at 0°F. Oils without the suffix are based on measurements taken at 210°F. The *W* suffix merely indicates a winter-grade oil. Also, the higher the SAE number, the more viscous is the oil for the same temperature.

SAE 10W-50 oil is an all-weather, or multigrade, oil that has essentially the same viscosity on a cold day as SAE 10W oil and about the same viscosity at 210°F as SAE 50 oil. This characteristic is shown approximately by the dashed line in Figure 3.18. Relatively speaking, the viscosity of SAE 10W-50 oil does not change as dramatically as SAE 10W or SAE 50 oil with temperature changes. Thus it is a desirable all-weather oil. If an oil with just the 10W classification were used in the summertime when engine oil temperatures could reach 210°F, the oil would become too thin. This results in increased oil consumption because of excessive leaks past bearing and piston ring seals. On the other hand, SAE 50 oil would be too sluggish on a cold (−20°F) day, and the oil pump might not deliver the oil to the bearings during engine warmup. This situation

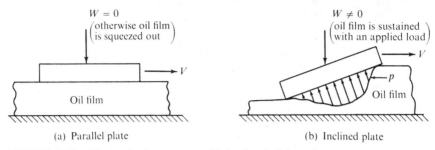

FIGURE 3.19   Hydroplaning occurs with inclined plate only

would result in metal-to-metal contact, which could cause bearing surface damage and increase the load on the starter motor and battery.

3.6   HYDRODYNAMIC ACTION OF JOURNAL BEARINGS ▬▬▬

When a shaft is rotating in a journal bearing in the presence of a lubricant, a phenomenon called *hydroplaning* occurs that can actually cause the shaft to lift off the surface of the bearing. To understand the cause of hydroplaning, let us first examine Figure 3.19*a*. In this figure see a horizontal flat plate moving at velocity *V* and separated from a stationary plate by a thick film of oil. If a load *W* is applied to the plate as shown, the plate lowers, squeezing the oil out until the two surfaces make contact. Thus the plate cannot sustain a load without destroying the oil film.

On the other hand, the plate can be inclined at some angle, as shown in Figure 3.19*b*. Even if a load *W* is applied, the oil film is sustained as a result of the pressure buildup in the film. This pressure, which opposes the load, is produced by the wedging action of the oil. As the inclined plate moves, the oil is forced to flow through a continually decreasing area. This requires pressure. The greater the plate velocity, the faster must be the flow of oil through the wedge-shaped area and hence the greater the pressure buildup and load-carrying capacity. This phenomenon of hydroplaning is a serious concern when an automobile is traveling on a wet road. If the speed is sufficient (usually over 50 mph), the tires can actually lift off the road surface, causing complete loss of steering control.

A journal bearing can be designed so that the shaft hydroplanes at sufficiently high speeds. Figure 3.20 shows a typical curve made during the testing of such a bearing. The vertical axis represents the coefficient of friction μ. The horizontal axis provides values of Zn/p,

where Z = the absolute viscosity in centipoise

n = the shaft speed in revolutions per minute

p = the average bearing pressure in pounds per square inch

The curve is plotted for a given bearing-shaft material combination and r/c ratio, where r is the shaft radius and c is the radial clearance between the shaft and bearing.

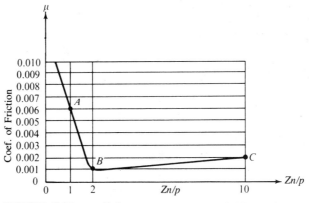

FIGURE 3.20   μ-*Zn/p* curve

The *radial clearance* is defined as one-half the diametral clearance (see Figure 3.2*b*). For hydrodynamic journal bearings, *r/c* usually varies between 500 and 1000. Thus, for a 1 in. diameter shaft, the radial clearance would be between 0.0005 and 0.001 in. Proper clearance is important. As clearance increases, the film thickness increases. Although this will result in a decrease in the coefficient of friction, the film's load-carrying capacity is reduced. On the other hand, too thin a film may cause high spots on the journal and shaft to rub. The *Zn/p* curve in Figure 3.20 is for a steel shaft running in a copper-lead bearing having an *r/c* ratio of approximately 675.

For discussion purposes, let us assume that *Z* and *p* are constant (temperature and load are constant) and that the shaft speed is increased from zero to some maximum value. As a result, the *Zn/p* parameter increases directly with speed. At zero speed, μ equals approximately 0.1 (from Figure 3.20). This situation is depicted in Figure 3.21*a,* where metal-to-metal contact exists. As the shaft starts to turn counterclockwise, it tries to roll up the left side of the journal bearing, as seen in Figure 3.21*b*. As the speed increases, the bearing starts to behave as a small pump because the shaft drags oil that clings to its surface into the narrowing wedge. This produces pressure, which, at low speeds, produces a very small amount of shaft lift-off. This corresponds approximately to point *A* in Figure 3.20. Note in Figure 3.21*b* that only a small pressure buildup opposes the bearing load *R*. A further increase in speed causes increased shaft lift-off, until μ decreases to a minimum value of 0.001, at point *B* in Figure 3.20 (note that *Zn/p* = 2). Finally, the design speed of the shaft is reached at point *C,* where μ = 0.002 and *Zn/p* = 10. The corresponding shaft-bearing relationship is shown in Figure 3.21*c*.

The following observations should be noted at this time.

1. Values given for the μ versus *Zn/p* curve will vary with different material combinations and *r/c* ratios.

2. The high value of μ to the left of point *A* is due to interlocking of surface irregularities, since the oil film may be only a few molecules thick. The principal

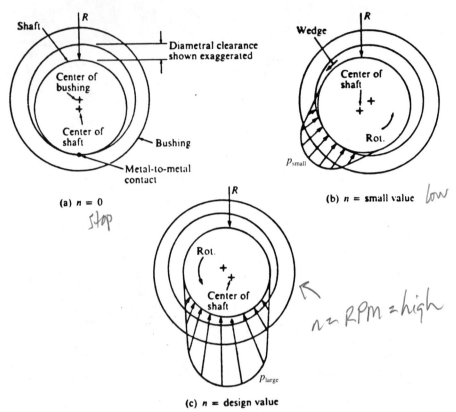

FIGURE 3.21   Hydrodynamic action of journal bearings

cause of friction to the right of point A is the viscosity of the oil, since there is essentially no metal-to-metal contact.

3. Point B is where μ is at a minimum. However, this is a very unstable operating point, since any sudden increase in load or temperature would drop the Zn/p value to a dangerously low level where bearing seizure might occur. A factor of safety of at least 5 is considered good practice. Thus point C would be the minimum design point for continuous operation. Please bear in mind that Figure 3.20 applies to a specific bearing and that other bearings and applications can have vastly different Zn/p values for a given friction coefficient. Safety factors up to 100 would not increase the friction very much.

4. Operation to the left of point B is generally called *boundary, thin-film,* or *imperfect lubrication* because the film of oil that exists is very thin and unstable. Perfect lubrication is experienced to the right of point B because of the existence of a thick, stable oil film. This region is called *thick-film lubrication.*

5. The shape of the Zn/p curve from B to C indicates a gradual increase in μ.

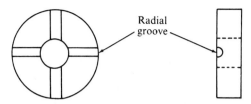

FIGURE 3.22   Washer-type thrust bearing

6. Perfect lubrication is enhanced by the following:
— Low temperature (therefore high viscosity)
— Low loads (For example, sudden acceleration from a standstill is undesirable for the main bearings of the automotive engine because it hastens wear and produces premature bearing knock due to an out-of-round journal.)
— Adequate supply of oil
— High speed

The *Zn/p* curve is used primarily to develop data tables recommending pressures and viscosities, as in Table 3.2. Problem 8 at the end of this chapter asks you to use the data in Table 3.2 to find the probable *Zn/p* values for various applications.

3.7 SLIDING-CONTACT THRUST BEARINGS

The simplest type of sliding-contact thrust bearing is a flat, cylindrical washer, as shown in Figure 3.22. The purpose of the radial grooves is to distribute the lubricant uniformly over the mating surfaces between the shaft and thrust bearing.

Thrust bearings are typically placed at the end of a shaft to absorb axial or thrust loads, as shown in Figure 3.23a. Another arrangement is to transmit the axial force through collars attached to the shaft, as illustrated in Figure 3.23b.

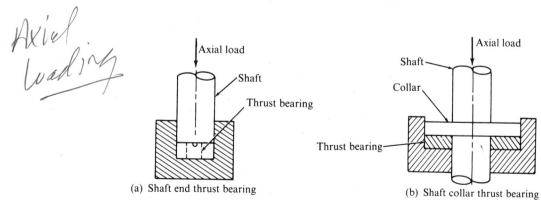

(a) Shaft end thrust bearing   (b) Shaft collar thrust bearing

FIGURE 3.23   Two thrust bearing arrangements

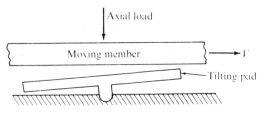

FIGURE 3.24   Tilting-pad thrust bearing

Since the flat washer-type thrust bearing does not produce hydrodynamic action, a thick film of oil is not produced. Therefore, this type of thrust bearing experiences metal-to-metal contact and is limited to low speeds and pressures.

Figure 3.24 shows a hydrodynamic type of thrust bearing called a *tilting-pad bearing*. Between the moving and stationary members of the bearing is a pad that can pivot to provide the desired wedging action for proper hydroplaning. This type of thrust bearing enjoys thick-film lubrication and thus can operate at high speeds and loads. It is used widely on steam and water turbines, as well as on centrifugal pumps.

An air thrust bearing is illustrated in Figure 3.25. As shown, an air line is attached to the bottom of the bearing. The air, under pressure, flows between the end of the shaft and the bearing and out to the atmosphere through the radial clearance. The film of pressurized air acts as a lubricant and separates the shaft from the bearing. The following facts should be taken into account when using air bearings:

1.   The air must be free of dirt, which could damage the bearing or shaft surface finish.
2.   Failure of the air supply system eliminates the supply of air and results in dangerous metal-to-metal contact.

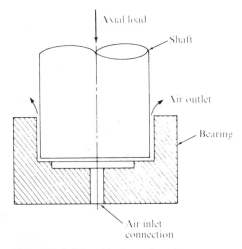

FIGURE 3.25   Air thrust bearing

3. Air is compressible, and therefore, an air bearing cannot sustain heavy loads.

If a liquid such as oil is used instead of air, very large load-carrying capacities can be developed because of the incompressibility of the oil and the high oil-pressure capabilities of positive-displacement fluid power pumps. Thrust bearings and journal bearings that employ oil supplied at a high pressure are called *hydrostatic bearings*. These bearing systems are more expensive than the other bearings mentioned here because of the oil pump, high-pressure piping, and leak-free drain systems required. However, a big advantage of such a system is that sudden increases in load can be sustained. As the load increases, the axial clearance starts to decrease. This action, however, increases the resistance to oil flow through the bearing. Thus the oil pressure builds up until the higher load is overcome. Even though sudden increases in load do reduce the film thickness, proper design can prevent metal-to-metal contact from occurring. Because of the high load-carrying capacities and low friction, hydrostatic oil bearings are finding increasing applications in large numerical control (N/C) machining centers.

3.8 POROUS BEARINGS

Porous bearings are self-lubricating bearings that are made by the powdered-metal process. Powdered metal, usually bronze, is heavily compressed in a die cavity until solid shapes are formed. These solid parts are then sintered in a temperature-controlled furnace. Tolerances of 0.001 in. can be obtained without performing additional machining operations. The finished bearing is porous, containing about 20% air pores and 80% metal by volume. The air pores are then filled with a lubricating oil. When a shaft starts to rotate in such a bearing, the metal-to-metal contact heats the bearing and the oil. This heating causes the oil to expand and ooze out of the pores onto the shaft. In this way, the entire bearing area of the shaft is lubricated. When the shaft stops turning, the bearing cools down. As the bearing cools, the oil cools and is drawn back into the bearing by capillary action.

Usually, the oil contained in the bearing will last for the life of the bearing. However, in some cases, provision is made to add oil to the bearing when necessary.

Porous bearings are not as strong as bearings made of solid metal. Hence they are normally used in light-load applications as either radial or thrust bearings. Two important advantages of porous bearings are as follows:

1. They eliminate oil drip, which could contaminate the product (e.g., food and clothing).
2. They eliminate oil leakage, which could be a fire hazard in applications where heat or steam is nearby.

3.9 SELF-ALIGNING JOURNAL BEARINGS

As mentioned earlier, shafts that transmit power absorb bending loads that cause shaft deflection between bearing supports. This misalignment can load a rigidly supported

FIGURE 3.26   Unibal spherical bearing *(Courtesy of Heim Universal Division, North American Rockwell)*

bearing with an extremely large peak pressure (see Figure 3.13). Under such a condition, the bearing is not able to develop a sufficiently thick oil film, thereby causing premature wear and seizure. The problem can be reduced by adding more closely spaced bearings and increasing the diameter of the shaft. However, this solution is very expensive. A better approach is to use a self-aligning bearing, as illustrated by Figure 3.26. This figure shows a Unibal spherical bearing (full and cutaway views). Observe the oil hole for adding the lubricant. A lubricating groove (not shown) extends all the way around the outer surface of the movable ball. Angular misalignments of up to ± 30° can be readily accepted by the Unibal spherical bearing.

3.10 HEAT GENERATION IN JOURNAL BEARINGS ▰▰▰▰▰

Even with perfect lubrication, friction is always present in journal bearings, and thus heat is generated. If this heat is not removed from the bearing, the temperatures can become excessive. In general, bearing temperature should not exceed approximately 180°F for the following reasons:

1. The oil viscosity becomes too low, a condition that increases oil leakage past seals and also lowers the Zn/p parameter below the thick-film region.
2. The oil may oxidize, damaging its lubrication features.
3. The r/c ratio may be adversely affected, since the coefficient of thermal expansion is generally not the same for the shaft and bearing materials.

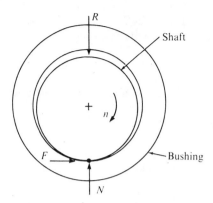

FIGURE 3.27   Parameters affecting heat generation

A set of useful equations for determining the effects of heat generation will now be developed by referring to Figure 3.27 and observing the following symbols and their definitions,

where

R = the bearing load in pounds

N = the reaction force that is equal and opposite to R

n = the shaft speed in revolutions per minute

D = the shaft diameter in inches

L = the bearing length in inches

μ = the coefficient of friction

F = the frictional force in pounds

T_L = the torque loss in pound·inches

hp_L = the horsepower loss = 33,000 ft-lb/min

Q_T = the total heat generation rate in Btu per minute

Q_R = the heat radiation rate in Btu per minute

t_o = the bearing temperature or temperature of hot oil in degrees F

t_{amb} = the ambient temperature in degrees F

k = an empirically determined constant

Q_{oil} = the heat absorption rate of oil in Btu per minute

m = the flow rate of oil in pounds per minute

m' = the flow rate of oil in gallons per minute (GPM)

C_P = the specific heat of oil (Btu/lb°F) in Btu per pound-degree F

ΔT = the temperature change of oil in degrees F

t_{co} = the temperature of cool oil in degrees F

The frictional force F is defined by $F = \mu N$. When we apply this general formula to journal bearing problems, the formula becomes

$$F = \mu R \tag{3.7}$$

Since torque equals force multiplied by moment arm, the torque loss becomes

$$T_L = \mu R \frac{D}{2} \tag{3.8}$$

Now the horsepower loss can be found:

$$\text{hp}_L = \frac{T_L n}{63,000} = \frac{\mu R (D/2) n}{63,000} = \frac{\mu R D n}{126,000} \tag{3.9}$$

The heat generation rate Q_T is equal to the horsepower loss, so to get the horsepower loss back into units of footpounds per minute and then into units of Btu's per minute, we must multiply horsepower by 33,000 ft-lb/min-hp and divide by 778 ft-lb/Btu:

$$Q_T = \frac{\mu R D n}{126,000} \times \frac{33,000}{778} = \frac{\mu R D n}{2970} \tag{3.10}$$

When a bearing temperature rises above the ambient temperature, the bearing radiates heat to the environment. This natural method of cooling bearings may be adequate to prevent excessive temperatures. The following gives an empirical relationship for the heat radiation rate Q_R of a journal bearing:

$$Q_R = \frac{(t_0 - t_{amb} + 33)^2 LD}{778k} \tag{3.11}$$

where $k = 31$ for heavy-duty bearings with good ventilation

$k = 55$ for light-duty bearings with poor ventilation

and Q_R and Q_T have the same units (Btu/min).

When the heat-radiation rate equals the heat-generation rate ($Q_R = Q_T$), we have a steady-state condition in which the bearing temperature reaches its highest value. This can be seen in Figure 3.28. As the bearing temperature increases, the heat-radiation rate (Q_R) increases parabolically. However, the heat-generation rate (Q_T) is not a function of bearing temperature and hence is shown as a horizontal line. Where the two curves intersect is the steady-state bearing temperature (t_o).

By equating Q_T and Q_R, we arrive at an equation for the steady-state bearing temperature when oil is not used as a coolant:

$$Q_T = Q_R$$

$$\frac{\mu r D n}{2970} = \frac{(t_0 - t_{amb} + 33)^2 LD}{778k} \tag{3.12}$$

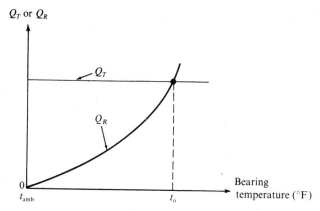

FIGURE 3.28   Relationship of Q_T versus Q_R

Solving for t_o,

$$t_0 - t_{amb} + 33 = \sqrt{\frac{778\mu RDnk}{2970LD}}$$

$$t_0 = \sqrt{\frac{0.262\mu Rnk}{L}} + t_{amb} - 33 \qquad \textbf{(3.13)}$$

If the steady-state temperature resulting from radiational cooling alone is excessive, the bearing temperature can be reduced to a safe level by providing excess oil circulation. Such a system is shown in Figure 3.29, where an abundant supply of cool oil is fed into the bearing in a closed recirculating system. As the oil flows through the bearing, it absorbs heat. The hot oil discharging from the bearing then goes to an oil cooler where the heat is extracted.

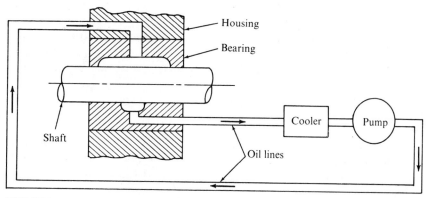

FIGURE 3.29   Oil-cooled bearing system

The heat-absorption rate of the oil can be found using

$$Q_{\text{oil}} = mC_P\Delta T \tag{3.14}$$

The steady-state temperature of an oil-cooled bearing can be found by the following two methods:

Method 1. *Oil cooling (ignoring the benefit of radiational cooling)* Ignoring the benefit of radiational cooling is a very conservative method, and it means that the heat-generation rate equals the heat-absorption rate of the oil for steady-state conditions:

$$Q_T = Q_{\text{oil}}$$

$$\frac{\mu RDn}{2970} = mC_P\Delta T \tag{3.15}$$

This equality is rewritten to solve for m:

$$m = \frac{\mu RDn}{2970}\frac{1}{C_P\Delta T} \tag{3.16}$$

Since 1 gallon of oil weighs about 7.4 lb, the pump flow rate in gallons per minute (GPM) can be found by dividing m by 7.4:

$$m' = \frac{m}{7.4} \tag{3.17}$$

If we assume that the temperature of the hot oil equals the bearing temperature (a conservative assumption), we have

$$\Delta T = t_0 - t_{\text{co}} \tag{3.18}$$

A solution for the bearing temperature is obtained by substituting Equation 3.18 into Equation 3.15:

$$t_0 = \frac{\mu RDn}{2970mC_P} + t_{\text{co}} \tag{3.19}$$

Method 2. *Oil cooling and radiational cooling.* Taking both into account means that the heat-radiation rate plus the heat-absorption rate of the oil for steady-state conditions must equal the total heat-generation rate. Using a consistent set of units of Btu's per minute, we have

$$Q_R + Q_{\text{oil}} = Q_T$$

$$\frac{(t_0 - t_{amb} = 33)^2 LD}{778k} + mC_P\Delta T = \frac{\mu RDn}{2970}$$

Rewritting to solve for m, we have

$$m = \frac{1}{C_P\Delta T}\left[\frac{\mu RDn}{2970} - \frac{(t_0 - t_{amb} + 33)^2 LD}{778k}\right] \qquad (3.20)$$

PROCEDURE

Solving Journal Bearing Heating Problems

Step 1. From the data given and questions asked, determine which of the equations is needed.

Step 2. Substitute the known values into the equation and solve for the unknown.

SAMPLE PROBLEM 3.3

Journal Bearing Temperature—Radiational Cooling

PROBLEM: A light-duty journal bearing operates in still air at an ambient temperature of 70°F and sustains a load of 1000 lb. A shaft rotates inside the bearing at 1000 rpm. $\mu = 0.0382$; $D = 2$ in.; $L = 5$ in. Find the steady-state bearing temperature.

Solution

Step 1   Equation 3.12 is the proper one because oil is not used to cool the bearing.

Step 2 $t_0 = \sqrt{\dfrac{0.262\mu Rnk}{L}} + t_{amb} - 33$

$\quad = \sqrt{\dfrac{0.262 \times 0.0382 \times 1000 \times 1000 \times 55}{5}} + 70 - 33$

$\quad = 332 + 70 - 33$

$\quad = 369°F$

SAMPLE PROBLEM 3.4

Journal Bearing Oil Flow Rate

PROBLEM: It is decided to provide excess oil circulation to the bearing in Sample Problem 3.3 in order to reduce the operating temperature to 140°F. Ignoring the benefit of radiational cooling, what oil flow rate must the pump produce? $C_p = 0.5$; $t_{co} = 70°F$.

Solution

Step 1   Looking through our choices, we see that Equations 3.15 and 3.16 are appropriate for this problem.

Step 2 $m = \dfrac{\mu R D n}{2970} \times \dfrac{1}{C_P \Delta T} = \dfrac{0.0382 \times 1000 \times 2 \times 1000}{2790} \times \dfrac{1}{0.5(140 - 70)}$

$m = 0.735 \text{ lb/min}$

$m' = \dfrac{0.735}{7.4} = 0.10 \text{ gal/min}$

3.11   SI UNITS ▰▰▰

Problems dealing with friction circles and journal bearing pressure can be readily handled in the SI system. However, we will not involve you with SI solutions to heat-generating problems because some empirical factors are used that may not apply in the SI system. Sample Problem 3.5 illustrates how a journal bearing pressure problem is handled.

SAMPLE PROBLEM 3.5

Journal Bearing Pressure Using SI Units

PROBLEM: Each crankpin on an air compressor supports a load of 50 kg. The bearing diameter is 20 mm. Use the maximum recommended *L/D* range in Table 3.2, and determine the pressure and the length of the bearing.

Solution

Note that we must convert kilograms to newtons and millimeters to meters.

$$50 \text{ kg} \times 9.81 = 490.5 \text{ N}$$

From Table 3.2 the maximum *L/D* ratio is 1.7.

$$L = D \times 1.7 = 20 \text{ mm} \times 1.7 = 34 \text{ mm}$$

Equation 3.2:

$$p = \frac{R}{LD} = \frac{490.5 \text{ N}}{0.034 \text{ m} \times 0.020 \text{ m}} = 721 \text{ kPa}$$

3.12   SUMMARY ▰▰▰

A journal bearing is simply a journal, or shaft, rotating in a sleeve. The bearing is classified as a sliding-contact bearing. The choice of bearing material involves compromises between many desired features.

Friction is always present, even with perfect lubrication. The friction analysis of bearings is based on the fundamental formula for the coefficient of friction:

$$\mu = \frac{F}{N}$$

Friction circles are most commonly used to determine the forces required to operate rocker arms and other linkage mechanisms employing journal bearings.

The journal bearing average pressure is equal to the bearing load divided by the projected bearing area:

$$p = \frac{R}{A_P}$$

The principal function of a lubricant in a journal bearing is to form a film between the bearing surfaces, preventing metal-to-metal contact. Common lubricants are oils, greases, and solid lubricants such as graphite.

Viscosity is a measure of sluggishness with which a fluid flows. The most common method of measuring viscosity is with a Saybolt viscosimeter, and the unit of measure is the Saybolt second universal (SSU). The metric system unit of viscosity is the centistoke (cS). Virtually all types of oils become more viscous when the temperature drops, and vice versa.

Most journal bearings operate at low coefficients of friction because they rely on the hydrodynamic action present when oil or another fluid is the lubricant. This action causes the oil to squeeze between the journal and the bearing, eliminating metal-to-metal contact. However, some journal bearings that operate at low speeds or have a reciprocating action do not obtain hydrodynamic action, and the forces involved are frequently determined using friction circle analysis.

The friction in a bearing, even with perfect lubrication, generates heat. This heat must be dissipated or the lubricant and bearing may be damaged. Bearings lubricated with oil should not be allowed to heat in excess of 180°F.

Thrust bearings are used to absorb axial or thrust loads. Most thrust bearings and some journal bearings do not develop hydrodynamic action. Therefore, oil under pressure must be pumped into the bearing to provide a film between the rubbing surfaces. Bearings that employ oil supplied at a high pressure are called *hydrostatic bearings.*

Porous bearings are impregnated with oil and are self-lubricating. They are most commonly used in fractional horsepower motors and other light machinery.

Self-aligning journal bearings help reduce shaft and bearing wear resulting from excessive shaft deflection or misalignment.

3.13 QUESTIONS AND PROBLEMS ▬▬▬▬▬▬

Questions

1. What is the difference between conformability and embedability relative to bearing materials?
2. What are the two general types of bearings in existence?
3. When would rubber normally be used as a bearing material?
4. Name one disadvantage of a long bearing.
5. What is the principal function of a lubricant in sliding-contact bearings?

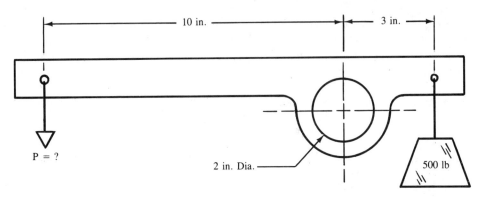

FIGURE 3.30 Sketch for problem 1

6. When would grease be used as a lubricant rather than oil?
7. Under what conditions would graphite be used as a lubricant?
8. Explain what is meant by the designation SAE 10W-40 as a motor oil.
9. Name three factors that enhance perfect hydrodynamic lubrication.
10. What causes the shaft of a hydrodynamic bearing to lift off its bearing support surface?
11. Why is it that the coefficient of friction in a journal bearing can vary approximately from a low of 0.001 to a high of 0.1?
12. Refer to Figure 3.20. Why shouldn't the bearing be designed to operate at the minimum coefficient of friction—at point *B?*
13. What is the purpose of a thrust bearing?
14. What desirable feature does a tilting-pad thrust bearing have?
15. What is the essential difference in operation between a hydrostatic and a hydrodynamic bearing?
16. What is an air bearing? Name one disadvantage of an air bearing.
17. Name two advantages of porous bearings.
18. What is a self-aligning journal bearing, and when would it normally be used?
19. Give three reasons for limiting the maximum bearing temperature to approximately 180°F.
20. If a bearing runs too hot, what can be done to reduce its operating temperature?

Problems (* indicates problems that may be solved with the computer program)

1. We are given the information in Figure 3.30, and we know that the coefficient of friction is 0.3 and that counterclockwise motion is impending. Find *P* to overcome friction and start lifting the weight.
2. Use the setup in Sample Problem 3.1, but make the following changes: the coefficient of friction is 0.25; the bearings are 1 in. in diameter, and the applied load is 150 lb. Solve for the force *P.*
3. Refer to Figure 3.31. The rocker arm is supported by a 1 in. diameter journal bearing having a coefficient of friction equal to 0.3. What value of applied force *P* is required such that clockwise rotation of the rocker arm takes place?

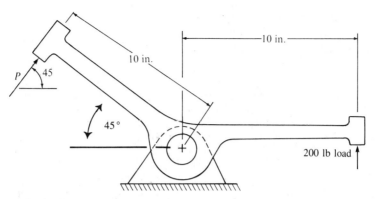

FIGURE 3.31   Sketch for problem 3

4.   Solve Sample Problem 3.2, except that for this problem the shaft has rotated 45° from the position shown.
5.   The journal bearings in Figure 3.32 are to be identical. The allowable bearing pressure is 100 psi, and the L/D ratio is 2.
   (a) Find the length and diameter of the two identical bearings.
   (b) Find the pressure of the lightly loaded bearing. Give one reason for using identical bearings.
6.   An oil has a viscosity of 150 SSU. What is its viscosity in centipoise?
7.   Refer to Figure 3.18, and calculate the kinematic viscosity (in centistokes) and *SAE 20W* the absolute viscosity (in centipoise) of SAE 20 oil at a temperature of (a) 220°F and (b) 20°F.
8.   Obtain the required data from Table 3–2, and calculate the Zn/p values (at 100°F) for electric motors, pumps, the main bearings for automobiles and air compressors. Use the maximum pressures and the following typical speeds: electric motors, 1750 rpm; pumps, 500 rpm; automobiles, 2500 rpm; air compressors, 320 rpm.
* 9.   Solve Sample Problem 3.3, except that for this problem the ambient temperature is 60°F and the shaft rotates at 600 rpm.
*10.   A heavy-duty bearing has good ventilation and, with radiational cooling only, has an operating temperature of 280°F when the ambient temperature is 85°F. The bearing has a diameter of 1.5 in. and a length of 2.5 in. It is decided to cool the bearing using cool excess circulating oil whose cool temperature is 70°F and specific heat is 0.47. What flow rate of oil (GPM) must be supplied to reduce the

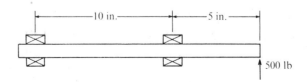

FIGURE 3.32   Sketch for problem 5

bearing temperature to 140°F if (a) the benefit of radiation cooling is ignored, and (b) radiational cooling is taken into account?

*11.   A light-duty bearing operates in still air at an air temperature of 70°F and sustains a load of 1000 lb. A shaft rotates inside the bearing at 1000 rpm. The coefficient of friction is 0.04, the shaft diameter is 1 in., and the bearing length is 4 in. It is decided to provide excess circulating oil to the bearing to reduce its operating temperature to 170°F. The specific heat of the oil is 0.47, and the temperature of the cool oil is 70°F. Assuming no heat loss due to radiation, what flow rate (GPM) of oil is required?

12.   Find the horsepower loss for the bearing in Problem 11.

SI Unit Problems

13.   Solve Student Problem 1 above except use the following data: P is 200 mm from the center of the journal and a 200 kg load is applied 80 mm from the journal. The journal has a diameter of 45 mm. The coefficient of friction is 0.25.

14.   Solve Sample Problem 3.1 except use the following data: The coefficient of friction is 0.16, the load is 130 kg, and all bearings have diameters of 40 mm. The load is 350 mm from the fulcrum point, and P is 200 mm from the fulcrum point.

15.   Solve Sample Problem 3.2 except make the following changes in data:
Cam diameter = 150 mm.
Distance from shaft center to cam center = 40 mm
$\mu = 0.28$
Shaft supplies a torque of 60 N·m.

16.   Use the information in Sample Problem 3.4 to determine the total heat-generation rate Q_T in Btu's per minute and then convert this to joules per minute.

17.   Give the answers to Problem 10 above in liters per minute.

4
Antifriction Bearings

Objectives

After completing this chapter, you will

- Be able to calculate the rolling resistance of bearings and wheels.
- Have an understanding of the various types of antifriction bearings and know their advantages and disadvantages.
- Be able to apply the appropriate equations when selecting bearings for your application.

4.1   INTRODUCTION

Antifriction bearings operate with rolling elements (either balls or rollers), and hence rolling resistance, rather than sliding friction, predominates. The cause of rolling resistance is the deformation of mating surfaces of the rolling element and the raceway on which it rolls. The three basic types of antifriction bearings are

—   Ball bearings
—   Roller bearings
—   Needle bearings

Ball and roller bearings can be designed to absorb thrust loads in addition to radial loads, whereas needle bearings are limited to radial load applications only.

One of the principal advantages of antifriction (rolling contact) bearings is the almost complete elimination of friction. Therefore, the main function of antifriction bearing lubricants is to help prevent corrosion and dirt contamination. Since some friction does exist, especially between the rolling elements and their separator, the lubricant also must remove the heat caused by this sliding action. As a result, although there is minimal friction present, a lubricant is still absolutely essential.

Fatigue, which is the cause of failure of a properly lubricated antifriction bearing, is a stress-reversal phenomenon that takes place on the contacting surfaces of the raceways and rolling elements. Hence the life of an antifriction bearing is measured by the number of shaft revolutions that occur prior to fatigue failure. The greater the bearing load, the shorter will be the life of the bearing in revolutions. Also, the greater the shaft speed, the shorter will be the bearing life. Since fatigue is a statistical occurrence, the percent probability of failure also must be considered.

It is essential to know the advantages and disadvantages of the various types of bearings, as well as their salient features. For example, ball and roller bearings can be of the self-aligning type to accommodate shaft misalignment and deflections. In addition, shields and seals are available to protect bearings from foreign contaminants. Some antifriction bearings contain a permanently sealed lubricant that is administered by the manufacturer and lasts for the life of the bearing.

Manufacturers of antifriction bearings produce a multitude of bearing types and sizes. Handbooks are available that tabulate the various types and sizes along with recommended speeds and loads. Thus the principal problem is not the design of an antifriction bearing, but the selection of the proper bearing for a given application.

4.2   ROLLING RESISTANCE

A rolling element experiences opposition to motion just as if sliding friction were present. However, there is essentially no sliding friction. Rolling resistance is in reality a result of deformation.

Figure 4.1 shows a ball or cylinder with a downward load resting on a horizontal flat surface. The load is represented by the vector P in Figure 4.1. To start rolling action, a force Q is required; the magnitude of Q is the amount needed to overcome the

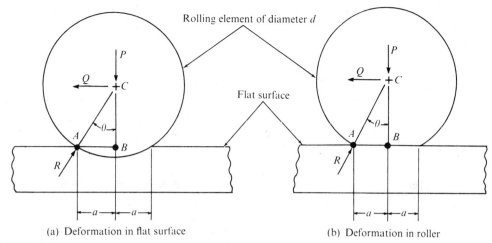

(a) Deformation in flat surface   (b) Deformation in roller

FIGURE 4.1   Example of rolling resistance

rolling resistance. Figure 4.1a shows the flat surface being indented by a perfectly rigid rolling element. Figure 4.1b shows all the deformation occurring in the rolling element. This is the case when the ball or cylinder is rolling on a perfectly rigid surface. An example is an automobile tire rolling on a concrete road. Most of the deformation occurs in the rubber tire.

In the case of bearings, the deformation occurs to varying degrees in both the roller and the surface on which it rolls (called a *raceway*). For example, in ball and roller bearings, the deformation in the rolling element is about equal to the deformation in the inner and outer raceways, usually less than 0.001 in., which explains why the rolling resistance is so small. Of course, the actual deformation increases as the load increases. To minimize the deformation, the rolling elements and the raceways are typically made of a high carbon-chromium steel (1% carbon, 1.5% chromium). The elements are then hardened to approximately Rockwell C58 and ground to a 6 μin. finish. This high hardness corresponds to an ultimate strength of 300,000 lb/in.^{2}

Using Figure 4.1a as a model, let us determine the mathematical relationship that defines the magnitude of the force Q. In order to get a mental picture of the action, imagine that you are pulling the wheel of a wagon over a curb stone and point A is at the edge of the curb. Similarly, the ball or roller bearing must be pulled out of the hole it is in. The reaction R between the roller and flat surface must be inclined such that its line of action goes through point C, the center of the roller. This is so because we have a three-force system (P, Q, and R) and three nonparallel forces acting on an object (in equilibrium) must be concurrent. The lines of action of the three forces intersect at point C. Rolling resistance exists because the roller must move against the horizontal component of the reaction force R. If there were no deformation, R would act vertically at point B and there would be no rolling resistance.

The laws of statics tell us that if an object is in equilibrium, the sum of all the moments about any point equals zero. Let us take moments about point A so that the unknown reaction force R is not involved:

$$-\Sigma M_A = 0 = Pa - Q\left(\frac{d}{2}\right)\cos\theta$$

$$Q = \frac{2aP}{d\cos\theta} \tag{4.1}$$

Since θ is very small for hardened steel bearings, $\cos\theta$ very nearly equals unity and therefore can be disregarded. Thus we have

$$0 = Pa - Q\frac{d}{2} \quad \text{or} \quad Q = \frac{2a}{d}P \tag{4.2}$$

where
- Q = the force to overcome rolling resistance in pounds
- $2a$ = the width of contact area in inches
- d = the roller diameter in inches
- P = the applied load in pounds
- a = the coefficient of rolling resistance (by definition)
- θ = the angle between the line of action of R and the vertical

Let us use this information to analyze an industrial application such as hand trucks. We wish to reduce the rolling resistance. Equation 4.2 indicates what should be done to accomplish this: Use large wheels made of hard materials to reduce deformation.

PROCEDURE
Determining the Force Required to Overcome Rolling Resistance

Sum the moments or apply Equation 4.1 for relatively soft materials such as rubber tires. Apply Equation 4.2 for hardened steel bearings.

SAMPLE PROBLEM 4.1
Rolling Resistance

PROBLEM: A 2000 lb machine is to be pushed into a room by rolling it on four pipes equally spaced between the steel bottom of the machine and a concrete floor. The coefficient rolling resistance for each pipe is 0.005 in. on top and 0.010 in. on the bottom. How much force would be required to move the machine if the pipe diameter is 1 in.?

Solution

$$Q = \frac{2a}{d}P_{\text{top}} + \frac{2a}{d}P_{\text{bottom}}$$

$$= \frac{0.010}{1}2000 + \frac{0.020}{1}2000 = 60 \text{ lb}$$

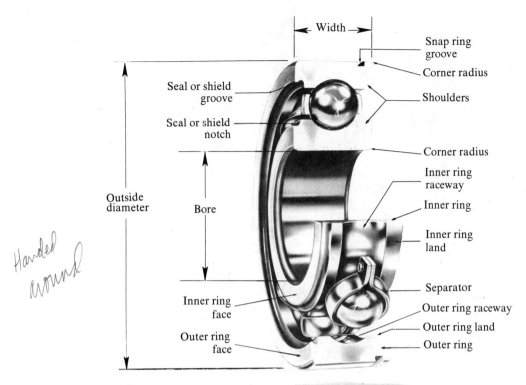

Handed
around

FIGURE 4.2   Terminology of typical single-row ball bearing *(Courtesy of New Departure Hyatt Bearings, Division of General Motors Corporation)*

4.3 BALL BEARINGS

The design of and terminology applicable to a typical single-row ball bearing is illustrated in Figure 4.2. Observe that there are essentially four different components:

— Outer ring, which contains the outer raceway
— Inner ring, which contains the inner raceway
— Complement of balls
— Two-piece separator (also called the *cage* or *retainer*)

This design is called a *Conrad bearing,* and the bearing is assembled by initially moving the inner ring into an eccentric position as illustrated in Figure 4.3*a*. Then as many balls as possible are inserted into the space between the inner and outer rings. (Fig. 4.3*b*). Next, the inner ring is returned to its concentric position and the balls are equally spaced, as shown in Figure 4.3*c*. Finally, the two halves of the separator are positioned from each side and fastened using rivets (Fig. 4.3*d*). The Conrad-type bearing is primarily a radial bearing because only very small thrust loads can be sustained. In addition, the radial capacity is somewhat limited because of the relatively small

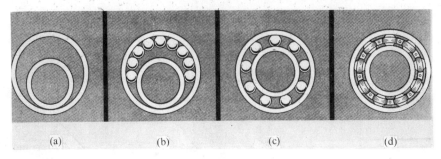

(a)   (b)   (c)   (d)

FIGURE 4.3   Assembly of Conrad-type ball bearing *(Courtesy of New Departure Hyatt Bearings, Division of General Motors Corporation*

number of balls. For example, only slightly more than half the annular space between
the inner and outer rings can be filled in a Conrad design. It should be noted that the
terms *ring* and *race* are used interchangeably.

Figure 4.4 is a second cutaway view of a Conrad bearing, this time without nomenclature details. Some of the dimensional terms are depicted in Figure 4.5 and are
defined as follows;

where W = the width of the bearing

D = the outside diameter (supported in housing)

B = the bore (support for shaft)

r = the corner radius of the inner and outer rings

In Figure 4.6, we see a comparison of the curvatures of the ball and the inner
raceway. There is an exaggerated difference for illustration purposes. If the radii of the
ball and raceway were made exactly equal, there would be maximum area of contact
and hence greater load capacity. However, this would increase bearing friction. Therefore, the radius of the ball is made slightly smaller than the radius of the raceway.

FIGURE 4.4   Cutaway view of Conrad ball bearing *(Courtesy of New Departure Hyatt Bearings, Division of General Motors Corporation)*

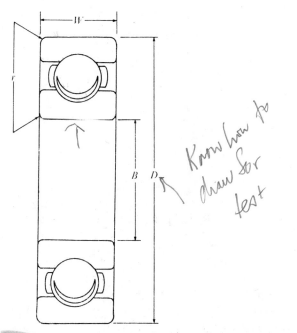

FIGURE 4.5   Dimensional nomenclature of ball bearing
(Courtesy of New Departure Hyatt Bearings, Division of General Motors Corporation)

A second type of ball bearing, the *maximum-capacity design,* is readily recognized by its two filling notches, or grooves, as shown in Figure 4.7. A maximum number of balls can be inserted in this type of bearing because of the filling notches. Initially,

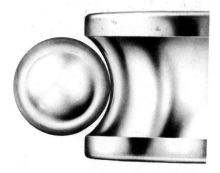

FIGURE 4.6   Conformity of ball and race *(Courtesy of New Departure Hyatt Bearings, Division of General Motors Corporation)*

FIGURE 4.7   Maximum-capacity ball bearing with filling
notches *(Courtesy of New Departure Hyatt
Bearings, Division of General Motors Corporation)*

some of the balls are inserted with the rings positioned eccentrically; then the remainder are inserted through the filling notches with the rings located in a concentric position. The maximum capacity ball bearing has a greater radial load-carrying capacity
than the Conrad bearing because of the greater number of balls.

The shape of the two filling notches is illustrated in Figure 4.8. Observe that the
loading grooves do not extend to the opposite side of the raceway. Therefore, this bearing can carry a small but significant amount of thrust load.

A ball bearing also can be supplied with shields, as illustrated in Figure 4.9. Notice the shield groove in the outer ring and the shield notch in the inner ring. The
subject of shields and seals will be discussed in Section 4.8.

In Figure 4.10 we see an *angular-contact bearing.* Observe that one of the shoulders of the outer ring has been machined thin. The balls are usually inserted by thermally expanding the outer ring, and hence, a full complement of balls can be

Loading grooves

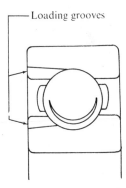

FIGURE 4.8   Loading grooves of maximum capacity ball
bearing

FIGURE 4.9   Ball bearing with shields *(Courtesy of New Departure Hyatt Bearings, Division of General Motors Corporation)*

assembled. After cooling, the bearing cannot be taken apart. The single-row angular contact bearing can take a significantly greater thrust load than Conrad or maximum capacity ball bearings, but in one direction only. Figure 4.11 shows that the thrust comes into the heavier shoulder of the outer ring and then goes through the ball along the contact angle and out the bearing through the opposite-side inner-ring shoulder.

In Figure 4.12 we see a double-row angular contact bearing, which can absorb thrust in both directions. As illustrated, this bearing has two ball raceways, two rows of balls, and two distinct separators. This design essentially represents two opposed angular contact bearings in a single assembly, as depicted in Figure 4.13. This type of bearing can support combinations of heavy radial and thrust loads.

Figures 4.14 and 4.15 illustrate a two-piece inner-ring ball bearing. It combines all the outstanding features of the Conrad, maximum capacity, and angular contact bearing types. Observe the one-piece symmetrical separator that provides good

FIGURE 4.10   Angular-contact ball bearing *(Courtesy of New Departure Hyatt Bearings, Division of General Motors Corporation)*

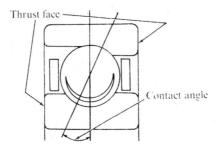

FIGURE 4.11   Single-thrust   direction   of   single-row
angular contact bearing *(Courtesy of New
Departure Hyatt Bearings, Division of
General Motors Corporation)*

FIGURE 4.12   Double-row angular contact bearing *(Courtesy of New Departure Hyatt Bearings, Division of General Motors Corporation)*

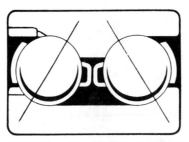

FIGURE 4.13   Double-thrust directions of a double-row
angular contact bearing *(Courtesy of New
Departure Hyatt Bearings, Division of
General Motors Corporation)*

FIGURE 4.14   Two-piece inner-ring ball bearing *(Courtesy of New Departure Hyatt Bearings, Division of General Motors Corporation)*

dynamic balance with great strength and rigidity. Because of the two-piece inner-ring design, the following advantages are obtained:

1.   The maximum complement of balls makes for higher load-carrying capacity.

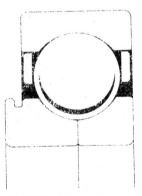

FIGURE 4.15   Cross section of two-piece inner-ring ball bearing *(Courtesy of New Departure Hyatt Bearings, Division of General Motors Corporation)*

2.   The very deep ball raceway shoulder makes for high thrust-load capacity.

3.   Thrust loads can be handled in both directions.

4.4   ROLLER BEARINGS

When severe shock or heavy loads are encountered, roller bearings are frequently employed instead of ball bearings. Roller bearings can be classified into three basic types depending on which of the following roller shapes they possess:

—   Cylindrical
—   Tapered
—   Spherical

Roller bearings can sustain much greater loads than ball bearings because roller bearings experience line contact, whereas ball bearings have point contact. As a result, roller bearings have more rolling resistance. For example, the average coefficient of rolling resistance is 0.0008 for a ball bearing and 0.0015 for a roller bearing.

The cylindrical roller bearing is designed to sustain primarily radial loads. Figure 4.16 illustrates a single-row cylindrical roller bearing. The rollers are cold-formed from triple-alloy carburized steel and then are hardened, ground, and surface-finished to about 4 μin. The separator is of a rigid segmented design so that it provides high strength and positive roller spacing. Moreover, although the rollers are guided by a flanged outer race, they are free to slide axially on the inner race (see Fig. 4.17a. This means that the design shown cannot absorb any thrust. On the other hand, if flanges are included in both races (Fig. 4.17b), a thrust load of about 15% of the radial load

FIGURE 4.16   Single-row cylindrical roller bearing *(Courtesy of Link-Belt Bearing Division, FMC Corporation)*

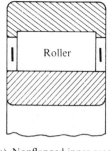

(a) Nonflanged inner race

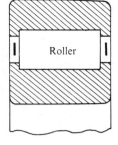

(b) Flanged inner race

FIGURE 4.17   Cylindrical roller bearings

can be sustained. Observe, however, that this thrust load is absorbed by sliding friction, which increases resistance to motion and generates heat.

Tapered roller bearings can simultaneously sustain both large radial and large thrust loads. Figure 4.18 shows the tapered roller and gives the nomenclature specifically used for tapered roller bearings. Observe that a tapered roller bearing can be readily disassembled. Various values for the conical angle can be obtained. As the conical angle increases, the thrust-load capacity likewise increases. However, an increased conical angle reduces the radial load capacity. One very popular application of tapered roller bearings is wheel bearings in automobiles.

Spherical roller bearings use spherical-shaped rollers and hence are self-aligning. As a result, they can accommodate a small amount of misalignment between the shaft

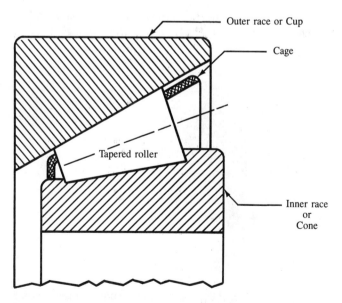

FIGURE 4.18   Nomenclature of tapered roller bearing

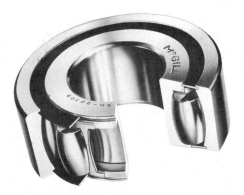

FIGURE 4.19   Single-row spherical roller bearing *(Courtesy of McGill Manufacturing Company, Inc.)*

and housing. Figure 4.19 illustrates a single-row spherical roller bearing with internal seals. This design can accommodate misalignments up to $\pm 3°$ with unsealed versions and $\pm 2°$ with sealed versions. Figure 4.20 illustrates a double-row spherical roller bearing that provides extra heavy capacity as well as the self-aligning feature. This bearing type is used extensively on earth-moving equipment, steel mill equipment, printing presses, and torque converters.

4.5   NEEDLE BEARINGS

Needle bearings use rollers and hence are very similar to roller bearings. The following is a listing of the principal features of needle bearings:

FIGURE 4.20   Double-row spherical roller bearing *(Courtesy of Link-Belt Bearing Division, FMC Corporation)*

FIGURE 4.21   Drawn-cup needle bearing *(Courtesy of the Torrington Company)*

1. Needle bearings use a large number of small-diameter rollers.

2. In needle bearings, there are no spaces between adjacent rollers, and thus no cage or separator is needed.

3. The outer ring (cup) of a needle bearing is sometimes made of a drawn, case-hardened, thin steel shell (see Fig. 4.21). The cylindrical shape of the outer ring is maintained by a rigid supporting housing. The outer ring of a drawn-cup needle bearing is not as thick as that of a machined heavy-duty type. Therefore, the drawn-cup needle bearing cannot sustain as much load. The drawn-cup design, however, is less costly.

4. Some needle bearings do not contain an inner ring, as illustrated in Figure 4.21. In this case, the rollers make direct contact with the shaft. For proper operation, the shaft must be hardened to a recommended value of Rockwell C58. Figure 4.22 shows how rapidly the load-carrying capacity decreases as the shaft hardness is reduced below Rockwell C58.

5. Needle bearings have a large load-capacity/size ratio. The main advantage of this characteristic is the small amount of radial space required. As a result, needle bearings are widely used in roller-type cam followers. Such a design is illustrated in Figure 4.23. As the figure indicates, the outside diameter of the roller can be kept small with the use of needle bearings. Since the roller makes continuous contact with an irregularly shaped cam, it is important that the roller size be minimized so that proper motion is transmitted to the follower. The subject of cams is discussed in Chapter 6.

6. Needle bearings have a higher coefficient of friction than roller bearings. Two contributing factors are (a) the small diameter of the rollers (refer to Equation 4.2) and (b) the fact that the rollers rub against each other. A typical value for needle bearings is 0.0025, as compared with 0.0015 for roller bearings. Consequently, needle bearings have a lower maximum speed limit.

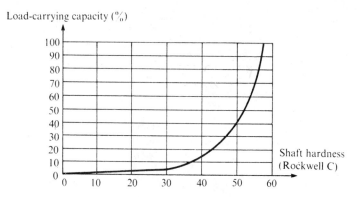

FIGURE 4.22   Load-carrying capacity versus shaft hardness for
a needle bearing running directly on a steel shaft

4.6   THRUST BEARINGS

Thrust bearings are designed essentially to carry only thrust loads. Figure 4.24 shows a ball-type thrust bearing that cannot absorb any radial load. As a result, two radial bearings must be used in conjunction with a pure thrust bearing. It should be noted that each ball in the bearing in Figure 4.24 carries an equal share of the load only if the load is applied along the axis of the bore.

The thrust capacity of a ball-type thrust bearing is very limited as a result of the small contact area. Roller thrust bearings have a much greater thrust capacity. Figure 4.25 illustrates a cylindrical roller thrust bearing that can absorb only thrust loads. The disadvantage of this design is that substantial sliding occurs because the outer ends of the rollers are forced to travel along a larger circle than the inner ends. If tapered rollers are used, this sliding action is replaced by completely pure rolling. Such a bearing is called a *tapered-roller thrust bearing,* and it can absorb only thrust loads. On the other hand, if the rollers are tapered and arranged along a conical surface, the result is

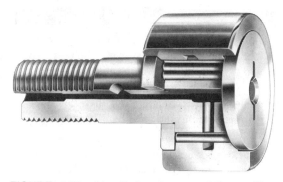

FIGURE 4.23   Needle bearing cam follower *(Courtesy of
The Torrington Company)*

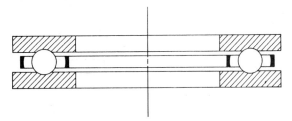

FIGURE 4.24   Ball thrust bearing

a conical thrust bearing, as depicted in Figure 4.26. Such a bearing can absorb some radial load and is commonly used as wheel bearings in automobiles and for supporting the steering columns of automobiles.

4.7 LIFE OF ANTIFRICTION BEARINGS ▬▬▬

Antifriction bearings are subjected to repeated stress cycles that can ultimately lead to fatigue failure. Contact surfaces on the races and rolling elements are loaded in

FIGURE 4.25   Cylindrical roller thrust bearing *(Courtesy of The Torrington Company)*

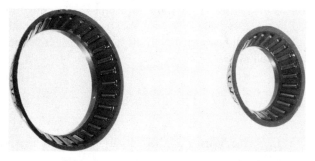

FIGURE 4.26   Conical thrust bearing *(Courtesy of The Torrington Company)*

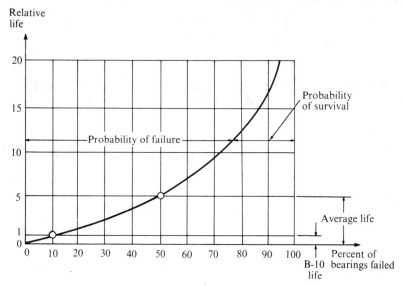

FIGURE 4.27   Typical bearing life-expectancy curve

compressive stress cycles that vary from zero to some maximum value. Fatigue failure is said to occur when a tiny flake of metal is removed from one of the races or the rolling element and is usually preceded by very tiny surface cracks that develop during repeated stress applications.

B-10 Life and Average Life.

The life of a bearing is the number of revolutions or the number of hours at a fixed speed that the bearing will run before fatigue begins. If a large number of identical bearings are tested to failure under specific operating conditions, a life-expectancy curve can be established, as shown in Figure 4.27. Here, the *B-10 life* is defined as the number of hours that 90% of the bearings tested will exceed. Thus, for a large number of bearings tested to failure, the probability of failure is 10% at the B-10 life. The *average life* is defined as the number of hours that 50% of the bearings will exceed. Figure 4.27 shows that the average life is *five times* the B-10 life. Since the bearing contact stresses increase with load, the fatigue life decreases as the load increases. Specifically, the life in revolutions or cycles is proportional to the reciprocal of the load raised to the third power for ball bearings and the reciprocal of the load raised to the 10/3 power for roller bearings. Thus doubling the load reduces the life by 8 times for ball bearings and by 10 times for roller bearings. This relationship is represented as follows:

$$\frac{L_1}{L_2} = \left(\frac{R_2}{R_1}\right)^b \tag{4.3}$$

where L_1 = the bearing life for condition 1 (in millions of revolutions)

             L_2 = the bearing life for condition 2 (in millions of revolutions)

Double the load reduces life 8x

TABLE 4.1   Data for Typical Size Conrad-Type Ball Bearings

Bearing Size Number	Bore Diameter B	Outside Diameter D	Width W	Basic Dynamic Capacity* (lb)	Static Strength Capacity (lb)
1	0.1250	0.3750	0.1562	18	17
2	0.1250	0.5000	0.1719	18	17
3	0.1875	0.5000	0.1562	44	45
4	0.2500	0.6250	0.1960	48	58
5	0.2500	0.7500	0.2188	93	117
6	0.3750	0.8750	0.2188	124	185
7	0.5000	1.1250	0.2500	235	460
8	0.6250	1.3750	0.2812	275	620
9	0.7500	1.6250	0.3125	475	960
10	0.8750	1.8750	0.3750	510	1120
11	1.0000	2.0000	0.3750	510	1120
12	1.1250	2.2150	0.3750	700	1600
13	1.2500	2.2500	0.3750	740	1760
14	1.3750	2.5000	0.4375	870	1980
15	1.5000	2.6250	0.4375	930	2200

*At 90×10^6 revolutions.

R = the load for condition 1 or 2 (lb)

b = 3 for ball bearings, 10/3 for roller bearings

Basic Dynamic Capacity.   Bearing manufacturers must supply sufficient data in catalogs so that technicians and engineers can use Equation 4.3 to select the appropriate bearing for their application. The bearing manufacturers do this by listing the maximum load their bearings can support for a specific number of revolutions. This load is called the *basic dynamic capacity C* of the bearing and is defined (by many bearing manufacturers) as the radial load a bearing can sustain for 1 million revolutions before failure occurs. A statement also must be made as to what percent of the bearings will reach the revolutions specified. Usually, this is the B-10 life. Note that in Table 4.1 the basic dynamic capacity is based on a B-10 life of 90 million revolutions (that is 3000 hours at 500 rpm). The reason why the basic dynamic capacity is based on revolutions rather than hours is that many machines such as the automobile are variable-speed

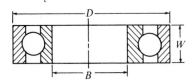

FIGURE 4.28   Bearing dimensions used in Table 4.1.

systems. The phenomenon of fatigue relates uniquely to the number of stress cycles (not to time) because of speed variability.

Equation 4.3 can be rewritten to account for the bearing manufacturer's data:

Conversion to your revolutions formula

$$\frac{L_1}{L_2} = \left(\frac{C}{R}\right)^b$$

(4.4)

where

L_1 = the number of millions of revolutions of bearing life desired

L_2 = the number of millions of revolutions stated in the manufacturer's basic dynamic capacity data

R = the actual bearing load in pounds

C = the basic dynamic capacity in pounds

Specify load and RPMs

The basic dynamic capacity of a bearing depends on whether the inner race or the outer race rotates. Normally, the inner race turns with the shaft. In this case, the same point on the outer race is continuously loaded. However, the stress on the outer race is less than on the inner race because a smaller area of contact exists for the inner race. This difference results from the smaller radius of the inner race and the fact that the curvature of the inner race is opposite to the curvature of the rolling element. Consequently, the effect of a single loaded spot is essentially canceled by the lower stress level. On the other hand, if the outer race rotates, the load is continuously applied to the same point on the higher-stressed inner race. In this case, a correction factor of 1.2 is applied to the actual radial bearing load. For example, a 100 lb load would be considered as a 120 lb load.

In addition, some bearings are subjected to a combination of radial and thrust loads. An equivalent radial load must be determined that has the same effect on bearing life as the combination of radial and thrust loads. An approximate relationship for the equivalent radial load is given by

$$P_E = 0.5R + 1.7T$$

(4.5)

where

P_E = equivalent radial load in pounds

R = actual radial load in pounds

T = actual thrust load in pounds

In the event that Equation 4.5 yields an equivalent load that is less than the radial load, the radial load should be used.

Static Strength Capacity.   Another important parameter is the *static strength capacity*, which is defined as the maximum load that can be applied without bearing damage when neither race is moving. Bearing damage in this case is the amount of permanent deformation of the races and rolling elements (commonly referred to as *brinneling*) that would cause noisy operation when the bearing is subsequently operated. The requirement that noisy operation not occur when a bearing is subsequently operated is normally satisfied if the total permanent deformation does not exceed 0.0001 in. per inch of diameter of the rolling element. If a bearing is rotated very slowly, it can sustain a load several times greater that the static strength capacity because brinneling is much more difficult to produce.

Table 4.1 provides the basic dynamic capacity and the static strength capacity for typical size Conrad-type ball bearings. These data are to be used for sample problems and text exercises only. Equations 4.4 and 4.5 will be used to assist in the selection of bearings from the table. However, it should be noted that when applying data from manufacturers' catalogs, the specific method proposed by the manufacturer must be used. Following the manufacturer's specific instructions is important because calculation methods vary from manufacturer to manufacturer. There are several reasons for this variation:

1. Equation 4.5 is approximate. Different manufacturers use various formulas that more precisely apply to their specific bearings and applications.
2. The factor 1.2 used to correct for outer race rotation is approximate.
3. The tabulated values of basic dynamic capacity are not all based on the same conditions. Most are based on 1 million revolutions (500 h B-10 life at 33⅓ rpm). However, as noted for Table 4.1, other conditions may apply.

PROCEDURE

Selecting Bearings Knowing the Bearing Life and Load

Step 1 If a thrust load is involved, find the equivalent radial load with Equation 4.5.

Step 2 Apply the correction factor of 1.2 to the radial load if the outer race rotates.

Step 3 Apply Equation 4.4, and solve for the basic dynamic capacity C that is required.

Step 4 Enter the data table with this information, and choose the appropriate bearing.

SAMPLE PROBLEM 4.2

Bearing Selection

PROBLEM: Select a ball bearing from Table 4.1 to operate with a 3000 lb radial load and have a B-10 life of 1 million revolutions.

Solution

$L_1 = 1$ (for 1 million revolutions)

$L_2 = 90$ (for 90,000,000 revolutions the basis for the basic dynamic capacity in the table)

C = the basic dynamic capacity desired

R = the required bearing load

Handwritten annotations (left):

load RPms Bearing Size

↑   ↑   ∼   ↑
↓   ↑   →
↑   ↓   →
↓   ↓   ↓

$$\frac{L_1}{L_2} = \left(\frac{C}{R}\right)^3$$

$$\frac{1}{90} = \left(\frac{C}{3000}\right)^3$$

$$C = 669\ lb$$

Handwritten annotations (right):

Bigger Bearing

$$\frac{1}{90} = \left(\frac{700}{3000}\right)^3$$

$$\frac{1}{90} = \left(\frac{C}{4000}\right)^3 \frac{1}{10} = \left(\frac{892}{4000}\right)^3$$

Answer: Use a number 12 bearing.

SAMPLE PROBLEM 4.3

Bearing Selection with Thrust Load Added

PROBLEM: Select a bearing from Table 4.1 to operate with a radial load of 1000 lb and a thrust load of 500 lb. The B-10 life is to be 100 h at 3600 rpm.

Solution:

From Equation 4.5 we obtain the equivalent radial load:

$$P_E = 0.5R + 1.7T$$

$$= 0.5(1000) + 1.7(500) = 1350\ lb$$

(handwritten: $10/3$)

The B-10 life in revolutions = 100 h x 3600 rpm x 60 m/h:

$$B\text{-}10\ life = 21,600,000\ revolutions$$

Using Equation 4.4, we have

$$\frac{L_1}{L_2} = \left(\frac{C}{P_E}\right)^3$$

$$\frac{21.6}{90} = \left(\frac{C}{1350}\right)^3$$

$$C = 1350\left(\frac{21.6}{90}\right)^{1/3} = 839\ lb$$

Answer: Use a number 14 bearing.

The following useful facts about the life of antifriction bearings should be clearly understood at this time:

1.   The load-carrying capacity of an antifriction bearing is roughly proportional to the square of the diameter of the rolling elements. Hence, if a bearing is doubled in size throughout, it can carry four times as much load.

2.   Roller bearings have much greater load-carrying capacities than ball bearings. This is due to line contact in contrast to point contact.

3. Antifriction bearings are subjected to repeated stress cycles with maximum compressive stress values of approximately 150,000 lb/in.^{2} The cause of failure is usually fatigue.

4. If the load on a ball bearing is doubled, its fatigue life is reduced by a factor of 8.

5. If the load on a roller bearing is doubled, its fatigue life is reduced by a factor of 10.

6. Deformation is the cause of rolling resistance. Hence, to eliminate almost all resistance to motion, the surfaces of the rolling elements and raceways are hardened to about Rockwell C58.

7. If the wheel bearings of an automobile are designed for a B-10 life of 100,000 miles, 10% of the bearings may fail before 100,000 miles are reached.

8. For properly loaded bearings, the following approximate forces are required to pull a 1 ton load:

— Ball bearing: 1.6 lb
— Roller bearing: 3.0 lb

4.8 LUBRICATION OF ANTIFRICTION BEARINGS ▬▬▬▬▬

The function of a lubricant for antifriction bearings is quite different from that for journal bearings. In a journal bearing, the load is supported by an oil film, and maintenance of this oil film under varying loads, speeds, and temperatures is the prime consideration. On the other hand, antifriction bearings operate with extremely high contact pressures between the rolling elements and the race. This prevents the formation of a continuous oil film; thus metal-to-metal rolling contact occurs.

The ability of the rolling elements to carry heavy loads is due primarily to their minimal deformation. Any heat generated as a result deformation must, however, be dissipated by the lubricant. Since there is some sliding contact between the rolling elements and the separator, lubrication is required to reduce friction and heat at these areas of contact.

Basically, an antifriction bearing lubricant must fulfill the following functions:

— Provide a lubricating film between the rolling elements and the separator at the areas of contact
— Dissipate heat caused by deformation of the rolling elements and raceways as well as heat caused by the sliding contact between the rolling elements and the separator
— Prevent corrosion of bearing components
— Aid in preventing dirt and other contaminants from entering the critical areas of the bearing where sliding or rolling contact takes place

Either oil or grease can be used as a lubricant for antifriction bearings. Because of its high viscosity, grease requires the least amount of attention as far as lubricant

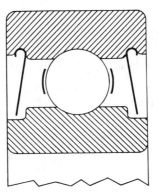

FIGURE 4.29   Ball bearing with shields

leakage is concerned. In addition, grease is generally used when extremely high temperatures are not encountered so that excessive leakage does not occur as a result of the
reduced viscosity caused by heat. Oil is more efficient, and for high-speed applications,
it is a better lubricant. However, intricate oiling devices are normally required.

Some bearings are supplied with shields or seals to keep foreign particles from
entering the inside of the bearing. The simplest form of closure is the *shield,* which is
a formed metal disk. Normally, shields are attached to the outer race and fit in a notch
in the inner race, as shown in Figure 4.29. There is a small running clearance between
the shield and the inner race. Hence shields help to retain grease lubricants and to
prevent the entrance of chips or large particles into the bearing. A *seal* differs from a
shield in that there is no running clearance between a seal and the inner ring. Since a
seal rubs on the inner race, it keeps out all foreign particles. It is also the function of a
seal to retain the lubricant inside a permanently prelubricated bearing. Such a bearing
does not have to be relubricated, because the lubricant is retained for the life of the
bearing. Figure 4.30 shows a ball bearing with a steel disk to which a molded synthetic
rubber seal is bonded. Since the seal rubs on the inner race, sliding friction exists. This
generates heat that must be dissipated so that excessive temperatures do not occur at
high speed operation.

4.9 INSTALLATION OF ANTIFRICTION BEARINGS ▬▬▬

In most bearing installations, the inner ring rotates and the outer ring is stationary. The
inner ring is usually press-fitted to the shaft, while the outer ring is push-fitted. A
press-fitted ring will normally not slip on a rotating shaft, but since the outer ring has
a push fit, it will creep rotationally in the supporting housing. This creep occurs very
slowly and is highly desirable because it does not allow prolonged stressing of any one
area of the raceway. Any wear due to normal creep is negligible.

Suitable shoulders must be provided in an installation design to hold the bearings
against axial movement, and if such shoulders are provided, it is sometimes not necessary to clamp the inner or outer rings. However, in most cases, axial forces that must
be resisted by one of the bearings on the shaft are present. Under these conditions, it is

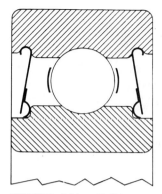

FIGURE 4.30   Ball bearing with seals

necessary to clamp the bearing in place to sustain the thrust loads. This is especially true if the direction of the axial loads is subject to reversals. Figure 4.31 shows an arrangement in which the inner ring is clamped against a shoulder on the shaft using a lock nut. This method requires that the end of the shaft be threaded.

In some applications, such as machine-tool spindles, there can be no looseness in the radial or axial directions at a bearing support. In these cases, two bearings can be preloaded against each other to remove all the play from each bearing. Preloading of

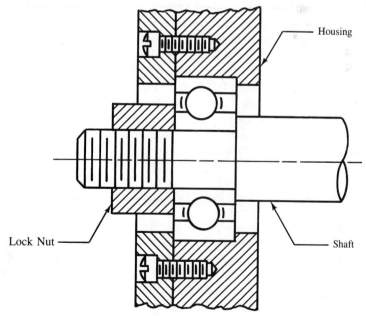

FIGURE 4.31   Common bearing installation

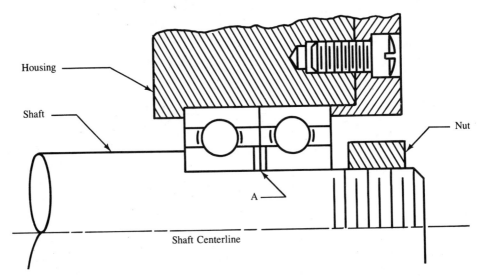

FIGURE 4.32   One arrangement for preloaded bearings

two ball bearings can be accomplished as shown in Figure 4.32. The widths of the
outer races are slightly greater than the widths of the inner races by a precise and
controlled amount. Thus there is initially a small but precise axial gap between the two
inner races at location A in Figure 4.32. As the nut is tightened, the inner races are
brought together by the force of the nut. As a result, the balls are displaced such that
all the internal looseness inside the bearings is eliminated. Of course, the outer race of
the bearing at the other end of the shaft must be allowed to float freely along the axial
direction. Care must be taken not to overstress the bearing by excessive preloading.
Therefore, the maximum axial gap at location A must be limited. Generally, a total
preload of 0.002 in. (0.001 in. per bearing) is adequate to remove all the looseness
from each bearing. Hence the total initial gap at location A would be 0.002 in. Preloaded bearings also result in a stiffer system. Thus, when an external load is applied,
preloaded bearings will deflect by a smaller amount. Increased bearing stiffness is especially important on computer numerical control (CNC) machines, where high machining precision is required.

4.10   PREMOUNTED BEARINGS

A *premounted bearing* is an entire bearing assembly that can be purchased as an off-
the-shelf item from most bearing manufacturers. The complete assembly contains the
actual bearing, provisions for administering the lubricant, seals or shields, and the
housing. Figure 4.33 illustrates a single-row ball bearing unit with internal seals. It
contains a one-piece iron housing with provisions for administering a lubricant. Observe that the outer surface of the outer ring is spherical and mates with a spherical

FIGURE 4.33   Self-aligning premounted ball bearing unit
*(Courtesy of Link-Belt Bearing Division,
FMC Corporation)*

surface inside the housing. As a result, this unit is self-aligning to allow for minor
misalignment and shaft deflection. Also notice the thick collar around the front end of
the wide inner ring, which permits locking the shaft into place by the use of a setscrew
located inside the collar.

Figure 4.34 illustrates a double-row spherical roller bearing unit. Like the single-
row unit, the double-row spherical roller bearing has integral seals and provisions for
lubrication. In addition to being self-aligning, it is designed for heavy-duty operation
and can absorb substantial thrust loads. It contains a one-piece iron housing with a
positive shaft locking collar.

Three principal advantages of premounted bearings are

—   Low cost
—   Availability as off-the-shelf units
—   Easy installation

FIGURE 4.34   Self-aligning premounted spherical roller
bearing unit *(Courtesy of Link-Belt Bear-
ing Division, FMC Corporation)*

Table 4.2   Journal Bearings versus Antifriction Bearings

Journal Bearings Advantages	Antifriction Bearings Advantages
Require little radial space	Have less friction
Run quietly	Require no wearing-in period
Have a longer life span	Require less axial space
Are less sensitive to contamination	Can run at higher speeds
Are less costly	Have fewer maintenance problems
Can better sustain shock loads	Have fewer lubrication difficulties
Require less precise mounting	Allow for considerable misalignment
Are available in split halves	Are easy to replace
(one application: automotive	Allow for greater precision
crankshaft bearings)	Are readily available in a large variety of types
	and sizes

4.11   JOURNAL BEARINGS VERSUS ANTIFRICTION BEARINGS ▰

Rotating shafts or reciprocating stems must be supported by properly designed bearings. For each application, a decision must be made between using journal bearings or antifriction bearings. Before a particular bearing is selected, the following factors should be considered:

— Shaft speed
— Type and magnitude of loads
— Consequences of failure
— Required life
— Sensitivity to contamination and corrosion
— Sensitivity to temperature
— Required space
— Shaft misalignment and deflections
— Costs
— Required lubrication provisions
— Required maintenance
— Bearing stiffness requirements

Table 4.2 lists the salient advantages of journal bearings over antifriction bearings, and vice versa. Each advantage should be considered when selecting a bearing for a particular application.

4.12   SUMMARY ▬▬▬▬▬▬▬▬▬

Antifriction bearings operate with rolling elements. There are three basic types: ball, roller, and needle bearings. The cause of rolling resistance is the deformation of the mating surfaces of the rolling elements and the raceways. The balls make point contact with their raceways. The balls in a ball bearing roll between two raceways and have a separator to keep them spread apart evenly. A raceway will have a different radius of curvature from the ball. Balls may be loaded into their retaining rings by (1) placing the rings eccentric to each other and loading the balls through the large opening, (2) pushing them through loading grooves in the rings, (3) thermally expanding the outer ring, and (4) using a two-piece inner ring.

Because of their line contact, roller bearings can take larger loads than ball bearings. There are three basic shapes for roller bearings: (1) cylindrical, (2) tapered and (3) spherical. Cylindrical bearings are not designed to take large thrust loads. Tapered bearings can be designed to take large thrust and radial loads. Spherical roller bearings are self-aligning.

Needle bearings require very little radial space because of the small diameters of the rolling elements. Needle bearings have no separators, so the rollers can fill the entire space between the raceways. In some cases, needle bearings do not have an inner ring; the rollers are placed directly on the shaft. This further reduces the radial space required for the bearing.

Those tapered roller bearings which have their rollers arranged on a conical surface are used in applications where both radial and thrust loads are present.

Properly maintained bearings usually fail as a result of fatigue because of the repeated stress cycles the contacting surfaces receive. Bearing manufacturers supply the technician and engineer with the life expectancy of a bearing for a specified load. This life expectancy is usually called the B-10 life and means that no more than 10% of the bearings will fail at the stated life and load. The technician or engineer then has to apply these data to his or her application through appropriate equations. The average life of a bearing is usually five times the B-10 life. The life of a ball bearing is inversely proportional to the cube of the load. The load capacity of an antifriction bearing is roughly proportional to the square of the diameter of the rolling element.

Lubricants for antifriction bearings do not form a film of lubricant between the rolling elements and their raceways. They do (1) provide a film between the rolling element and the separator, (2) dissipate heat, (3) prevent corrosion, and (4) keep out dirt and other contaminants. Bearings may be obtained with shields or seals to aid in keeping out contaminants.

The inner ring of a bearing is usually press-fitted to its shaft to eliminate slip between the shaft and inner ring. However, it is desirable to have some slip, or creep, between the outer ring and its housing. Premounted bearings are entire bearing assemblies, including the housing, that are stocked for immediate shipment to a customer. The sizes of premounted bearings are standardized to reduce the cost.

4.13 QUESTIONS AND PROBLEMS ▬▬▬▬▬▬▬▬

Questions

1. What is the cause of rolling resistance? How can it be minimized?

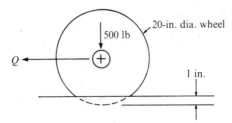

FIGURE 4.35   Sketch for Problem 1

2.   Name the four main components of a ball bearing.
3.   What is the difference between a Conrad ball bearing and a maximum capacity ball bearing?
4.   Discuss the concept of conformity of a ball and its raceway.
5.   What is the difference between a seal and a shield?
7.   Name the three basic types of roller bearings, and give an application for each.
8.   Why do roller bearings have much greater load-carrying capacities than ball bearings?
9.   What is meant by a self-aligning bearing?
10. Some needle bearings do not have inner rings, and thus the rollers have direct contact with the shaft. Discuss the hardness requirements of the shaft under these conditions.
11. What is a thrust bearing?
12. Describe the disadvantage of the roller thrust bearing shown in Figure 4.25.
13. Why does an antifriction bearing normally fail as a result of fatigue?
14. What is meant by the B-10 life of an antifriction bearing?
15. Define the bearing parameter *basic dynamic capacity*.
16. What is the significance of the term *static strength capacity*.
17. Name three functions that an antifriction bearing lubricant must fulfill.
18. Why are bearings sometimes preloaded?
19. What is a premounted bearing, and when would it normally be used?
20. Give four advantages and four disadvantages of journal bearings as compared to antifriction bearings.

Problems

1.   A 20 in. diameter rubber wheel is deformed 1 in. on a concrete road, as shown in Figure 4.35. If its axle is assumed to be frictionless, how much force is required to start the wheel rolling against a 500 lb load?
2.   A 500 lb slide rests on six steel rollers. The coefficient of rolling resistance is 0.020 between the slide and the rollers and 0.015 between the rollers and the concrete supporting floor. If the roller diameter is 1 in., what horizontal force is required to start the slide moving?
3.   Select a ball bearing from Table 4.1 to operate with a 2000 lb radial load and have a B-10 life of 2 million revolutions.

Thrust →

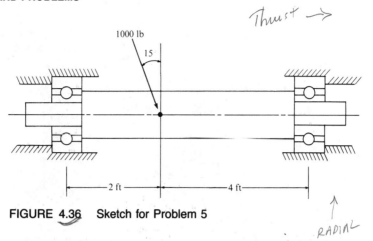

FIGURE 4.36   Sketch for Problem 5

RADIAL

4. Select a ball bearing from Table 4.1 to operate with a radial load of 500 lb and a thrust load of 250 lb. The B-10 life is to be 100 h at 1800 rpm.

5. Figure 4.36 shows a shaft absorbing a 1000 lb load. Decide which bearing should take the full thrust load, and select a ball bearing for each end of the shaft from Table 4.1 based on the following requirements:
    (a) The desired B-10 life is 1 year at 40 h per week.
    (b) The shaft rotates at 500 rpm.

$L_1 = 40 hr \times 52 weeks \times 60 \times 500$
$L_2 = 90 \times 10^6$

SI Unit Problems

6. Solve Problem 1 above, except use these data: The rubber wheel is 350 mm in diameter, the deformation is 18 mm, and the load it supports is 225 kg.

7. A 1500 kg machine part is placed on four steel rollers that roll on a steel way. The rollers are 80 mm in diameter with the coefficient of rolling resistance (for both top and bottom contact surfaces = 0.0015 mm. What horizontal force is required to start the machine part moving?

1000 lb

Ball B.   15° 1000 cos 15   Ball B.   By

1000 sin 15   Bx

Ax

moment around
point A & B

find Ax by moments.

$-6 Ax + 1000 \cos 15 \times 4 ft = 0$

5

Shafts and Couplings

Objectives

After completing this chapter, you will be able to

- Find stresses resulting from combined forces applied to shafts at different angles (say, one force vertically down and another horizontal), as well as stresses resulting from a combination of torques and forces.
- Explain the application and construction of flexible shafts and flexible couplings.
- Determine the critical shaft speed for a number of different types of loadings.
- Apply the appropriate equations to select appropriate keys and splines to transmit power from shafts to gears, pulleys, wheels, and so forth.
- Analyze the forces on rigid couplings and on torque-limiting couplings.

5.1 INTRODUCTION

Virtually all machines contain shafts. The most common shape for shafts is circular, and the cross section can be either solid or hollow. Hollow shafts can result in weight savings, but they are more common on machines that require some other component to pass through the shaft. For instance, drive spindles on lathes are usually hollow to allow bar stock to pass through, and drive spindles on vertical milling machines are hollow to allow a long bolt to pass through to fasten the tool chuck (or collet) in place. Square shafts are sometimes used, as in screwdriver blades, socket wrenches, and control-knob stems.

A shaft must have adequate torsional strength to transmit torque and not be over-stressed. It also must be torsionally stiff enough that one mounted component does not deviate excessively from its original angular position relative to a second component mounted on the same shaft. Generally speaking, the angle of twist should not exceed 1° in a shaft length equal to 20 diameters.

Shafts are mounted on bearings and transmit power through such devices as gears, pulleys, cams, and clutches. These devices introduce forces that tend to bend the shaft; hence the shaft must be rigid enough to prevent overloading of the supporting bearings. In general, the bending deflection of a shaft should not exceed 0.01 in. per foot of length between bearing supports. However, if the bearings are self-aligning, a greater deflection may be acceptable.

In addition, a shaft must be able to sustain a combination of bending and torsional loads. Thus an equivalent load must be considered that takes into account both torsion and bending. In addition, the allowable stress must contain a factor of safety that includes fatigue, since torsional and bending stress reversals occur.

For diameters less than 3 in., the usual shaft material is cold-rolled steel containing about 0.4% carbon. Shafts are either cold-rolled or forged in sizes from 3 in. to 5 in. For sizes above 5 in., shafts are forged and machined to size. Plastic shafts are used widely for light load applications. One advantage of using plastic is safety in electrical applications, since plastic is a poor conductor of electricity.

In selecting a shaft diameter, the calculated size is considered the minimum value. A standard size that is the smallest standard size exceeding the calculated value should be selected. Standard nominal sizes are given to Table 5.1

TABLE 5.1 Standard Diameters for Shafts

Diameter (in.)	Diameter Increments (in.)
Up to 3	1/16
3–5	1/8
5–8	1/4

Components such as gears and pulleys are commonly mounted on shafts by means of keys. The design of the key and the corresponding keyway in the shaft must be properly evaluated. For example, stress concentrations occur in shafts as a result of keyways, and the material removed to form the keyway further weakens the shaft.

Another important aspect of shaft design is the method of directly connecting one shaft to another. This is accomplished by devices such as rigid and flexible couplings.

If shafts are run at critical speeds, severe vibrations can occur that can seriously damage a machine. It is important to know the magnitude of these critical speeds so that they can be avoided. As a general rule of thumb, the difference between the operating speed and the critical speed should be at least 20%.

5.2 HORSEPOWER-TORQUE EQUATION

A common equation used in this chapter and in others is one that converts torque to horsepower, or vice versa. It is helpful to understand how this equation is obtained. The basic relationships we must start with are

$$\text{Work} = \text{force} \times \text{distance (in.-lb)}$$
$$\text{Torque} = \text{force} \times \text{perpendicular distance to center of rotation (lb-in.)}$$
$$\text{Horsepower} = 33,000 \text{ ft-lb/min}$$
$$= 396,000 \text{ in.-lb/min}$$

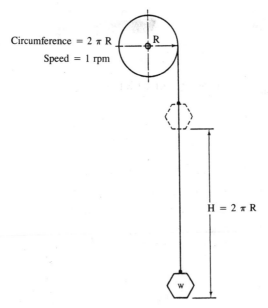

Circumference = $2 \pi R$

Speed = 1 rpm

$H = 2 \pi R$

FIGURE 5.1 Sketch of hoist to explain horsepower-torque relation

Also, Figure 5.1 will aid our discussion. It is a schematic diagram of a hoist. The hoist is lifting the weight by means of a drum with a cable wrapped around it. An electric motor is turning the drum. The question is: At constant speed, how much power is required to lift the weight? We will specify that the rotational speed of the drum is *1 rpm* in order to make the explanation clearer. The figure indicates that the weight is raised a height *H* in 1 min. *H* is equal to the circumference of the drum. Therefore,

$$\text{Power} = \frac{\text{work}}{\text{time}} = \frac{W \times H}{t} = \frac{W \times 2\pi \times R}{1 \text{ min}}$$

where W = the load that must be raised in pounds

R = the radius of the drum in inches

If the numerator is rearranged so that W is placed next to R, it becomes clear that $W \times R$ is torque. Therefore, $W \times R$ is replaced by the symbol T. Moreover, if the speed of the drum were 10 rpm, the height would simply be multiplied by 10, which, in turn, means that the numerator is multiplied by 10. Therefore, the power formula can simply be multiplied by the drum speed (symbol n). The time remains at 1 min and is not shown in the final equation because it is included in the drum speed (rpm). We now have

$$\text{Power} = 2\pi T n$$

$P = \frac{Torque}{min}$

The units are inch-pounds per minute. To obtain the answer in horsepower, we must divide by 396,000 in.-lb/min:

$$\text{Horsepower} = \frac{2\pi T n}{396,000}$$

Divide the numerator and denominator by 2π:

$$\text{Horsepower} = \frac{T n}{63,000} \tag{5.1}$$

where T = the torque in pound-inches

n = the rotational speed in revolutions per minute

5.3 TORSION OF CIRCULAR SHAFTS ▬▬▬▬▬▬▬

A shaft experiences torsion when it transmits torque. This twisting action produces torsional stresses and torsional deflections. To evaluate these stresses and deflections, let us first refer to Figure 5.2. In Figure 5.2*a* we see an unloaded shaft with line *AB* marked on its side. Rectangle *ABCD* represents an imaginary cutting plane penetrating the shaft up to its centerline. Figure 5.2*b* shows the same shaft loaded with torque *T*.

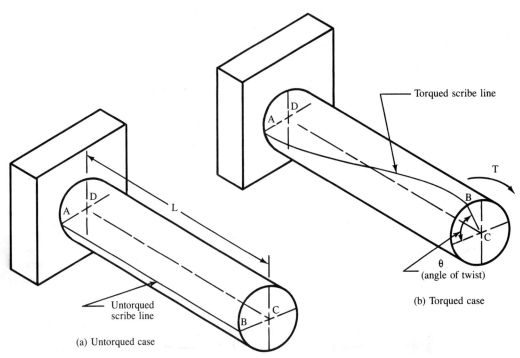

FIGURE 5.2   Untorqued versus torqued shaft configuration

Note that the front end of the shaft twists through an angle θ, the angle of twist, which is the *torsional deflection*. The original flat area *ABCD* becomes warped as the scribe line assumes a helical path. The *angle of twist* can be calculated from

$$\theta = \frac{TL}{GJ} \qquad Material\ Dependent.$$ (5.2)

where   θ = the angle of twist in radians

T = the applied torque in pound·inches

L = the shaft length in inches

J = polar moment of inertia of the shaft cross section in inches^4

G = the shear modulus of elasticity of the shaft material, also called *modulus of rigidity* in pounds per inch^2

The concept of angle of twist can be clarified by referring to a side view of the shaft (Fig. 5.3). Imagine the shaft to consist of a number of thin circular disks (a total of 10 is chosen here for discussion purposes). Notice that the untorqued scribe line is horizontal when the applied torque equals zero. However, when torque is applied, each disk slides rotationally a small amount relative to its neighbor. Disk no. 1 is fixed and

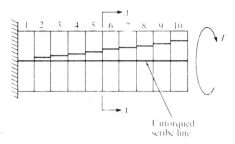

FIGURE 5.3   Shaft angle of twist

hence is in the zero rotation position. The total angle of twist is the summation of each individual disk's relative rotation. Thus, a long shaft produces a greater angle of twist than does a short shaft, as is confirmed by Equation 5.2. Observe that the series of short line segments represents an approximation to the actual torqued scribe line. If we let the number of disks become very large, each disk thickness becomes very small and each short line segment approaches a point. The curve through all these points gives the torqued scribe line depicted in Figure 5.2b.

Let us now evaluate the torsional stresses by passing a cutting plane as shown in Figure 5.3 and examining cross-sectional view A–A (Fig. 5.4). The size of the cross section has been enlarged for illustration purposes. In order to have rotational equilibrium, there must be stresses acting on the cross-sectional area of the shaft, as shown by the vectors marked with half an arrowhead. These are shear stresses because they are parallel to the surface over which they act. For discussion purposes, five annular ring areas are depicted in Figure 5.4. The stress on each annular ring multiplied by the ring area gives the force acting on each ring. The product of the force and the radius of each ring equals the ring internal resistive torque opposing the applied torque T. The summation of all the internal resistive torques for all the ring areas of the cross section must equal the applied torque to satisfy equilibrium. Figure 5.2b illustrates that the stress value is small near the center of the shaft and is at a maximum at the outer diameter. The reason that torsional stress is at a maximum at the shaft's surface is

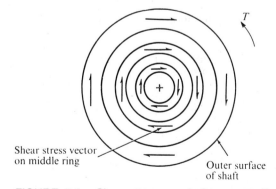

FIGURE 5.4   Shear stress on shaft cross section A–A

because point B on the surface is forced farther out of position than particles closer to the shaft's center. Specifically, the stress increases linearly with radius, starting with zero stress at the center of the shaft. We are interested only in the maximum torsional shear stress on the outer shaft surface, which can be calculated using the following equation:

$$J = \frac{Tc}{\tau_{max}}$$
$$T = \frac{\tau_{max} J}{c}$$

$$\tau_{max} = \frac{Tc}{J}$$

(5.3)

where τ_{max} = maximum torsional shear stress in pounds per inch^2

T = the applied torque in pound-inches

c = the radius of the shaft in inches

J = the polar moment of inertia of the shaft cross section in inches^4

Refer to Table 5.2 to find J.

SAMPLE PROBLEM 5.1

Shear Stress and Angle of Twist

$$T = \frac{\theta G J}{L}$$

PROBLEM: A 1 in. diameter solid steel shaft has an allowable shear stress of 5000 lb/in.^{2} and an allowable angle of twist of 0.1° per foot of length. $G = 12 \times 10^6$ lb/in.^{2} Find (a) the maximum value of torque that can be applied, and (b) how much horsepower can be delivered at 1700 rpm.

Solution

(a) We must check for both stress and stiffness and select the torque for the limiting case.

$$J = \frac{\pi d^4}{32} = \frac{3.14 \times 1}{32} = 0.098 \; in.^4$$

Torque based on stress:

$$T = \frac{\tau_{max} J}{c} = \frac{5000 \times 0.098}{0.5} = 980 \; lb-in.$$

Torque based on stiffness:

$$\frac{0.1^0}{57.3^{0/rad}} = 0.001,475 \; rad$$

$$T = \frac{GJ\theta}{L} = \frac{12 \times 10^6 \times 0.098 \times 0.001,745}{12 \; in.} = 171 \; lb-in.$$

The angle of twist is the limiting case.

(b) $Power = \dfrac{Tn}{63,000} = \dfrac{171 \times 1700}{63,000} = 4.61 \; hp$

It was stated earlier that the material near the center of a shaft in torsion is not as highly stressed as the material near the periphery. This statement brings us to the question, Why not leave out the material near the center of the shaft? If we did leave it out, the result would be a hollow shaft. Of course, a hollow shaft is not as strong as a solid shaft of the same diameter, but a hollow shaft is more efficient than a solid shaft because the "lazy" material near the center has been removed. Thus a weight savings can be realized when using a hollow shaft of equal torsional strength. Equations 5.2 and 5.3 can be used for hollow shafts, but the equation for the polar moment of inertia J must be adjusted (see Table 5.2).

5.4 BENDING OF CIRCULAR SHAFTS

Shafts transmit power through the use of devices such as gears and pulleys. In addition to torque loads, these devices introduce forces that tend to bend the shaft as it rotates. The bending stresses and deflections must be evaluated along with torsional stresses to ensure adequate strength and rigidity. Actually, we will not find bending stresses separately. The approach used in this text will be to find the bending moment and then combine it with the torque to reach an appropriate shaft design.

The problem here is to find the maximum bending moment on a shaft. In your strength of materials course, you have already done this for loads and forces acting in the same plane along the shaft. Now you will learn how to determine the maximum bending moment when forces acting on the shaft are not in the same longitudinal plane.

TABLE 5.2 I and J Relationships for Circular Cross-Sectional Areas

Area Shape	I	J
T5.2(1) Solid circle	$\dfrac{\pi d^4}{64}$	$\dfrac{\pi d^4}{32}$
T5.2(2) Hollow circle	$\dfrac{\pi(D_o^4 - D_i^4)}{64}$	$\dfrac{\pi(D_o^4 - D_i^4)}{32}$

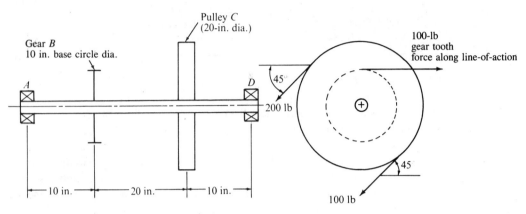

FIGURE 5.5   Power transmission shaft

Notice the direction of the forces on the gear and pulley in Figure 5.5. What must be done is to construct the shear and moment diagrams for both the horizontal and vertical directions (assuming the shaft is horizontal). From these diagrams, the location of the maximum moment is determined. If the horizontal and vertical moments at this location are considered to be vectors, then the Pythagorean theorem can be used to determine the maximum combined moment acting on the shaft. The procedure used in Sample Problem 5.2 is as follows.

PROCEDURE

Finding the Maximum Combined Moment

Step 1 Determine the horizontal and vertical component of each force acting on the shaft.

Step 2 Calculate the horizontal and vertical reactions at the bearing supports.

Step 3 Draw shear and moment diagrams for horizontal and vertical directions.

Step 4 Determine the location of the maximum combined moment (several locations along the shaft may have to be investigated). Use the Pythagorean theorem to find its magnitude. *p 121*

SAMPLE PROBLEM 5.2

Repeat

Combined Loads on a Shaft

PROBLEM: In the system in Figure 5.5, power is received on pulley *C*, whose belt pulls are at 45°, as shown. On the same shaft is gear *B*, which experiences a tooth thrust force of 100 lb horizontally by delivering power to a second shaft. (Note: the *base circle* of a gear is explained in Chapter 7.) Neglect all weights. Find the maximum bending moment acting on the shaft.

Solution

Steps 1 & 2   Analyze the vertical forces and reactions first. The gear has no force in the vertical direction. The vertical downward force on pulley C is

$$F_C = 300 \cos 45° = 212 \text{ lb}$$

$$\sum M_A = 0 = 212 \times 30 - 40 R_D \quad (R_D = 159 \text{ lb})$$

$$\sum F = 0 = -212 + 159 + R_A \quad (R_A = 53 \text{ lb})$$

Now analyze the horizontal forces and reactions.

$$F_C = -300 \sin 45° = -212 \text{ lb}$$

$$F_B = +100 \text{ lb}$$

$$\sum M_A = 0 = -100 \times 10 + 212 \times 30 - 40 R_D \quad (R_D = 134 \text{ lb})$$

$$\sum F = 0 = +100 - 212 + 134 - R_A \quad (R_A = 22 \text{ lb})$$

Step 3   Figure 5.6 shows the vertical FBD of the shaft and the vertical shear and moment diagrams. The horizontal FBD of the shaft, shear diagram, and moment diagram are shown in Figure 5.7.

Step 4   In this case we can determine by inspection of the moment diagrams that the combined maximum moment will be at the location of the pulley (see Fig. 5.8):

$$M_{combined} = \sqrt{M_V^2 + M_H^2} = \sqrt{1590^2 + 1340^2} = 2080 \text{ lb–in.}$$

5.5 COMBINED TORSION AND BENDING ▬▬▬▬▬▬

Recall that bending stresses act parallel to a shaft's longitudinal axis and that the shear stresses resulting from torque act in planes perpendicular to the axis of a shaft. Stresses are vectors because they are essentially unit forces. And since the bending stresses and torsional stresses are perpendicular to each other, the Pythagorean theorem must be used to find the maximum combined stress. There are several proposed methods for properly combining the effects of torsion and bending. This text will combine the moments and torques rather than the bending and shear stresses.

Methods for Finding Combined Stresses

The two methods we will use are called the *equivalent torque method* and the *equivalent bending moment method*. In general, a shaft is analyzed by both methods, and the greater calculated shaft size is used.

Want Maximums numbers

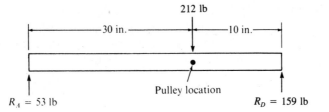

(a) Loading diagram

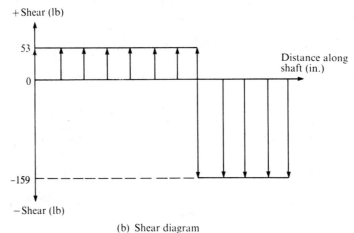

(b) Shear diagram

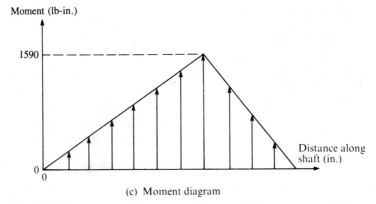

(c) Moment diagram

FIGURE 5.6   Shaft vertical loading diagrams

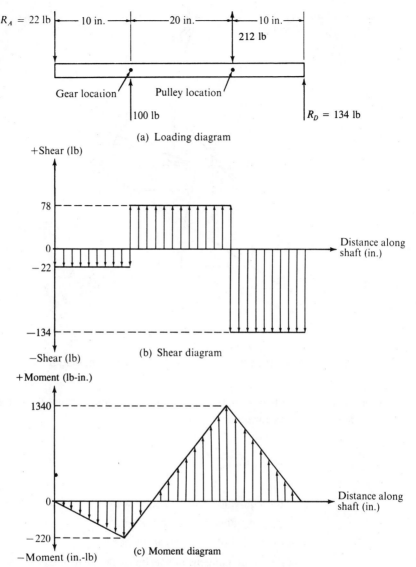

(a) Loading diagram

(b) Shear diagram

(c) Moment diagram

FIGURE 5.7 Shaft horizontal loading diagrams

Equivalent Torque Method. An equivalent torque T_E is defined by the equation

$$T_E = \sqrt{M^2 + T^2} \tag{5.4}$$

M and T represent the moment and torque values at the location determined to have the maximum value for the equivalent torque. Since the moment and torque values are considered perpendicular to each other (as are their stresses), the Pythagorean theorem

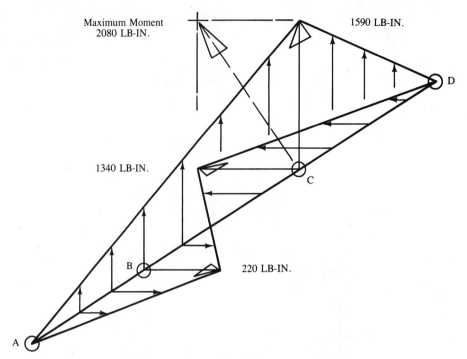

Maximum Moment
2080 LB-IN.

1590 LB-IN.

1340 LB-IN.

D

C

B

220 LB-IN.

A

FIGURE 5.8   Combined horizontal and vertical moment diagrams

is used as indicated in Equation 5.4. The equivalent torque T_E is then used in the torsional shear stress formula:

$$\tau_E = \frac{T_E c}{J} \tag{5.5}$$

where τ_E = the equivalent shear stress, which takes into account the effects of both bending and torsion (lb/in.^{2})

            c = the radius of the shaft in inches.

           J = the polar moment of inertia of the shaft cross section in inches^4

           Refer to Table 5.2 to find J.

Equation 5.5 is used to select a shaft size such that the value of τ_E is less than the allowable shear stress. In actual practice, it is easier to let the equivalent shear stress equal the allowable shear stress and solve for the appropriate shaft diameter. The steps used is rewriting Equation 5.5 to find a *solid* shaft diameter is as follows:

$$\tau_E = \frac{T_E(d/2)}{\pi(d^4/32)}$$

$$d = \sqrt[3]{\frac{16T_E}{\pi\tau}}$$

(5.6)

Equivalent Bending Moment Method. The equivalent moment M_E is arrived at by averaging the bending moment M and the equivalent torque T_E (which is found by using Equation 5.4):

$$M_E = \frac{M + T_E}{2}$$

(5.7)

Now that the equivalent bending moment has been found, the bending stress formula is used to find an equivalent bending stress S_E:

$$S_E = \frac{M_E \times c}{I}$$

(5.8)

where S_E = the equivalent tensile or compressive bending stress in pounds per inch^2

M_E = the equivalent bending moment in pound-inches

c = the shaft outside radius in inches

I = the moment of inertia of the shaft cross section in inches^4

Again, when solving for a solid shaft diameter, it is easier to rewrite Equation 5.8 using the value for I shown in Table 5.2 and letting S_E equal the allowable bending stress:

$$S_E = \frac{M_E \times (d/2)}{\pi \times (d^4/64)}$$

$$d = \sqrt[3]{\frac{32M_E}{\pi S}}$$

(5.9)

PROCEDURE

Finding Solid Shaft Size under Combined Torsion and Bending

Use the maximum moment and torque values for the location on the shaft that will have the maximum combined value. Solve for the shaft size using both the equivalent torque method and the equivalent moment method. Choose the largest shaft size.

Note: Remember that the allowable shear stress is used for the torsional shear stress formula and the allowable bending stress (tension or compression) is used for the bending stress formula.

SAMPLE PROBLEM 5.3

Shaft Design

PROBLEM: Refer to the shaft assembly in Figure 5.5 and the data supplied in Sample Problem 5.2. In addition, we are given that the allowable shear stress is 5000 lb/in.^{2} and the allowable tensile stress is 7500 lb/in.^{2} Find a suitable solid shaft diameter.

Solution (Equivalent Torque Method)

First, the shaft diameter will be determined by the equivalent torque method, which makes use of the allowable shear stress. Moment (at pulley) $M = 2080$ lb-in. (from Sample Problem 5.2). Torque (at pulley) $T = (200$ lb - 100 lb$) \times 10$ in. $= 1000$ lb-in.

$$T_E = \sqrt{M^2 + T^2} = \sqrt{2080^2 + 1000^2}$$

$$= 2308 \; lb - in.$$

From Equation 5.6,

$$d = \sqrt[3]{\frac{16 T_E}{\pi \tau}}$$

$$= \sqrt[3]{\frac{16 \times 2308}{3.14 \times 5000}} = 1.33 \text{ in.}$$

Solution (Equivalent Bending Moment Method)

The required shaft diameter is again determined by the equivalent bending method, which makes use of the allowable tensile strength.

$$M_E = \frac{(M + T_E)}{2} = \frac{2080 + 2308}{2} = 2194 \text{ lb-in.}$$

From Equation 5.9,

$$d = \sqrt[3]{\frac{32 M_E}{\pi S}}$$

$$= \sqrt[3]{\frac{32 \times 2194}{3.14 \times 7500}} = 1.44 \text{ in.}$$

Answer The larger calculated value of 1.44 in. is the minimum acceptable shaft diameter. Using Table 5.1, we select a standard size shaft diameter of 1.50 in.

5.6 CRITICAL SPEEDS OF SHAFTS

A shaft supported between bearings deflects under its own weight. Hence the center of gravity of a shaft does not coincide with the axis of rotation, which is the centerline of

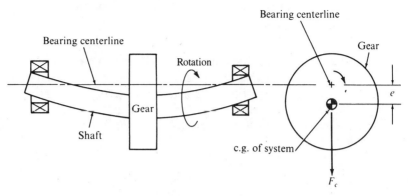

FIGURE 5.9 Eccentricity in a rotating shaft

the bearings. The eccentricity becomes even greater when the shaft carries components such as gears, pulleys, cams, and flywheels. Figure 5.9 shows a rotating shaft supported by two bearings and carrying a single gear. Observe that e is the eccentricity of the center of gravity (c.g.) of the system measured from the axis of rotation. The eccentricity is exaggerated for illustration purposes.

Very little vibration occurs if the center of gravity of the system remains in its position as the shaft and gear rotate. Vibration becomes a serious problem if the system operates at such a speed (called the *critical speed*) that the center of gravity rotates with the shaft along a circle of radius e whose center is the centerline of the bearings. The motion of the center of gravity is analogous to the motion of the middle of a jump rope being twirled by two children. This action produces a centrifugal force F_c that rotates with the shaft and is therefore an unbalanced force. This action causes forced vibrations.

At critical speeds, the shaft system can become dynamically unstable. This occurs when the shaft speed equals the natural frequency of the system. The *natural frequency* is the frequency of oscillation at which the system wants to vibrate. For example, consider a shaft between two bearings but not rotating. If a bending force is applied and suddenly removed, the shaft will vibrate at its natural frequency until all the dynamic energy is removed as a result of friction. At critical speeds, the shaft deflections become so large that severe damage to the entire machine can occur. This phenomenon is commonly called *shaft whirl*. It is therefore necessary to calculate these critical speeds so that they can be avoided. A general rule of thumb is that the value of the operating speed should be at least 20% away from the value of the critical speed.

The basic equation for the vibration frequency of a body vibrating because of an elastic restoring force is as follows:

$$f = \frac{1}{2\pi} \sqrt{\frac{kg}{W}}$$

where $\qquad$ f = the frequency in cycles per second

$\qquad\qquad$ k = the force per inch of deflection and is called the *force constant*

g = the acceleration due to gravity (386.4 in./s^2)

W = the weight in pounds

This equation, when adjusted, will provide us with the first or lowest critical speed for a simply supported shaft carrying a single concentrated load. The *lowest critical speed* is equivalent to the first harmonic of a vibrating shaft. There are other critical speeds, but these are higher and are less important than the first. The frequency in cycles per second (cps) must be changed to revolutions per minute by multiplying by 60 s/min:

$$N_c = 60 \times f$$

where N_c is the *critical* shaft speed in revolutions per minute (rpm).

The force constant k is equal to the weight of the object W divided by the static deflection of the shaft y (y is at the position of W):

$$k = \frac{W}{y}$$

The equation now becomes

$$N_c = \frac{60}{2\pi} \sqrt{\frac{W \cdot g}{W \cdot y}}$$

Since g = 386.4 in./s^2 and π = 3.14, we get the following equation for a simply supported shaft with a single concentrated load (shaft weight is neglected).

$$N_c = 187.7 \sqrt{\frac{1}{y}} \tag{5.10}$$

For shafts supporting n concentrated loads, Rayleigh's equation is used (the weight of the shaft is considered negligible):

$$N_c = 187.7 \sqrt{\frac{W_1y_1 + W_2y_2 + \cdots + W_ny_n}{W_1y_1^2 + W_2y_2^2 + \cdots W_ny_n^2}} \tag{5.11}$$

where W_1 = the weight of object number one in pounds

y_1 = the static deflection at the position of W_1, in inches

N_c = the first critical speed in revolutions per minute

Table 5.3 shows the formulas for the deflections of simply supported shafts with concentrated loads.

TABLE 5.3 Bending Moments and Deflections for Simply Supported Shafts

	Bending Moment at Load	Shaft Deflection
Concentrated Load at Shaft Midpoint	$\dfrac{WL}{4}$	$y = \dfrac{WL^3}{48EI}$ (at midspan)
Concentrated Load at any Position on Shaft	$\dfrac{Wab}{L}$	$y = \dfrac{Wa^2b^2}{3EIL}$ (at load) $y_1 = \dfrac{Wbx}{6EIL}(L^2 - b^2 - x^2)$ (at any distance x to the left of the load) $y_2 = \dfrac{Wa(L - x)}{6EIL}[L^2 - a^2 - (L - x)^2]$ (at any distance x to the right of the load)

PROCEDURE

Finding the Critical Speed of a Shaft

Step 1 Refer to Table 5.3 for the appropriate deflection formula.

Step 2 Solve for the deflection at each load.

Step 3 Substitute the deflection(s) and load(s) into either Equation 5.10 or Equation 5.11 and solve for the critical speed.

SAMPLE PROBLEM 5.4

Critical Speed

PROBLEM: Figure 5.10 shows a 1 in. diameter steel shaft mounted in bearings 30 in. apart. Disk A weighs 7 lb and is mounted 10 in. from the left bearing. Disk B weights 15 lb and is mounted 12 in. from the right bearing. Find the critical speed for the shaft (a) with disk A alone and (b) with disk A and B in place.

Solution

Steps 1 and 2

(a) Deflection at A: Use the equation for the deflection at the load.

$$I = \pi \frac{d^4}{64} = \pi \frac{1}{64} = 0.049087 \ in.^4$$

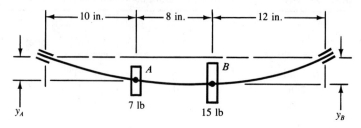

FIGURE 5.10   Steel shaft setup

$$y = \frac{Wa^2b^2}{3EIL} = \frac{7 \times 10^2 \times 20^2}{3(30 \times 10^6 \times 0.049,087 \times 30)}$$

$$= 0.002,112,67 \text{ in.}$$

Note: $EIL = \mathbf{44.178 \times 10^6}$

Step 3   Use Equation 5.10

$$N_c = 187.7 \sqrt{\frac{1}{y}} = 187.7 \sqrt{\frac{1}{0.002,112,67}}$$

$$= 4080 \; rpm$$

(b)   Steps 1 and 2 Deflection at B from load B:

$$y = \frac{Wa^2b^2}{3EIL} = \frac{15 \times 18^2 \times 12^2}{3(44.178 \times 10^6)} = 0.0052805 \text{ in.}$$

Deflection at B from load at A:

$$y_2 = \frac{Wa(L - x)}{6EIL}[L^2 - a^2 - (L - x)^2]$$

$$= \frac{7 \times 10(30 - 18)}{6(44.178 \times 10^6)}[30^2 - 10^2 - (30 - 18)^2]$$

$$= 0.002,078,86 \text{ in.}$$

Total deflection at B = 0.005,280,5 + 0.002,078,86 = 0.007,359,36 in.
Deflection at A due to load B:

$$y_1 = \frac{Wbx}{6EIL}(L^2 - b^2 - x^2) = \frac{15 \times 12 \times 10}{6(44.178 \times 10^6)}(30^2 - 12^2 - 10^2)$$

$$= 0.004,454,7 \; in.$$

Total deflection at A = 0.002,112,67 + 0.004,454,7 = 0.006,567,37 in.

Step 3 Use Equation 5.11.

$$N_c = 187.7 \sqrt{\frac{W_A y_A + W_B y_B}{W_A y_A^2 + W_B y_B^2}}$$

$$= 187.7 \sqrt{\frac{7(0.006,567,37) + 15(0.007,359,36)}{7(0.006,567,37)^2 + 15(0.007,359,36)^2}}$$

$$= 2223 \text{ rpm} \quad (\text{round to 2220 rpm})$$

5.7 FLEXIBLE SHAFTS

A *flexible shaft* is a self-contained mechanical device that permits the transmission of power between any two points regardless of the relative positions of the two points or of the obstacles in the path between them. Figure 5.11a illustrates how flexible shafting performs the functions of gearing or solid shafting with couplings or universal joints. The versatility of flexible shafting in transmitting power around corners is shown in Figure 5.11b. Another important application of flexible shafts is in remote control, where it is necessary to operate a device manually or mechanically from a distance. Examples of flexible shaft applications are (1) speedometer cables, (2) power screwdrivers, (3) surgical instruments (see Fig. 1.3), (4) steering mechanisms, (5) power-driven automobile seats, (6) spotlight controls, and (7) antennas.

It should be realized that a flexible shaft consists of a helical winding of a small-diameter wire. Hence, for power-drive applications, the torque must be applied in the same direction as the helical winding so that the helix tends to tighten. The torque-carrying capacity can be reduced by as much as 50% if the shaft is rotated in the wrong direction. Design data such as the following can be readily obtained from manufacturers' catalogs: (1) torque capacity, (2) minimum operating radius of curves, (3) lubrication requirements, and (4) maximum permissible shaft length.

The torque capacity of a flexible shaft is greatly reduced as its operating radius of curvature is reduced. Also, to transmit the greatest amount of power, flexible shafts should be run at the highest possible speed, because torque is usually the limiting parameter when sizable horsepower is required. Speeds of 1800 to 3600 rpm are very typical. However, speeds of 20,000 rpm and higher can be accommodated with small shaft diameters of approximately 1/4 in. or less.

Flexible shafts are enclosed in casings that also must be flexible to ensure satisfactory operation. The functions of the casing are

— To provide a continuous guide for the flexible shaft so it can rotate smoothly
— To prevent the shaft from tending to loop under torsional load
— To reduce the transverse bending of the flexible shaft
— To protect the shaft from moisture, dirt, and damage
— To retain the lubricant

Figure 5.12 illustrates a typical construction of a metallic casing that consists of two separate wires. This type of casing is strong and durable, but not oil-tight. Figure 5.13 illustrates a plastic-covered casing that consists of a flat inner steel liner

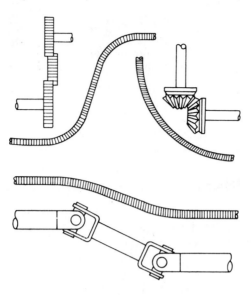

(a) Gear and universal joint functions

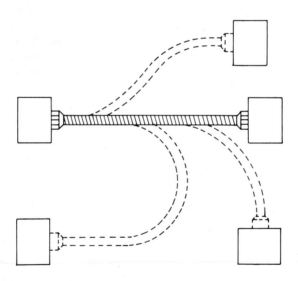

(b) Diversification of power transmission around corners

FIGURE 5.11   Functions and versatility of flexible shafting *(Courtesy of S.S. White Division of
Pennwalt Corporation)*

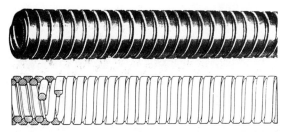

FIGURE 5.12 Typical construction of a metallic casing *(Courtesy of S.S. White Division of Pennwalt Corporation)*

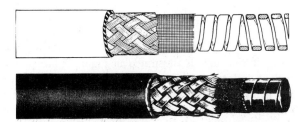

FIGURE 5.13 Typical construction of a plastic-covered casing *(Courtesy of S.S. White Division of Pennwalt Corporation)*

reinforced with layers of cotton duck, steel braid, and plastic. This type of casing is both oil- and water-tight.

Flexible shafts have special end fittings for attachment to rotating shafts. One method commonly used is to make the two ends of the flexible shaft square to fit a square opening. The casing ends are fastened by means of separate fittings that are threaded to the housings. Bicycle speedometer cables are designed in this way. There are also fittings that permit quick attachment to the power supply and driven members. These fittings consist of a shaft and casing combination with a special coupling at each end. To readily install this design, all that is necessary is to slide the couplings of proper sizes over the corresponding spindles at each end. A setscrew in each coupling is then tightened, and the flexible shaft is ready for use. No provisions are necessary for attaching the casing.

5.8 KEYS

A *key* is a device that mechanically connects a member such as a gear to a shaft. As such, a key transmits power from a shaft into the hub of a mounted component, and vice versa.

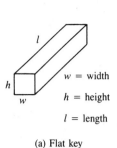

w = width

h = height

l = length

(a) Flat key

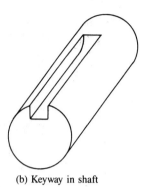

(b) Keyway in shaft

FIGURE 5.14   Key design

Flat Keys

The most common type of key is the *flat key*, as shown in Figure 5.14a. The corresponding keyway in the shaft is illustrated in Figure 5.14b.

Let us make a failure analysis of a flat key assuming that the force acting on the side of the key is uniform. This analysis also includes square keys. We will use Figure 5.15, which shows a gear keyed to a shaft. The torque T is transmitted from the shaft through the key into the hub. The torque then travels through the web and out the gear teeth by producing the output gear tooth force F_g. In transmitting the torque, the key absorbs a force P as defined by

$$P = \frac{T}{r}$$ (5.12)

where r = the radius of the shaft in inches

P = the force in pounds

T = the torque in pound-inches

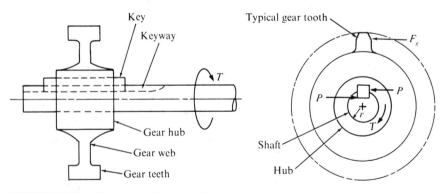

FIGURE 5.15   Gear keyed to a shaft

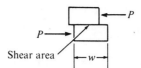

FIGURE 5.16   Shear deformation of key

There are two possible modes of failure: *shear failure* and *crushing failure*.

Shear Failure of Flat Keys.   Figure 5.16 shows that the keyways of the shaft and hub exert equal and opposite forces of value P; these forces attempt to shear the key at the radius of the shaft. The shear stress equals the applied force divided by the shear area

$$\tau = \frac{P}{w\ell}$$

(5.13)

where τ = the shear stress in pounds per inch^2
            P = the force in pounds
            w = the width of the key
            ℓ = the length of the key

Combining Equations 5.12 and 5.13 and rearranging to solve for key length, we have

$$\ell = \frac{T}{wr\tau}$$

(5.14)

Thus, if the torque, key width, shaft radius, and allowable shear stress are known, the required key length can be found based on shear strength.

Crushing Failure of Flat Keys.   Figure 5.17 shows that the key can suffer permanent compressive deformation where it contacts the keyways in the shaft and hub. The crushing stress is actually a compressive stress that equals the applied force divided by the bearing area of contact:

$$S = \frac{P}{\ell \frac{h}{2}}$$

(5.15)

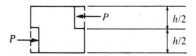

FIGURE 5.17   Crushing deformation of key

where S = the compressive stress pounds per inch^2

h = the total height of the key in inches

Combining Equations 5.12 and 5.15 and solving for key length yields

$$\ell = \frac{2T}{Shr} \tag{5.16}$$

Hence, if the torque, key height, radius of the shaft, and allowable compressive stress are known, the required key length based on crushing strength can be determined. The greater of the two values obtained from Equations 5.14 and 5.16 is then selected.

Flat and Square Key Selection. A hub must have a length equal to or greater than the required key length. It is possible to select a key cross section ($w \times h$) that is so large that the shaft is weakened excessively. On the other hand, if the selected key cross section is too small, the required key length becomes impractical.

Table 5.4 lists standard key sizes for various shaft diameters. Note that one column is for square keys and one for flat keys. Most metallic materials have a much larger allowable compressive stress than allowable shear stress, so flat keys are usually satisfactory. However, if the two allowable stresses are close together, a square key (which has a larger compressive area) may be required.

PROCEDURE

Selecting a Key

Step 1 Refer to Table 5.4, and select the appropriate key size.

TABLE 5.4 Standard Size Square and Flat Keys

Shaft Diameter	Square Key Size, *w* or *h*	Flat Key Size, *w* × *h*
1/2–9/16	1/8	1/8 × 3/32
5/8–7/8	3/16	3/16 × 1/8
15/16–1 1/4	1/4	1/4 × 3/16
1 5/16–1 3/8	5/16	5/16 × 1/4
1 7/16–1 3/4	3/8	3/8 × 1/4
1 13/16–2 1/4	1/2	1/2 × 3/8
2 5/16–2 3/4	5/8	5/8 × 7/16
2 7/8–3 1/4	3/4	3/4 × 1/2
3 3/8–3 3/4	7/8	7/8 × 5/8
3 7/8–4 1/2	1	1 × 3/4
4 3/4–5 1/2	1 1/4	1 1/4 × 7/8
5 3/4–6 /	1 1/2	1 1/2 × 1

Step 2 Solve Equations 5.14 and 5.16, and use the longer length.

SAMPLE PROBLEM 5.5

Key Length

PROBLEM: A standard size flat key is required to transmit 100 hp at 1800 rpm. A 1 1/2 in. shaft is used, and the allowable stresses are 5000 lb/in.^{2} in shear and 12,000 lb/in.^{2} in compression. Find the key length.

[handwritten: $\ell = \dfrac{T}{w \cdot \tau}$]

Solution

*[handwritten margin notes:
1. DIA of shaft
2. Kind of material
3. Pick size of key]*

Step 1 Table 5.4 indicates we should use a flat key 3/8 in. wide by 1/4 in. high.

Step 2 Before applying Equations 5.14 and 5.16, we must find the torque:

$$T = \frac{63{,}000 \times \text{hp}}{n} = \frac{63{,}000 \times 100}{1800}$$

[handwritten: $T = 3500$]

$$= 3500 \ lb\text{-}in.$$

Equation 5.14:

$$\ell = \frac{T}{w r \tau} = \frac{3500}{0.375 \times 0.75 \times 5000} = 2.5 \ \text{in.}$$

[handwritten: shear]

Equation 5.16:

$$\ell = \frac{2T}{S h r} = \frac{2 \times 3500}{12{,}000 \times 0.25 \times 0.75}$$

$$= 3.1 \ in.$$

[handwritten: compression]

Answer Thus the required key length is 3.1 in. Normally, the key is made of a weaker material than the shaft and hub because it is easier and less expensive to replace the key.

Woodruff Keys

Another type of key is the *Woodruff key*, which resembles a segment of a circle, as shown in Figure 5.18a (diameter is d and thickness is w). The keyway in the shaft is machined in the shape of a circular slot by using a milling cutter of diameter d. The key protrudes outside the shaft by one-half the thickness of the key (see Fig. 5.18b). It is common practice to make the cutter diameter d equal to four times the key thickness. The proper shaft diameter should be about five times the key thickness. Woodruff keys are suitable only for light torque applications such as hand wheels on lathes and milling machines. Woodruff keys are easy to remove from the shaft. Once the hub is removed from the shaft, the key can be removed by a light tap with a hammer.

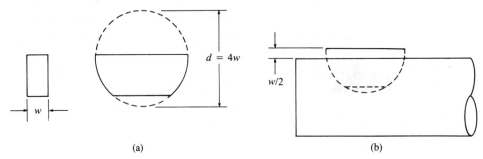

FIGURE 5.18   Woodruff key configuration

If high torque

Splined Shafts.   If the torque capacity of a key is not adequate, a splined shaft is frequently used. As shown in Figure 5.19, a *splined shaft* is basically a shaft that has a number of integral keys machined from the outer periphery. The hub of the mating member has matching keyways to accommodate the keys of the splined shaft. Splines can be used as fixed connectors or may be designed to permit relative axial movement between the shaft and hub. One common application of the sliding design is the shifting of gears in manual automotive transmissions.

The two basic spline contours are as follows.

1. *Parallel side contour:* The splines are actually rectangular in shape, like a flat key. Although only four teeth are shown in Figure 5.19, other standard sizes provide 6, 10, and 16 teeth.

2. *Involute contour:* The splines are shaped similar to teeth of involute gears. This provides greater strength.

It should be noted that the manufacturing cost of splined shafts is much higher than that for keys. Therefore, splines are used only if the extra torque capacity, or hub-to-shaft axial sliding capability, is needed.

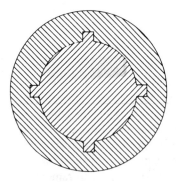

FIGURE 5.19   Four-tooth splined shaft with matching hub

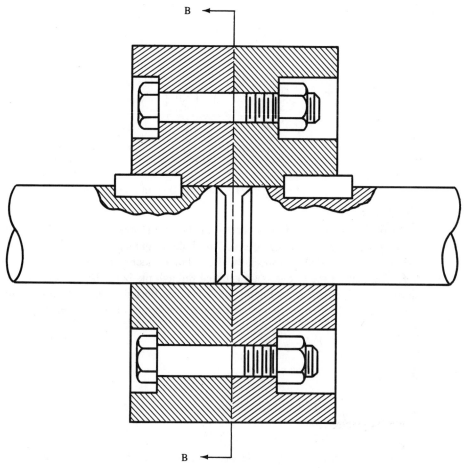

FIGURE 5.20   Rigid flanged coupling

5.9 RIGID COUPLINGS

A *coupling* is a device used to connect the end of one shaft to the end of a second. Rigid couplings do not provide for any misalignment (linear or angular) between the two shafts being connected. Therefore, rigid couplings can be used only when the centerlines of the two shafts are collinear.

Flange Couplings.   Figure 5.20 shows a flange-type rigid coupling that is transmitting torque T. The two flanges are bolted together by a series of bolts arranged along a bolt circle, as shown in Figure 5.21 (actually section B–B of Fig. 5.20). The radius of the bolt circle is r, and there are N bolts ($N = 4$ in the specific case shown in Figure 5.21).

Let us analyze the flange coupling to find the required number of bolts for transmitting the applied torque T. It is assumed that none of the torque is transmitted as a

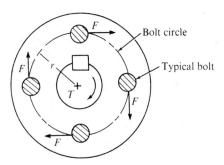

FIGURE 5.21   Section B–B of Figure 5.20

result of friction between the interface of the two flanges. This assumption is conservative because the clamping force of the bolts is quite large. However, the frictional forces are unpredictable, especially if the bolts undergo relaxation. We are therefore assuming that all the torque is transmitted through the bolts. Hence each bolt absorbs a shear force F as shown in Figure 5.21. For equilibrium, the summation of each bolt shear force multiplied by its moment arm to the center of rotation equals the torque T:

$$T = FrN$$

Rearranging,

$$N = \frac{T}{Fr}$$

The shear stress τ equals the shear force divided by the cross-sectional area A of each bolt:

$$\tau = \frac{F}{A}$$

Rearranging

$$F = \tau A = \tau \frac{\pi \times d^2}{4}$$

Substituting this into the previous equation results in

$$N = \frac{4T}{\tau \pi d^2 r} \qquad \qquad (5.17)$$

where N = the number of bolts

T = the applied torque in pound-inches

τ = the allowable shear stress in pounds per inch^2

r = the radius of the bolt circle in inches

d = the bolt diameter, usually the shank or maximum diameter in inches

SAMPLE PROBLEM 5.6

Number of Bolts in a Rigid Coupling

PROBLEM: A flanged coupling has a 4 in. diameter bolt circle designed to use 1/4 in. diameter bolts. If the applied torque is 5000 lb-in. and the allowable shear stress is 10,000 lb/in.^{2}, how many bolts are required?

Solution

$$N = \frac{4T}{\tau \times \pi \times r \times d^2}$$

$$= \frac{4 \times 5000}{10{,}000 \times 3.14 \times 2 \times 0.25^2} = 5.1$$

Answer: Use six bolts.

It should be noted that Equation 5.17 also assumes that the load is equally shared by each bolt. That is, each bolt must fit tightly in each hole without any radial clearance, which requires the following two manufacturing realities:

1. The holes must be drilled and reamed with both mating flanges assembled together, which, of course, means that interchangeability has been destroyed. For example, if only one flange needs to be replaced, a complete matching set (left- and right-hand flanges) would have to be installed.
2. The diameters of the bolts must be held to very tight tolerances to match the hole diameters.

Also observe the safety design feature of recessing the bolt heads and nuts inside the flanges. As a result, the sharp edges of the bolt heads and nuts are not exposed, eliminating the possibility of serious human injury occurring when the flange is rotating.

Sleeve Couplings.   A second type of rigid coupling is the *sleeve coupling,* which is illustrated in Figure 5.22. This coupling is readily assembled by sliding the sleeve over the ends of the two shafts until the two holes in the sleeve and shafts line up. Then the two pins are driven completely through the shaft as shown. The amount of torque T that can be transmitted depends on the shearing area of one of the drive pins, as shown in Figure 5.23 (which is section *B–B* of Fig. 5.22). Note that each pin experiences double shear because there are two shear areas absorbing the torque. Intuitively, we expect that a sleeve coupling cannot transmit a large amount of torque. However, a

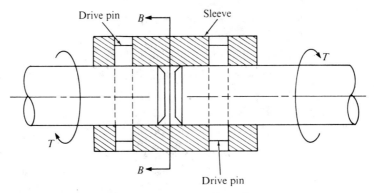

FIGURE 5.22   Sleeve coupling with drive pins

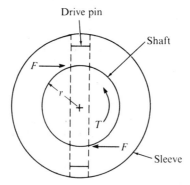

FIGURE 5.23   Section B–B of Figure 5.22

sleeve coupling is much less expensive than a flange coupling, and it can be used as a safety device by making the pin diameter small, as well as by using a weak pin material such as brass. An example of use as a safety device would be a sleeve coupling in the propeller shaft on an outboard motor. If the propeller hits a rock, for example, the resulting overload would quickly shear the pin. This would avoid damage to the rest of the mechanism. Such a device is analogous to a fuse in an electrical system.

5.10 FLEXIBLE COUPLINGS

It is almost impossible to obtain perfect alignment between two shafts. Misalignment can be attributed to the following:

— Lack of perfect collinearity of bearing support housings due to manufacturing tolerances
— Shaft bending deflections under load

FIGURE 5.24   Paraflex coupling. *(Courtesy of Dodge Manufacturing Division, Reliance Electric Company)*

— Use of two separately mounted units, such as coupling a motor shaft to a pump shaft

If a rigid coupling is used where there is any significant misalignment, the following serious consequences can result:

— Excessive shaft bending loads
— Excessive bearing loads
— Increased vibration and noise

All these problems can be solved by using *flexible couplings*, provided their misalignment capacities are not exceeded. Figure 5.24 shows one type of flexible coupling that not only accommodates for misalignment, but also cushions shock loads and reduces torsional vibrations. The connecting element is a flexible tire that consists of synthetic tension members bound together in natural rubber. The flexible tire can be replaced without moving either the driver or the drive shaft. Horsepower ratings of 500 or more are obtainable. Typical misalignment capacities are as follows:

— Angular: 4°
— Linear: 1/8 in.
— End float: 5/16 in.

Figure 5.25 illustrates a chain-type coupling that can permit enough relative movements between the two hubs to accommodate slight angular and linear misalignments. It provides a strong, positive, and compact direct coupling of shafts. Quick disconnection of the two shafts is accomplished by removing the one coupling pin and unwrapping the chain.

FIGURE 5.25   Chain coupling. *(Courtesy of Dodge Manufacturing Division, Reliance Electric Company)*

5.11   OLDHAM COUPLING

Sometimes two shafts are parallel but not collinear owing to a slight but unavoidable offset. An *Oldham coupling* can be used to connect two such shafts (see Fig. 5.26). This coupling consists of two flanges (parts 1 and 3), each of which is rigidly connected to the end of its shaft. Each flange contains a machined keyway. A third piece (part 2) fits between the two flanges. This central piece has two integral keys that are perpendicular to each other. Each key of the central piece fits loosely into a keyway of each flange. As the shafts rotate, the central piece slides in the keyways as a result of the shaft offset. Proper selection of materials is essential because of heat generation at the sliding surfaces.

5.12   UNIVERSAL JOINT

The *universal joint* is used to connect two nonparallel shafts whose centerlines intersect, as shown in Figure 5.27. A universal joint consists basically of two clevises (parts 1 and 2), each of which is attached to an end of a shaft. Two pins (parts 3 and 4) are used to connect the two clevises together at right angles. Thus the two clevis pins are perpendicular to each other, with the small pin (part 3) passing through a hole in the

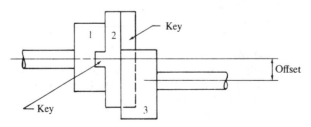

FIGURE 5.26   Oldham coupling

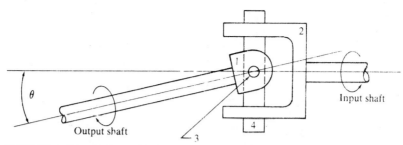

FIGURE 5.27   Universal joint

large pin (part 4). The angle between the shaft centerlines is shown as θ, which may be as large as 30°. The shaft centerlines intersect at the center of the two pins. Figure 5.28 is a photograph of an actual universal joint that can operate at speeds up to 1750 rpm depending on the angle of operation.

A single universal joint does not provide a constant speed ratio during each complete shaft revolution. The variation in the speed ratio depends on the misalignment angle θ. To eliminate the variation in speed ratio, a double universal joint can be used, as shown in Figure 5.29.

The angle θ between the input and middle shafts must equal the angle θ between the middle and output shafts. In this way, any speed variation existing for the middle shaft is canceled for the output shaft. This angular arrangement is in the automobile,

FIGURE 5.28   Single universal joint *(Courtesy of Lovejoy, Inc.)*

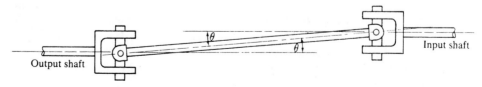

FIGURE 5.29   Double universal joint

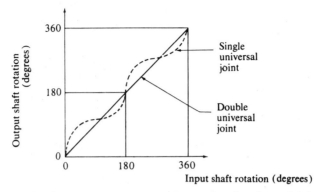

FIGURE 5.30   Angular displacements of universal joint output and input shafts

where the drive shaft is connected to the rear end by one universal joint and to the transmission by a second universal joint.

Figure 5.30 shows the relationship between the angular displacements of the input and output shafts for one revolution. With a double universal joint, the relationship is completely linear because the speed ratio is held constant.

5.13   TORQUE-LIMITING COUPLINGS

In Section 5.10 it was shown that a sleeve coupling could be used as a protective device to limit the amount of torque that could be transmitted to a load and thus prevent damage to the machine. At the maximum permissible torque, a pin would shear, causing the coupling to slip with no subsequent torque transmission.

There are other types of torque-limiting couplings in which there is no resulting damage when the coupling slips. The coupling merely slips until the excessive torque is removed. After the cause of the overload has been eliminated, the torque-limiting coupling is immediately ready for operation without any need for maintenance, such as resetting or part replacement.

Tools such as power screwdrivers and wrenches have devices that limit the torque transmitted to a screw or bolt. If excessive torque is applied by the driving member (which is connected to the screwdriver or torque wrench), the torque-limiting device prevents the driven member from receiving any additional torque beyond the desired value. Hence the driving member will continue to rotate, but the driven member will remain stationary. A related device is employed in the toe release of some ski bindings.

Figure 5.31 shows a device used to limit torque. The driving shaft is engaged with the driven member by means of two spring-loaded balls contacting holes in the driving shaft (Figs. 5.31 and 5.32). When the torque reaches the torque-limiting value, the balls are forced out of the holes against the spring, thus disengaging the two rotational parts. Various torque-limiting values can be obtained by changing the setting of the adjusting screws.

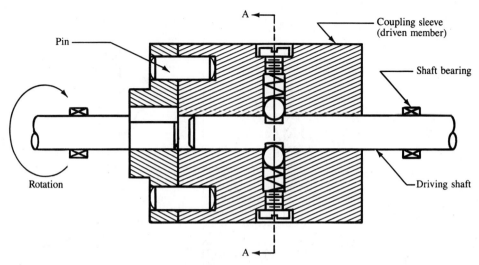

FIGURE 5.31   Torque-limiting coupling

Note that when the torque-limiting value is reached, a clicking noise is generated as a result of the balls clicking into and out of the holes. This alerts the operator.

In order to determine the spring force needed to obtain a specified torque, we need to solve a moment equation and relate the torque to the force trying to unseat the ball.

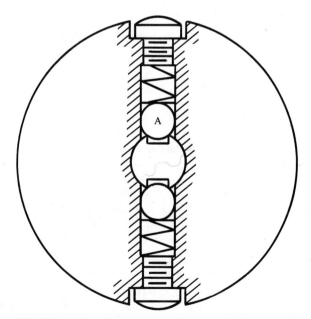

FIGURE 5.32   Section A–A of Figure 5.31

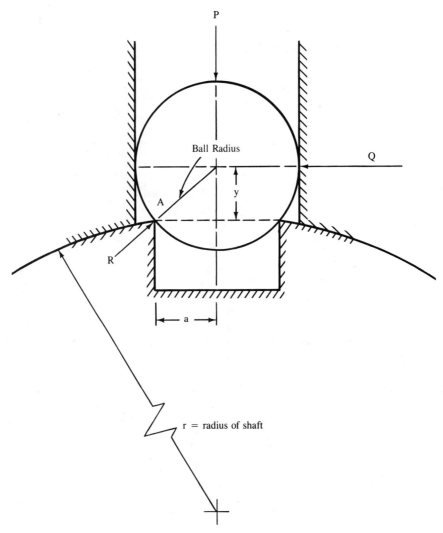

FIGURE 5.33   Forces acting on Ball *A* of Figure 5.31

The moment equation is the one we started with back in Chapter 4, Section 4.2, to develop Equation 4.1. Refer to Figure 5.33. *P* is the spring force; *Q* represents the force just necessary to push the ball out of the hole, and it is the result of the torque applied. Taking moments about point *A*, we have

$$\sum M_A = 0 = Pa - Qy \qquad\qquad \textbf{(5.18)}$$

(From the geometry of the setup, the distance *y* can be obtained.)

For balls housed in the sleeve, add y to the shaft radius to get the torque arm of Q. Thus,

$$\text{Torque} = T = Q(r + y) \tag{5.19}$$

Where r is the radius of the shaft, not the ball.

If the balls are housed in the shaft and the ball-seat holes are in the sleeve, then y must be subtracted from r. Thus,

$$T = Q(r - y) \tag{5.20}$$

Equations 5.18 and 5.19 or 5.20 can be used to find torque if the spring force is known, as well as to find spring force if torque is known:

P = the spring force in pounds

a = half the diameter of the seat hole in inches

Q = the force necessary to unseat the ball in pounds

y = the distance from the shaft surface to the line of action of the force Q in inches

r = the radius of the shaft in inches

T = torque in lb-in.

PROCEDURE

Solving Torque-Limiting Coupling Problems

Normally, you will be expected to find T or P.

Step 1 From the geometry of the setup, calculate y.

Step 2 Apply Equations 5.18 and 5.19 or 5.20 in the proper sequence for your problem.

Note: If there is more than one ball, each ball takes its fair share of the load.

SAMPLE PROBLEM 5.7

Torque-Limiting Coupling

PROBLEM: A coupling similar to the one in Figure 5.31 is to transmit a torque of 38 lb-in. The shaft is 1½ in. in diameter, the balls are each 7/16 in. in diameter, and the ball-seat holes in the shaft are each 11/32 in. in diameter. Find the spring force.

Solution

Step 1   From the preceding data,

$$a = 11/64 \text{ in.}   \text{Ball radius } r_b = 7/32 \text{ in.}$$

$$y = \sqrt{r_b^2 - a^2} = \sqrt{(7/32)^2 - (11/64)^2}$$

$$= 0.135 \text{ in.}$$

Step 2   Torque T transmitted by one ball is 38/2 = 19 lb-in. Thus, Equations 5.18 and 5.19 are rearranged:

$$Q = \frac{Pa}{y} = \frac{T}{r + y}$$

$$P = \frac{Ty}{a(r + y)} = \frac{19 \times 0.135}{0.172(0.75 + 0.135)}$$

Answer   Spring force $P = 16.8$ lb

5.14   SHAFT CALCULATIONS IN SI UNITS ▬▬▬▬▬▬

Sample Problem 5.8 below will demonstrate the use of SI units in solving a shaft problem. Remember that the basic units we must use in the SI system are the kilogram, newton, meter, pascal, and watt. The unit of work is the joule (J). One joule equals one newton-meter (1 J = 1 N-m). One watt equals one joule per second (1 W = 1 J/s). Torque is also in units of newton-meters. The equation for power is one we have discussed before:

$$P \text{ (watts)} = \frac{\text{work}}{t} = \frac{2\pi \times T \text{ (N-m)} \times n \text{ (rpm)}}{60 \text{ s/min.}} \qquad (5.21)$$

Note that since 1 W = 1 J/s, we must change the speed n from rpm to rps by dividing by 60 s/m.

SAMPLE PROBLEM 5.8 _____

SI Units

PROBLEM: A 35 mm diameter steel shaft is 500 mm long and must transmit 20 kW of power at 500 rpm. Find (**a**) the torque transmitted, (**b**) the maximum shear stress developed, and (**c**) the angle of twist.

Solution

(**a**)   Rearrange Equation 5.21 and solve for torque.

$$T = \frac{P \text{ (W)} \times 60 \text{ (s/min)}}{2\pi n} = \frac{20,000 \times 60}{2 \times 3.14 \times 500} = 382 \text{ N-m}$$

150 rpm

(b) Obtain the equation for J from Table 5.2 and then solve for Equation 5.3 for shear stress.

$$J = \frac{\pi \times d^4}{32} = \frac{3.14(0.035)^4}{32} = 147 \times 10^{-9}\,m^4$$

$$\tau = \frac{Tc}{J} = \frac{382 \times (0.035/2)}{147 \times 10^{-9}} = 45.5 \times 10^6\,Pa \quad (45.5\,MPa)$$

(c) Solve Equation 5.2 (obtain G from the Appendix).

$$\theta = \frac{TL}{GJ} = \frac{382\,N\text{-}m \times 0.5\,m}{83 \times 10^9\,Pa \times 147 \times 10^{-9}}$$

$$= 0.0157\,rad$$

$$= 0.0157\,rad \times 57.3^0/rad = 0.90°$$

5.15 SUMMARY

When solving for the bending moments on a shaft, resolve all the forces into vertical and horizontal components, construct the vertical and horizontal moment diagrams, determine the location where the maximum combined moment occurs, and then solve for it by considering the vertical and horizontal moments at that location as vectors used in the Pythagorean theorem.

Two methods are used in this text to determine the diameter of a shaft subjected to a combination of bending and torsional loads. The first method uses the Pythagorean theorem to obtain an equivalent torque from the moment and torque values. This equivalent torque value is then placed in the torsional shear stress formula. The second method averages the moment and equivalent torque values to obtain an equivalent moment. This value is then placed in the bending stress formula. The method that requires the largest shaft controls the design.

Shaft vibrations become dangerous when the rotational speed reaches a value equal to the natural vibration of the shaft. This point is called the *critical speed*. Good design requires that the operating speed of the shaft be at least 20% away from the critical speed.

A flexible shaft consists of a helical winding of a small-diameter wire. Such a shaft is employed where large bends are required or where the shaft must avoid some obstacle. For maximum torque-carrying capacity, flexible shafts should rotate in the same direction as the helical winding.

Keys are mechanical devices that transfer power between a shaft and the mounted component such as a gear or pulley. The keys themselves are subject to shear or compression failure. Therefore, both the shear and compressive critical areas must be investigated.

There are numerous types of couplings used to connect shafts to each other. A rigid coupling usually employs either flanges mounted on the shafts and bolted together or a sleeve that slips over the shafts and is held in place by pins driven through the sleeve and shafts. Rigid couplings require almost perfect alignment of shafts. Flexible couplings, on the other hand, allow some misalignment of the shafts. Again, there are numerous types. Some use a rubber and fabric combination to obtain flexibility, and others may employ chains (as shown in Fig. 5.25).

There are other types of couplings that allow for specific types of misalignment. The Oldham coupling is used when the shafts are parallel but are not collinear. The universal joint is used when the angle of misalignment is fairly large.

A common type of torque-limiting coupling employs spring-loaded balls that are seated in small holes. If the torque exceeds a specified value that is controlled by the spring compression, the balls will ride out of the holes and thus disengage the shafts.

5.16   QUESTIONS AND PROBLEMS ■■■■■■■■■■■■■■■

Questions

1. Name three applications for rectangular shafts.
2. Name one advantage of plastic shafts as compared to metallic shafts.
3. Name one advantage of a hollow shaft.
4. What is meant by the critical speed of a shaft? Why do critical speeds exist?
5. Name three applications of flexible shafts.
6. What is the purpose of the casing of a flexible shaft?
7. What is the significance of the direction of rotation of a flexible shaft?
8. Name one advantage of a Woodruff key.
9. What is a spline? When would a spline normally be used?
10. When a rigid coupling is used, what requirements must be met for each bolt to carry an equal share of the load?
11. Why are flexible couplings frequently used instead of rigid couplings?
12. What is an Oldham coupling, and when would it be used?
13. Give an application where a double universal joint is used.

Problems

1. A 1 in. diameter steel shaft rotates at 600 rpm and transmits 10 hp. Find the torque and the maximum shearing stress.
2. Solve Sample Problem 5.1, except make the shaft hollow with a 1¼ in. outside diameter and a 5/8 in. inside diameter.
3. Solve Sample Problem 5.2, except change the gear force to 212 lb.
4. A 30 in. long shaft is supported by bearings at each end. Two forces act at its midpoint. The force acting vertically down is 100 lb, and the horizontal force is 150 lb. Find the combined maximum bending moment.
5. Solve Sample Problem 5.3, except change the allowable tensile stress to 10,000 lb/in.^{2}
6. In the shaft system in Figure 5.34, power is received on pulley A, whose belt pulls are vertically down as shown. On the same shaft is gear B, which experi-

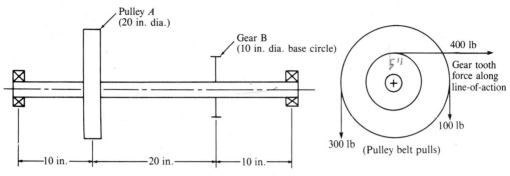

FIGURE 5.34   Sketch for Problem 6

ences a tooth force of 400 lb horizontally by delivering power to the second shaft. Find a suitable shaft diameter based on the equivalent torque method only. The allowable shear stress is 5000 lb/in.^{2} Neglect all weights.

7. Solve Sample Problem 5.4, but use a 1¼ in. diameter shaft.

8. A 1½ in. diameter steel shaft has a 100 lb disk mounted midway between two bearings that are 30 in. apart. Neglecting the weight of the shaft, find the system critical speed.

9. Solve Problem 8 above, except replace the 100 lb disk with two 50 lb disks. Each disk is placed 10 in. from a bearing.

10. A 2 in. diameter solid steel shaft that spans 40 in. between bearings has three 50 lb gears mounted on it. The gears are placed 10 in. from each other and from the bearings. At times, all the gears may be disengaged. Neglect the weight of the shaft, and find the system critical speed.

11. Solve Sample Problem 5.5, except double the speed.

12. A 1¾ in. diameter shaft rotates at 150 rpm and transmits 20 hp. It is keyed to a 9 in. diameter pulley. Find the size of square key required if the allowable shear stress is 5000 lb/in.^{2} and the allowable compressive stress is 15,000 lb/in.^{2}

13. Find the size of a flat key if it is used in Problem 12 above.

14. Solve Sample Problem 5.6, but make the bolt circle diameter 5 in.

15. A flange coupling made for eight 1/4 in. diameter bolts on a 4 in. diameter bolt circle was selected for a shaft connection but was not available. From a strength point of view, can a flange coupling with four 3/8 in. diameter bolts on a 6 in. diameter bolt circle be substituted?

16. Solve Sample Problem 5.7 using a shaft diameter of 1⅝ in.

17. The torque-limiting coupling in Figure 5.35 uses two spring-loaded balls of 5/8 in. diameter and a shaft of 2 in. diameter. If each ball-seat hole in the sleeve has a 1/2 in. diameter and the spring force is 15 lb, find the torque at which the coupling begins to slip.

18. Figure 5.36 is a plan view of a shaft power transmission system. A 50 hp motor drives the main shaft at 1000 rpm. Gear B pulls off 30 hp, and pulley C absorbs the remaining 20 hp, as shown in the plan view. It is assumed that the bearings are frictionless. Views X and Y are elevation views that show the

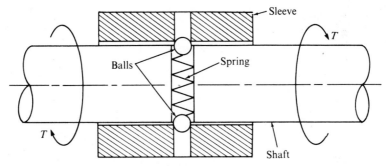

FIGURE 5.35   Sketch for Problem 17

force-transmission orientations for gear A and pulley D. Ignoring all weights, find the minimum diameter of the main shaft. The allowable shearing and tensile stresses are 5000 lb/in.^{2} and 10,000 lb/in.^{2}, respectively. The pulley belt tension ratio F_1/F_2 equals 3.

(Note: Gear A dia. is base circle Dia.)

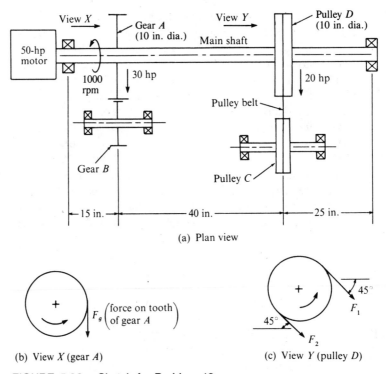

(a) Plan view

(b) View X (gear A)

(c) View Y (pulley D)

FIGURE 5.36   Sketch for Problem 18

SI Unit Problems

19. A 30 mm diameter shaft transmits 15 kW at 700 rpm. Find the maximum shearing stress in megapascals.
20. Solve Sample Problem 5.8, except change the shaft speed to 150 rpm.
21. A power transmission shaft is set up similar to Figure 5.5. The shaft has a span of 1500 mm between bearings. The gear, which has a base diameter of 250 mm, is placed 300 mm from bearing A and has a 500 N horizontal force applied to it. The pulley has a diameter of 600 mm, is placed 400 mm from bearing D, and has a total vertical force of 1500 N applied to it. Find the maximum combined moment in newton-meters.
22. We are given a key with these dimensions: Length, 70 mm, width 10 mm, and height 7 mm. The key is placed in a 40 mm diameter shaft that transmits 120 kW at 1800 rpm. Find the compressive and shear stresses developed in the key.

5.17 PROJECT ▬▬▬▬▬▬▬▬▬▬

Torque-Limiting Device

Tools such as power screwdrivers and wrenches have devices that limit the torque transmitted to a screw or bolt. If excessive torque is applied by the driving member (which is connected to the screwdriver or torque wrench), the torque-limiting device prevents the driven member from receiving any additional torque beyond the desired value. Hence the driving member will continue to rotate, but the driven member will remain stationary.

Design a torque-limiting device that will permit the torque to be applied in both the clockwise and counterclockwise directions and the torque limiter must accept either direction of rotation. Adjustments for varying the limiting torque values are to be available. The maximum amount of torque to be transmitted is 200 lb-in.

6

Analysis and Applications of Cams

Objectives

After completing this chapter, you will be able to

- Describe the most common types of cams and their functions and motions.
- Use the proper cam design terminology.
- Explain the reasons for each of the plate cam configurations and the applications of each.
- Design plate cam profiles and construct the related displacement diagrams.

6.1 INTRODUCTION ▬▬▬▬▬▬▬▬▬▬▬▬▬▬▬▬▬▬▬▬

Cams are among the most versatile mechanisms available. A *cam* is a simple two-member device. The input member is the cam itself, while the output member is called the *follower.* Through the use of cams, a simple input motion can be modified into almost any conceivable output motion that is desired. Some of the common applications of cams are

— Camshaft and distributor shaft of automotive engine
— Production machine tools
— Automatic record players
— Printing machines
— Automatic washing machines
— Automatic dishwashers

The contour of high-speed cams (cam speed in excess of 1000 rpm) must be determined mathematically. However, the vast majority of cams operate at low speeds (less than 500 rpm) or medium speeds (500–1000 rpm). The profiles of low- or medium-speed cams can be determined graphically using a large-scale layout. In general, the greater the cam speed and output load, the greater must be the precision with which the cam contour is machined.

6.2 TYPES OF CAM CONFIGURATIONS ▬▬▬▬▬▬▬▬▬▬▬▬▬

Plate Cams. This type of cam is the most popular type because it is easy to design and manufacture. Figure 6.1 shows a *plate cam.* Notice that the follower moves perpendicular to the axis of rotation of the camshaft. All cams operate on the principle that no two objects can occupy the same space at the same time. Thus, as the cam rotates (in this case, counterclockwise), the follower must either move upward or bind inside the guide. We will focus our attention on the prevention of binding and attainment of the desired output follower motion. The spring is required to maintain contact between the roller of the follower and the cam contour when the follower is moving downward. The roller is used to reduce friction and hence wear at the contact surface. For each revolution of the cam, the follower moves through two strokes—bottom dead center to top dead center (BDC to TDC) and TDC to BDC.

Figure 6.2 illustrates a plate cam with a *pointed follower.* Complex motions can be produced with this type of follower because the point can follow precisely any sudden changes in cam contour. However, this design is limited to applications in which the loads are very light; otherwise the contact point of both members will wear prematurely, with subsequent failure.

Two additional variations of the plate cam are the *pivoted follower* and the *offset sliding follower,* which are illustrated in Figure 6.3. A pivoted follower is used when rotary output motion is desired. Referring to the offset follower, note that the amount of offset used depends on such parameters as pressure angle and cam profile flatness, which will be covered later. A follower that has no offset is called an *in-line follower.*

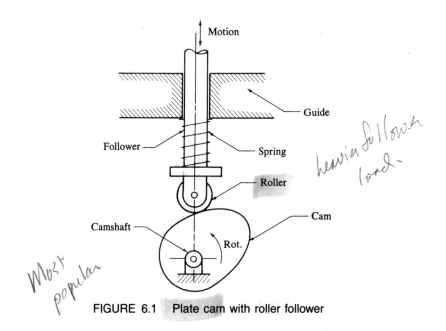

FIGURE 6.1   Plate cam with roller follower

Translation Cams.   Figure 6.4 depicts a *translation cam*. The follower slides up and down as the cam translates motion in the horizontal direction. Note that a pivoted follower can be used as well as a sliding-type follower. This type of action is used in certain production machines in which the pattern of the product is used as the cam. A variation on this design would be a three-dimensional cam that rotates as well as translates. For example, a hand-constructed rifle stock is placed in a special lathe. This stock is the pattern, and it performs the function of a cam. As it rotates and translates, the follower controls a tool bit that machines the production stock from a block of wood.

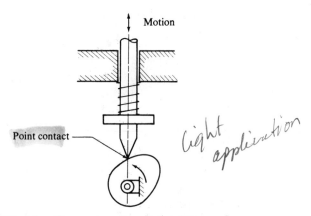

FIGURE 6.2   Plate cam with pointed follower

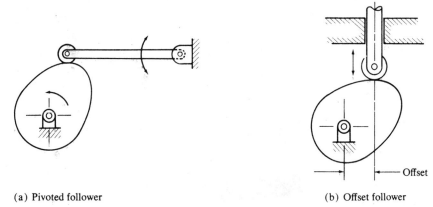

(a) Pivoted follower

(b) Offset follower

FIGURE 6.3   Variations of plate cam follower

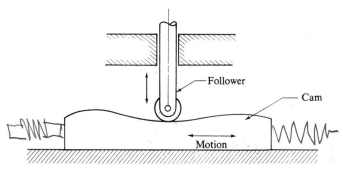

FIGURE 6.4   Translation cam

Positive-Motion Cams.   In the foregoing cam designs, the contact between the cam and the follower is ensured by the action of the spring forces during the return stroke. However, in high-speed cams, the spring force required to maintain contact may become excessive when added to the dynamic forces generated as a result of accelerations. This situation can result in unacceptably large stress at the contact surface, which in turn can result in premature wear. *Positive-motion cams* require no spring because the follower is forced to contact the cam in two directions. There are four basic types of positive-motion cams: the *cylindrical cam*, the *grooved-plate cam* (also called a *face cam*), the *matched-plate cam*, and the *Scotch yoke cam*.

Cylindrical Cam.   The cylindrical cam shown in Figure 6.5 produces reciprocating follower motion, whereas the one shown in Figure 6.6 illustrates the application of a pivoted follower. The cam groove can be designed such that several camshaft revolutions are required to produce one complete follower cycle.

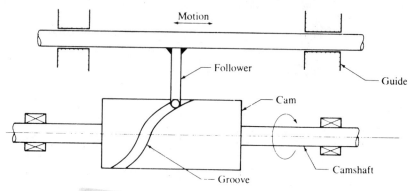

FIGURE 6.5 Cylindrical cam with reciprocating follower

FIGURE 6.6   Cylindrical cam with pivoted follower *(Courtesy of Commercial Cam Division, Emerson Electric Company)*

Grooved-Plate Cam.   Figure 6.7 illustrates a grooved-plate cam. The roller of the follower is guided inside a groove cut into the face of the cam. As the cam rotates, the follower translates.

Matched-Plate Cam.   In Figure 6.8 we see a matched-plate cam with a pivoted follower, although the design also can be used with a translation follower. Cams E and F rotate together about the camshaft B. Cam E is always in contact with roller C, while cam F maintains contact with roller D. Rollers C and D are mounted on a bell-crank lever, which is the follower oscillating about point A. Cam E is designed to provide the desired motion of roller C, while cam F provides the desired motion of roller D.

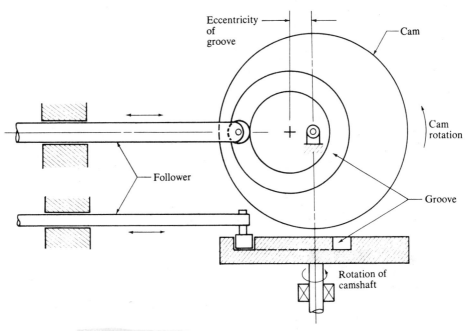

FIGURE 6.7   Grooved plate cam

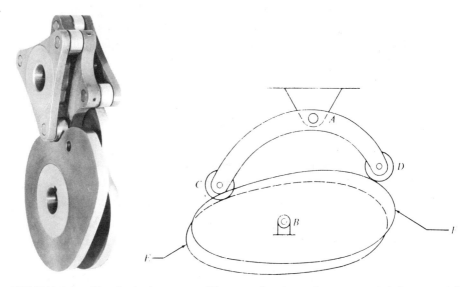

FIGURE 6.8   Matched plate cam *(Photo and schematic courtesy of Commercial Cam Division, Emerson Electric Company)*

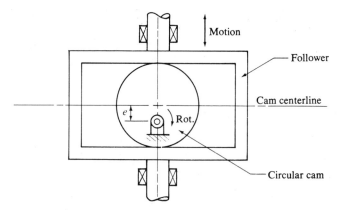

FIGURE 6.9   Scotch yoke cam

Scotch Yoke Cam.   This type of cam, which is depicted in Figure 6.9, consists of a *circular* cam mounted eccentrically on its camshaft. The stroke of the follower equals two times the eccentricity e of the cam. This cam produces simple harmonic motion with no dwell times. Refer to Section 6.8 for further discussion.

6.3   TYPES OF CAM FOLLOWERS

In Figure 6.10 we see the five most common types of cam followers. The simplest type is the sliding follower shown in Figure 6.10*a, b,* and *c.* Sliding followers can be either in-line or offset from the center of the cam. The three sliding followers show in Figure 6.10 make contact through either a point, a roller, or a flat surface. Figure 6.10*d* and *e* shows a pivot-type follower with either a roller or a flat face. It should be noted that the flat-face and knife-edge followers result in more wear than does the roller-type follower. Wear on flat-face followers can be minimized through proper design and

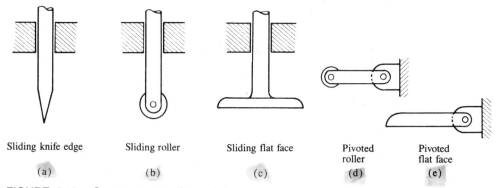

| Sliding knife edge | Sliding roller | Sliding flat face | Pivoted roller | Pivoted flat face |
| (a) | (b) | (c) | (d) | (e) |

FIGURE 6.10   Common types of cam followers

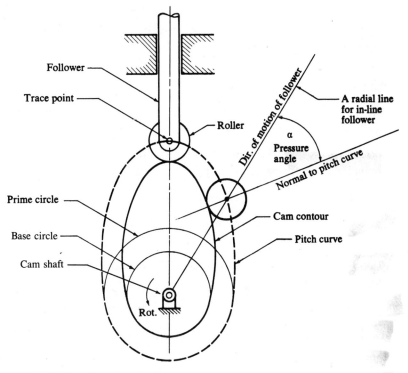

FIGURE 6.11   Cam terminology

lubrication. If the follower is used merely to actuate a switch, the flat-face and knife-edge followers present no problem. One example is the cam on the distributor shaft of an automobile engine. However, a roller-type follower or a properly designed and lubricated flat-face follower is needed in cams that operate heavily loaded mechanical components. An example is the camshaft in an automobile engine. Here the cams mounted on the camshaft operate push-rods (followers) to open and close spring-loaded valves in synchronization with the motion of the pistons.

6.4 CAM TERMINOLOGY

Before we become involved with the design of cams, it is desirable to know the various terms used to identify important cam design parameters. The following terms refer to Figure 6.11. The descriptions will be more understandable if you visualize the cam as stationary and the follower as moving around the cam.

Trace Point.   The end point of a knife-edge follower or the center of the roller of a roller-type follower.

Cam Contour.   The actual shape of the cam.

[margin note: Don't need to Know]

Base Circle.   The smallest circle that can be drawn tangent to the cam contour. Its center is also the center of the camshaft. The smallest radial size of the cam starts at the base circle.

Pitch Curve.   The path of the trace point, assuming the cam is stationary and the follower rotates about the cam.

Prime Circle.   The smallest circle that can be drawn tangent to the pitch curve. Its center is also the center of the camshaft.

Pressure Angle.   The angle between the direction of motion of the follower and the normal to the pitch curve at the point where the center of the roller lies.

Cam Profile.   Same as cam contour.

[margin note: Know]

BDC.   Bottom dead center, the position of the follower at its closest point to the cam hub.

[margin note: Know]

TDC.   Top dead center, the position of the follower at its farthest point from the cam hub.

[margin note: Raise maximum + 1 stroke]

Stroke.   The displacement of the follower in its travel between BDC and TDC.

Rise.   The displacement of the follower as it travels from BDC to TDC.

Return.   The displacement of the follower as it travels from TDC to BDC.

Dwell.   The action of the follower when it remains at a constant distance from the cam hub while the cam turns.

A clearer understanding of the significance of the pressure angle can be gained by referring to Figure 6.12. Here F_T is the total force acting on the roller. It must be normal to the surfaces at the contact point. Its direction is obviously not parallel to the direction of motion of the follower. Instead, it is indicated by the angle α, the pressure angle, measured from the line representing the direction of motion of the follower. Therefore, the force F_T has a horizontal component F_H and a vertical component F_V. The vertical component is the one that drives the follower upward and, therefore, neglecting guide friction, equals the follower F_{load}. The horizontal component has no useful purpose but is unavoidable. In fact, it attempts to bend the follower about its guide. This can damage the follower or cause it to bind inside its guide. Obviously, we want the pressure angle to be as small as possible to minimize the side thrust F_H. A practical rule of thumb is to design the cam contour so that the pressure angle does not exceed 30°. The pressure angle, in general, depends on the following four parameters:

— Size of base circle
— Amount of offset of follower
— Size of roller
— Flatness of cam contour (which depends on follower stroke and type of follower motion used)

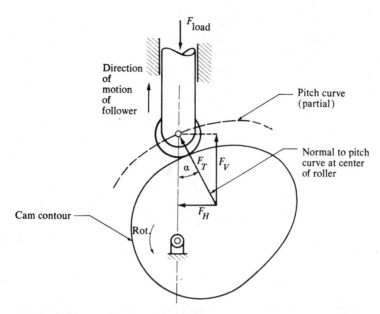

FIGURE 6.12   Cam pressure angle (α)

Some of the preceding parameters cannot be changed without altering the cam requirements, such as space limitations. After we have learned how to design a cam, we will discuss the various methods available to reduce the pressure angle.

6.5 TYPES OF FOLLOWER MOTIONS

For the high-speed automotive engine, the valves must be moved through stroke in approximately 0.01 s. It is a difficult problem to keep the follower accelerations low enough to avoid premature wear and yet achieve the proper valve movements. Another problem is a phenomenon called *jerk*, which is the time rate of change of acceleration:

$$\text{Jerk} = \frac{\Delta a}{\Delta t} \quad \text{(for very small values of } \Delta t\text{)}$$

where Δa = the change in acceleration

Δt = the time interval

Thus a rapid change in acceleration will result in a huge value of jerk, which causes the follower to receive an abrupt change in force. This induces a severe shock that contributes to noise, vibration, wear, and ultimate failure. It is important, therefore, that follower motion be carefully analyzed.

The four most widely used types of follower motions are *constant velocity motion*, *constant acceleration motion*, *simple harmonic motion*, and *cycloidal motion*. We will discuss these types of follower motions to determine their dynamic characteristics. Particular attention must be given to the beginnings and ends of each stroke of the follower. These are the locations where huge accelerations and jerks are most likely to occur. The reason for this is quite simple: The follower has to come to a dead stop at the end of each stroke and then, after a short rest period, start off the beginning of the next stroke in the opposite direction.

For convenience sake only, let us assume in each of the following explanations that the follower undergoes this cycle:

1.   It rises the full stroke (BDC to TDC) during the first 90° of cam rotation.
2.   It dwells during the second 90° of cam rotation (*dwells* means that the follower is stationary).
3.   It returns the full stroke (TDC to BDC) during the third 90° of cam rotation.
4.   It dwells during the final 90° of cam rotation.

This cycle repeats itself over and over for each revolution of the camshaft on which the cam is mounted.

6.6   CONSTANT VELOCITY MOTION

Constant velocity motion is characterized by the constant velocity of the follower during the rise and return strokes. Constant velocity motion is represented by the three diagrams in Figure 6.13. Figure 6.13a is called the *follower displacement diagram.* The coordinates are S for follower displacement and θ for cam rotation angle. By convention, θ equals 0° at BDC just after completing any dwell that might exist while the cam is at BDC. The straight-line relationship on the displacement diagram provides constant follower velocity because

$$\theta = \omega t$$

where ω (omega) is the constant angular velocity of the cam and t is time.

Note that we are now using the Greek letter ω for angular velocity, where previously n was the symbol. Angular velocity normally is measured in radians per second and the angle θ is measured in radians. However, there are times when rpms can be used for ω if we want θ in units of revolutions (see Sample Problem 6.1).

The equation above indicates that θ is essentially a time (t) axis. Also,

$$V = \frac{\Delta S}{\Delta t} \quad \text{and} \quad a = \frac{\Delta V}{\Delta t}$$

where V = the velocity of the follower

ΔV = the change in velocity of the follower

S = the displacement (stroke) of the follower

ΔS = the change in displacement of the follower

Δt = the time interval used

a = acceleration

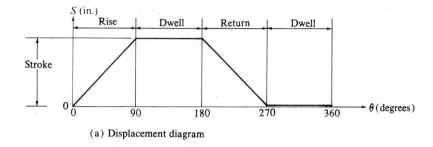

(a) Displacement diagram

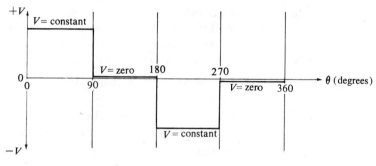

(b) Velocity diagram

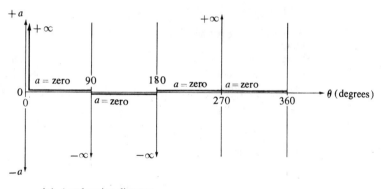

(c) Acceleration diagram

FIGURE 6.13 Constant velocity motion of a cam follower

At first glance, constant velocity (see Figure 6.13b) looks like a desirable type of motion. However, the velocity must increase from zero at 360° to some finite value at 0°. This increase theoretically requires an infinite acceleration spike at 0°.

Because 0° and 360° are actually the same point of the repeating cyclic curve, the velocity is expected to change over essentially a zero time change. The same phenomenon occurs at 90°, 180°, and 270°. Although we know that the acceleration will not actually reach infinity at these points, the value certainly will be too high for a medium- to high-speed cam. Thus constant velocity motion is reserved for low-speed cams. The actual accelerations do not become infinite at 0°, 90°, 180°, and 270° because of the elasticity of the system. This elasticity reduces the dynamic forces to a level that is still far too large for high-speed cams, causing component deflections and undesirable vibrations. Figure 6.13c shows the acceleration characteristics with the four theoretically, infinitely large acceleration spikes.

PROCEDURE

Determining Constant Velocity Motion of a Cam Follower

Step 1   Draw the S and θ axes and mark off the stroke on the S axis (see Figure 6.13a). On the θ axis, mark off one revolution (360°). Construct the displacement diagram from the data given.

Step 2   Calculate the time for one revolution and, in turn, the time for each required segment of the diagram. Use simple ratio-and proportion equations.

Step 3   Knowing the time and distance for the rise and return strokes, calculate the velocity and construct the velocity diagram.

Note: The acceleration is zero (except for the spikes) because the velocity is constant.

SAMPLE PROBLEM 6.1

Constant Velocity Motion

PROBLEM: A camshaft is rotating at 400 rpm, and the total rise of the follower is 2 in. The rise occurs in 72°, dwell for 108°, return in 90°, and dwell for 90°. Construct the three diagrams for displacement, velocity, and acceleration.

Solution

Step 1   See Figure 6.14.

Step 2   Time for one revolution ($\theta = \omega t$):

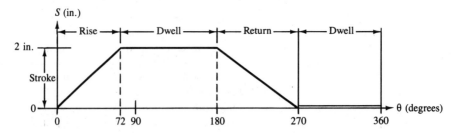

FIGURE 6.14   Displacement of a cam follower

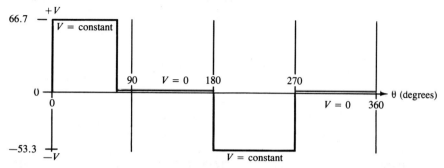

FIGURE 6.15 Velocity of a cam follower

$$t = \frac{1 \text{ rev.}}{400 \text{ rpm}} = 0.0025 \text{ min} \times 60 \text{ s/min} = 0.15 \text{ s}$$

Time for 72° (use ratio and proportion):

$$\frac{t}{0.15} = \frac{72°}{360°} \qquad t = 0.03 \text{ s}$$

Time for 90°:

$$\frac{t}{0.15} = \frac{90°}{360°} \qquad t = 0.0375 \text{ s}$$

Step 3 See Figure 6.15.

$$\text{Velocity of rise} = \frac{2 \text{ in.}}{0.03} = 66.7 \text{ in./s}$$

$$\text{Velocity of return} = \frac{-2}{0.0375} = -53.3 \text{ in./s}$$

Note: See Figure 6.16. The accelerations are zero except for spikes.

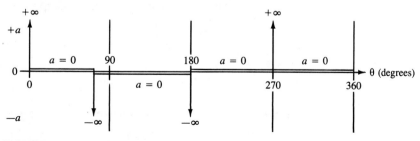

FIGURE 6.16 Acceleration of a cam follower

TABLE 6.1   Follower Displacement

Time t	Cam Angle θ	Follower displacement S	ΔS
0 units	0	0	
1 unit	1ω	$a/2$	$1(a/2)$
2 units	2ω	$4a/2$	$3(a/2)$
3 units	3ω	$9a/2$	$5(a/2)$
4 units	4ω	$16a/2$	$7(a/2)$
and so on			

6.7 CONSTANT ACCELERATION MOTION (PARABOLIC MOTION) ∎

A big improvement in dynamic characteristics will result if *parabolic motion* (constant acceleration) is used. For this type of motion, the follower will accelerate at a constant value during the first half of the rise and then decelerate at a constant value during the second half of the rise. Thus the maximum velocity is attained at the midstroke position. The same situation applies for the return stroke.

We can graphically generate the displacement diagram for parabolic motion by recalling that the motion equation is

$$S = \tfrac{1}{2} at^2$$

where a is the acceleration and is constant. By noting that $\theta = \omega t$, Table 6.1 can be derived, in which t is represented in arbitrary units of time.

The ΔS column in Table 6.1 indicates that the follower travels a distance of $a/2$ during the first unit of time, and during the second unit of time, it travels three times as far as it did during the first unit of time. It then travels five times as far during the third unit of time, seven times as far during the fourth unit of time, and so on. We can see that there is a specific progression in the distances traveled in each succeeding interval of time.

This observation leads us to a graphic means of deriving the follower displacement diagram. The steps are as follows.

PROCEDURE

Graphically Determining the Parabolic Motion Displacement of a Cam Follower

Step 1   Draw the S and θ axes and mark off the stroke AB along the S axis (see Figure 6.17a).

Step 2   Then establish six points on the displacement diagram for the rise stroke (for a more precise curve, additional points can be used). To do this, break the first 90° into six increments of 15° each.

Step 3   Draw an oblique line AC at any convenient angle from the S axis starting at $S = 0$.

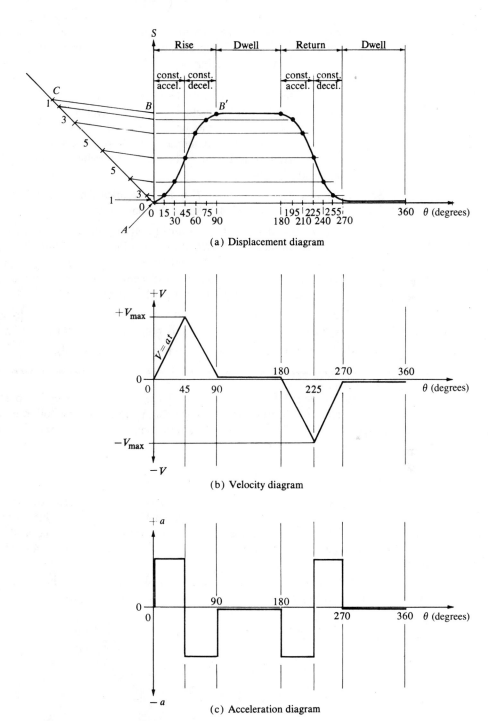

(a) Displacement diagram

(b) Velocity diagram

(c) Acceleration diagram

FIGURE 6.17   Parabolic motion of a cam follower

Step 4 Since you have selected six increments, break the line AC into length units of 1, 3, and 5 for acceleration. Then continue with length units of 5, 3, and 1 for deceleration. This gives you a total of six increments along line AC. A good approximate length of one unit is 1/4 in. because there are a total of 18 units of length along line AC. This gives you a full-scale length for line AC equal to $18 \times 1/4 = 4\ 1/2$ in. In order to obtain parabolic motion, an even number of increments (4, 6, 8, and so on) is required. This is so because the acceleration and deceleration phases must be identical parabolas in order to begin and end with dwells.

Step 5 Draw a straight line from C to B. Then draw a horizontal line from B until it reaches θ of 90°. Label this point B'.

Step 6 Draw the remaining lines from line AC to the S axis. Each of these lines is parallel to line CB. Project the lines horizontally in sequence to $\theta = 75°$, 60°, 45°, 30°, and finally, 15°. The six points just produced represent points on the displacement diagram. Draw a smooth curve through these points using a french curve.

Step 7 Repeat the process for the return stroke ($\theta = 180°$ to 270°) by extending the horizontal lines into this range at 15° increments.

Step 8 Add the two horizontal lines representing the two dwells.

An alternate method is to use the data in Table 6.2, which gives the percent of stroke for the percent of cam rotation for each rise or return of the follower. In this case, when laying out the displacement diagram, a convenient scale is one graduated to tenths of an inch or millimeters.

Parabolic Motion Velocity

The velocity diagram of the follower (see Figure 6.17b) consists of straight lines. During the first 45°, the velocity increases linearly until a maximum value is reached. From 45° to 90°, the velocity decreases linearly to zero, at which time a 90° dwell begins. The remainder of the velocity diagram follows a similar pattern for the return stroke.

Since the velocity curve is linear for the rise and return, the maximum velocity is twice the average velocity:

$$V_{\text{max}} = 2 \times V_{\text{avg}}$$

This is so because (with constant acceleration)

$$V_{\text{avg}} = \frac{V_0 + V_F}{2}$$

where V_0 = the initial velocity (which in this case is zero)

V_F = the final velocity (in this case V_{max})

TABLE 6.2   Relative Displacement Values for Cams

Percent of Cam Rotation Angle Achieved during Rise or Return	Percent of Stroke Achieved		
	Parabolic	SHM	Cycloidal
0	0	0	0
5	.5	.6	.08
10	2.0	2.4	.60
15	4.5	5.5	2.1
20	8.0	9.6	4.9
25	12.5	14.6	9.1
30	18.0	20.6	14.8
35	24.5	27.3	22.1
40	32.0	34.6	30.6
45	40.5	42.2	40.1
50	50.0	50.0	50.0
55	59.5	57.8	59.9
60	68.0	65.5	69.4
65	75.5	72.7	77.9
70	82.0	79.4	85.2
75	87.5	85.4	90.9
80	92.0	90.5	95.1
85	95.5	94.6	97.9
90	98.0	97.6	99.4
95	99.5	99.4	99.9
100	100	100	100

V_{avg} can be obtained from the distance and time data:

$$V_{avg} = \frac{\Delta S}{\Delta t}$$

Parabolic Motion Acceleration

The acceleration diagram (Figure 6.17c) shows that with the exception of key positions (θ = 0°, 45°, 90°, 180°, 225°, and 270°), the acceleration is constant as follows:

— Constant acceleration: θ = 0° to 45° and 225° to 270°
— Constant deceleration: θ = 45° to 90° and 180° to 225°
— Zero acceleration: θ = 90° to 180° and 270° to 360°

From an acceleration point of view, parabolic motion looks excellent. In fact, no other type of motion will yield as small a maximum value of acceleration when moving a follower through a given stroke during a given cam rotation angle. However, since jerk = $\Delta a/\Delta t$, we find that large, theoretically infinite values of jerk occur at θ = 0°, 45°, 90°, 180°, 225°, and 270°. Excessive jerk is responsible for noise and vibration problems, which hasten wear and ultimate failure. Thus parabolic motion is reserved for low- to medium-speed cams.

6.8   SIMPLE HARMONIC MOTION (SINUSOIDAL MOTION) ▬▬▬

In this section and the following one, calculus is used to explain the derivation of some equations. This is for the benefit of those students who are taking or have had a semester of calculus. Other students may disregard the derivations and simply apply the algebraic equations to the chapter problems.

Please refer to Figure 6.18. *Simple harmonic motion* (SHM) is the motion of the vertical projection of terminal point P of a rotating radius vector of a circle. The radius vector rotates at constant angular velocity ω_h, as shown in Figure 6.18. It is convenient to consider the angular displacement θ_h of the rotating radius to be zero when the radius vector is in its lowest position (bottom of circle). The linear displacement S of the follower becomes the vertical projection of point P, and hence S equals zero when P is located at the bottom of the circle rather than at the center of the circle. θ_h is used instead of θ to differentiate the radius vector angular displacement from the cam angular displacement. The radius of the circle is r. This circle is used to construct the displacement diagram for a simple harmonic motion cam, as shown in Figure 6.19a.

Let us just take a moment to list these symbols and terms and others that will be used in SHM equations.

S = follower displacement for any given cam angle (in.)

h = maximum rise of the follower (in.)

θ = angle of cam rotation (degrees).

γ = cam angle rotation to give maximum rise (or fall) of follower (radians).

r = length of radius vector (in.) Note that this equals $\frac{1}{2} h$.

θ_h = angle of radius vector rotation (degrees)

ω_h = rotational speed of radius vector (rad/s).

ω = rotational speed of cam (rad/s).

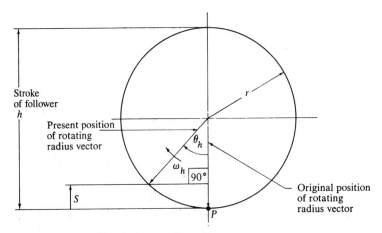

FIGURE 6.18   Simple harmonic motion

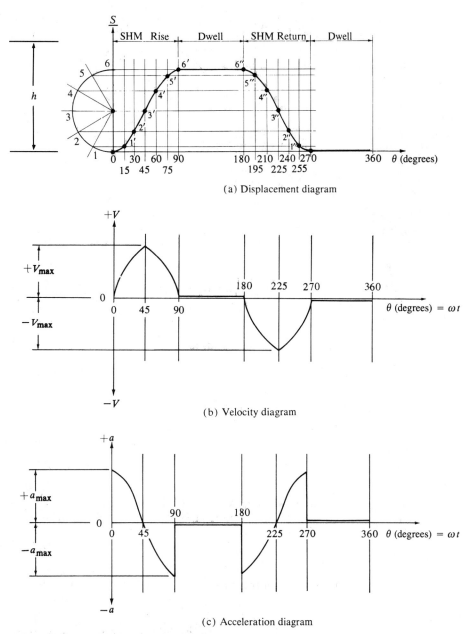

(a) Displacement diagram

(b) Velocity diagram

(c) Acceleration diagram

FIGURE 6.19 SHM of a cam follower

Notice that most of the angles have units of degrees. Strictly speaking, the units should be in radians. However, most of us are more familiar with finding trignometric functions of angles in degrees, so we will continue that practice in the following equations. For example, when you get to Equation 6.4 you must find the cosine of $\pi\theta/\gamma$. The angles θ and γ can be in degrees but you must convert π to $180°$ to be consistent. Again, Equation 6.4 includes the factors $\pi\omega/\gamma$. Since π/γ is a ratio of angles, γ can be in degrees if you convert π to degrees. Otherwise both should be in radians. Later on, in Equation 6.6, the angle γ can be in degrees when finding the cosine of "$2\pi\theta/\gamma$", but *must be in radians* when placed as a factor in $h\omega/\gamma$.

Note that in our example in Figure 6.19a, when point P has rotated $180°$ (that is, $\theta_h = 180°$), the angular displacement of the cam γ is just $90°$. The relationship between θ_h and γ is expressed by the proportion

$$\frac{\theta_h}{\theta} = \frac{180°}{\gamma}$$

which is usually written,

$$\theta_h = \frac{\pi}{\gamma}\theta \tag{6.1}$$

For the specific example represented by Figure 6.19, we have

$$\theta_h = \frac{180°}{90°}\theta = 2\theta$$

The follower displacement S at a radius vector rotation angle of θ_h can be determined as follows:

$$S = r - r\cos\theta_h = r(1 - \cos\theta_h)$$

where r is equal to one-half the maximum follower stroke.

It is desirable to have the preceding equation written in terms of h and θ. Therefore, by substituting $h/2$ for r and $(\pi/\gamma)\theta$ for θ_h, we get

$$S = \frac{h}{2}\left(1 - \cos\frac{\pi\theta}{\gamma}\right) \tag{6.2}$$

It can therefore be seen that the displacement diagram is basically a cosine type of wave. The velocity can be found by taking the derivative of S with respect to time:

$$V = \frac{ds}{dt} = \frac{d}{dr}(r - r\cos\theta_h) = r\sin\theta_h\left(\frac{d\theta_h}{dt}\right)$$

But

$$\frac{d\theta_h}{dt} = \omega_h = \text{angular velocity of rotating radius vector}$$

Substituting, we have

$$V = r\, \omega_h \sin \theta_h$$

Equation 6.1 tells us that for each degree the cam rotates, the radius vector rotates π/γ degrees. Moreover, when the cam rotates one revolution (2π), the radius vector rotates π/γ revolutions. Therefore,

$$\omega_h = \frac{\pi}{\gamma}\,\omega$$

With appropriate substitutions for r, ω_h, and θ_h, we obtain the general equation for velocity:

$$V = \frac{h}{2}\left(\frac{\pi}{\gamma}\,\omega\right)\sin\frac{\pi\theta}{\gamma} \tag{6.3}$$

For our specific example, we have

$$V = \frac{h}{2}\,2\omega \sin 2\theta = h\omega \sin 2\theta$$

Note that the velocity equation represents a pure sine wave. The acceleration can now be found by taking the derivative of V with respect to time:

$$a = \frac{dV}{dt} = r\omega_h \cos\theta_h\left(\frac{d\theta_h}{dt}\right)$$

$$= r(\omega_h^2)\cos\theta_h$$

Again, with appropriate substitutions, we obtain the general equation for acceleration:

$$a = \frac{h}{2}\left(\frac{\pi\omega}{\gamma}\right)^2 \cos\frac{\pi\theta}{\gamma} \tag{6.4}$$

Observe that in the preceding equation the acceleration changes as a cosine wave. Since a cosine wave is basically a sine wave displaced 90°, SHM is commonly referred to as *sinusoidal motion*. Also, the factors $r(\omega_h{}^2)$ represent the familiar centrifugal acceleration parameters.

When constructing a displacement diagram, you have three options: Use Equation 6.2, use the relative displacement data in Table 6.2, or use the graphic method shown in Figure 6.19a and explained here.

PROCEDURE

Graphically Determining the SHM Displacement of a Cam Follower

Step 1 Draw a semicircle whose diameter equals the follower's stroke, as shown in Figure 6.19a.

Step 2   Break the semicircle into equal sectors (six equal sectors of 30° are shown). Establish six points on the rise portion of the follower displacement diagram. Label the points on the semicircle as points 1, 2, 3, 4, 5, and 6.

Step 3   Break the cam rotation angle γ into the same number of equal increments as the semicircle. In our example, γ is 90° and the six increments are 15° each.

Step 4   Next, for our example, draw horizontal lines from points 1 through 6 to the corresponding points 1' through 6', which represent the 15° increments during the 90° rise. Draw a curve through the latter six points. This becomes the 90° SHM rise curve.

The velocity and acceleration curves require the use of Equations 6.3 and 6.4.

As can be seen from Figure 6.19, an SHM cam has good dynamic characteristics and thus is used for medium-speed applications. An SHM cam with a roller follower requires less power to operate than parabolic or cycloidal motion cams because the pressure angle will be smaller. An SHM cam with no dwell time is easy to design because it is simply an eccentrically mounted circular cam (see Figure 6.9, the Scotch Yoke cam). Figure 6.19c does indicate that a significant jerk (theoretically infinite) exists at $\theta = 0°$, 90°, 180°, and 270°, which accounts for the fact that SHM cams are not normally used at high speeds. Note that SHM does have a smoother acceleration curve than does parabolic motion.

6.9 CYCLOIDAL MOTION

Please refer to Figure 6.20. *Cycloidal motion* is experienced by a point Q on the periphery of a roller that is rolling on a flat surface. The total follower stroke h equals the circumference of the roller. θ_R equals the angle of rotation of the roller.

The cycloidal displacement diagram for the follower rise is produced by dividing the cycloidal path into equal increments, as shown in Figure 6.21a. Note that the roller rotates through 360° during the follower rise (which occurs in 90° of cam rotation in our example). Therefore, locate the roller at six incremental values of θ_R, which are 60°, 120°, 180°, 240°, 300°, and 360°.

At $\theta_R = 60°$, we have point Q_1 on the cycloidal curve. (See Figure 6.20 for a method of locating points on the cycloidal curve as a roller rotates through 360°.) Draw a horizontal line from Q_1 until it intersects θ of 15°. This intersection is point P_1. In a similar fashion, run lines Q_2P_2, Q_3P_3, and so on until reaching line Q_6P_6. The resulting follower cycloidal displacement curve goes through points P_0 through P_6.

Although Figure 6.21a does not show the displacement curve for the 90° return, all that is needed is to continue the six horizontal lines to the appropriate cam angles. For example, horizontal line Q_6P_6 would extend until in interseacts the 180° cam rotation angle. This point could be labeled P'_6, and the process could be continued until the six required points were located. Figures 6.21b and c shows the V-θ and a-θ curves, respectively, for the 90° rise.

degree rise.

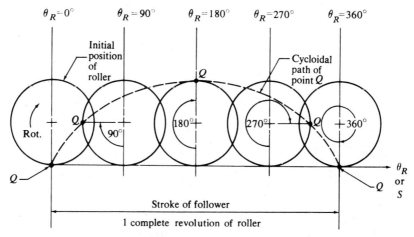

FIGURE 6.20   Cycloidal motion

The displacement of a follower having cycloidal motion can be represented by

$$S = \frac{h\theta}{\gamma} - \frac{h}{2\pi} \times \sin \frac{2\pi\theta}{\gamma} \tag{6.5}$$

where h = the total rise (stroke) of the follower in inches

S = the follower displacement in inches

θ = the cam rotation angle in degrees corresponding to follower displacement S.

γ = the angle in degrees the cam rotates while the follower is rising through its total stroke h.

The velocity equation is found by differentiating Equation 6.5 with respect to time:

$$V = \frac{dS}{dt} = \frac{h}{\gamma} \times \frac{d\theta}{dt} - \frac{h}{2\pi} \times \frac{2\pi}{\gamma} \times \frac{d\theta}{dt} \cos \frac{2\pi\theta}{\gamma}$$

Since $\omega = d\theta/dt$ = camshaft speed, the final result is

$$V = \frac{h\omega}{\gamma} - \frac{h\omega}{\gamma} \times \cos \frac{2\pi\theta}{\gamma} = \frac{h\omega}{\gamma}\left(1 - \cos \frac{2\pi\theta}{\gamma}\right) \tag{6.6}$$

Figures 6.21b and 6.22b show that the maximum velocity always occurs when

$$\theta = \frac{1}{2}\gamma$$

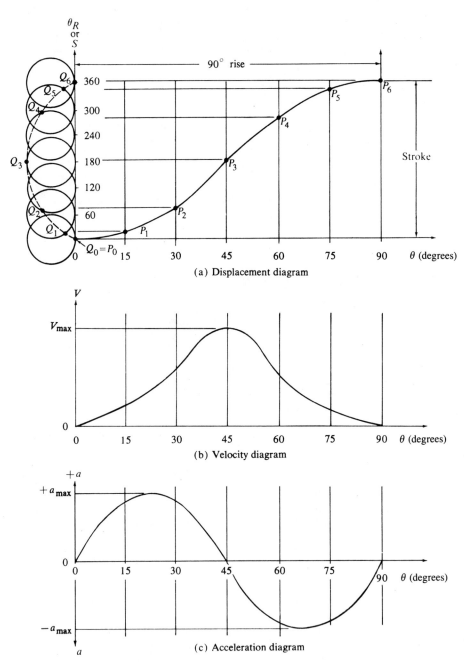

(a) Displacement diagram

(b) Velocity diagram

(c) Acceleration diagram

FIGURE 6.21   Cycloidal motion of a cam follower during rise time

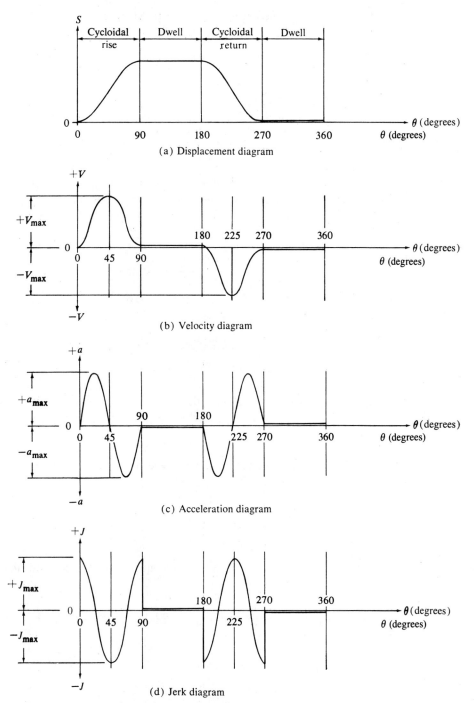

FIGURE 6.22   Cycloidal motion of a cam follower

Therefore, the cosine function in Equation 6.6 reduces to

$$\cos \frac{2\pi\gamma}{2\gamma} = \cos \pi = \cos 180° = -1$$

and Equation 6.6 becomes

$$V_{max} = \frac{2h\omega}{\gamma} \tag{6.7}$$

In Eq. 6.7 the units for θ, γ, and π will be radians, and ω will be in radians per second. The units for velocity will therefore be inches per second.

The acceleration equation can now be obtained by differentiating the velocity equation with respect to time:

$$a = \frac{dV}{dt} = 0 - \left(\frac{h\omega}{\gamma}\right)\left(\frac{2\pi d\theta}{\gamma\, dt}\right)\left(-\sin \frac{2\pi\theta}{\gamma}\right)$$

Simplifying,

$$a = \frac{2\pi h\omega^2}{\gamma^2} \times \sin \frac{2\pi\theta}{\gamma} \tag{6.8}$$

Figures 6.21c and 6.22c indicate that the maximum acceleration occurs when

$$\theta = ¼\, \gamma$$

Therefore, substituting in Equation 6.8 the sine function $= 1$ reduces the acceleration equation to

$$a_{max} = \frac{2\pi h\omega^2}{\gamma^2} \tag{6.9}$$

If we again differentiate the acceleration equation with respect to time, we obtain the equation for jerk J:

$$J = \frac{da}{dt} = \frac{2\pi h\omega^2}{\gamma^2} \frac{2\pi}{\gamma} \frac{d\theta}{dt} \cos \frac{2\pi\theta}{\gamma}$$

which reduces to

$$J = \frac{4\pi^2\omega^3 h}{\gamma^3} \times \cos \frac{2\pi\theta}{\gamma} \tag{6.10}$$

The equations for acceleration and jerk bring forth the following interesting facts concerning maximum accelerations and jerks for cycloidal motion:

1. *Increasing* the cam speed ω causes the maximum values of acceleration and jerk to *increase* in parabolic and cubic relationships, respectively.
2. For a constant value of γ, *increasing* the stroke h causes the maximum values of acceleration and jerk to *increase* linearly.
3. *Increasing* γ (the angle the cam rotates while the follower is rising through its total stroke) causes the maximum values of acceleration and jerk to *decrease* in parabolic and cubic relationships, respectively.
4. Not only is the maximum acceleration finite in value, this is also true for the maximum jerk. Therefore, no infinite accelerations or jerks will occur.

Figure 6.22 shows the motion curves for a full revolution of the cam. As can be seen, cycloidal motion is excellent for high-speed cams, since all infinite jerks have been completely eliminated. The elimination of jerks is accomplished because the acceleration curve is a pure sine wave, starting with zero acceleration and ending with zero acceleration. Thus, even with a dwell preceding and following the cycloidal rise, there would be no infinite jerks, as was the case with SHM. The manufacture of cycloidal cams requires the utmost dimensional precision if the excellent dynamic performance is to be realized. Therefore, in actual practice, the displacement diagram is mathematically calculated from an equation. Table 6.2 simplifies the calculation of points on the displacement diagram. The graphic method of Figure 6.21 is presented so that you will be able to gain a clearer understanding of cycloidal motion.

SAMPLE PROBLEM 6.3

Cam Displacement

PROBLEM: A follower is to have a rise of 2 in. for 100° of cam rotation. Find the displacement at 30° of cam rotation for **(a)** parabolic motion, **(b)** SHM, and **(c)** cycloidal motion.

Solution

Table 6.2 will be used. Thirty degrees is 30% of the cam rotation required for the rise. Find 30% in the left-hand column of the table and locate the corresponding percentage of stroke achieved for each type of motion:

(a) Parabolic motion displacement is 0.18 × 2 in. = 0.360 in.

(b) SHM displacement is 0.206 × 2 in. = 0.412 in.

(c) Cycloidal motion displacement is 0.148 × 2 in. = 0.296 in.

6.10 LAYOUT OF CAM CONTOUR

We have completed our discussion of the types of follower motions. Let us now develop the technique to design a cam contour that will impart the desired type of follower motion. Although the required points on a cam contour can be mathematically calcu-lated, we will use the graphic approach to provide you with a clearer understanding. For high-speed cams, however, the cam coordinates would be calculated instead of scaled from a large-scale layout.

The procedure for laying out the cam contour will now be presented in three sam-ple problems. In each sample problem, the follower displacement diagram will be iden-tical and will be assumed to have been derived previously.

To provide a clear picture of the cam layout process, 30° increments of cam rota-tion angle will be used. If greater accuracy is desired, smaller increments should be used. Normally, 10° or 15° increments are satisfactory. In each sample problem, the cam rotation is clockwise.

SAMPLE PROBLEM 6.4

Cam Design: SHM with an In-Line Roller Follower

PROBLEM: Make a layout of the cam profile of a plate cam having an in-line roller follower. The specifications are

> Base circle diameter = 2 in.
>
> Roller diameter = ⅝ in.
>
> Follower stroke = 1 in.
>
> Follower motion: SHM rise from θ = 0° to 180°
>
> Dwell from θ = 180° to 210°
>
> SHM return from θ = 210° to 330°
>
> Dwell from θ = 330° to 360°

Solution

Step 1   Mark off 30° increments along the θ axis of the follower displacement diagram, as shown in Figure 6.23a. The corresponding points on the displacement curve are calculated from the data in Table 6.2, rather than using the graphic method, and then are labeled 0, 1, 2, 3 through 12.

Step 2   Draw the vertical and horizontal centerlines of the camshaft (which in-tersect at point P), as shown in Figure 6.23b.

Step 3   Draw the base and prime circles with center at point P. The prime circle goes through the center of the roller when the follower is at BDC (roller at position Q_0). The base circle is tangent to the roller when the follower is at BDC.

Step 4   Divide the cam rotation angle in Figure 6.23b into 30° increments and label them from 0 through 12 in a counterclockwise direction, since the cam is rotating clockwise. Because we cannot rotate the cam clock-wise on the layout, we will assume that it is stationary and that the follower rotates about the cam in a counterclockwise direction. The numbers on the layout (Figure 6.23b) correspond to the identical num-bers (0 through 12) on the displacement diagram.

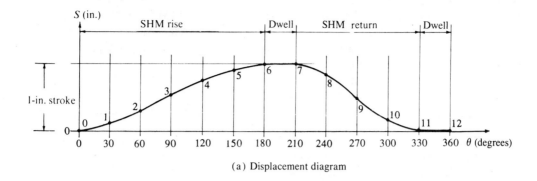

(a) Displacement diagram

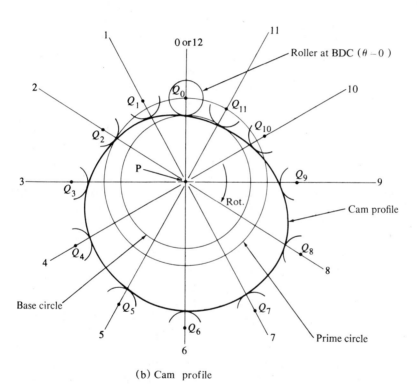

(b) Cam profile

FIGURE 6.23   Cam layout for in-line roller follower

Step 5   Using a pair of dividers, mark off the follower displacement at one of the points indicated on the displacement diagram. As an example, let us assume that this has been done for point 3 on the displacement diagram. Then put one needle point of the dividers on radial line 3 where the prime circle crosses (Figure 6.23*b*). Let the other needle point of the dividers fall radially outward on radial line 3. This point is labeled Q_3 and is the center of the roller at a cam rotation angle of 90°.

Step 6 Draw the bottom arc of the roller whose center is point Q_3.

Step 7 Repeat steps 5 and 6 for each of the remaining points.

Step 8 Using a french curve, draw a smooth curve that is tangent to the bottom arcs of all the rollers. Note that the cam profile during a dwell is a perfect arc of a circle if the follower is in-line. Therefore, use a compass to draw the cam contour during dwells.

It should be recognized that if the cam contour and roller size are given, it is possible to generate the corresponding follower displacement diagram. The procedure is merely the reverse of that used when making a cam layout.

SAMPLE PROBLEM 6.5

Cam Design: SHM Cam with a Pivoted Roller Follower

PROBLEM: The data are the same as for Sample Problem 6.4, except that we are to lay out a cam profile for a pivoted roller follower. The location of the follower pivot (point **A**) in relationship to the camshaft centerline is shown in Figure 6.24.

Solution

Steps 1 through 4 are the same as the corresponding steps in Sample Problem 6.4.

Step 5 Draw the pivot circle whose center is point P and which goes through the follower pivot point A.

Step 6 Draw a curved arc of radius AB (center at A) extending radially outward from the prime circle and terminating past the pivot circle, at point C.

Step 7 Break the pivot circle into twelve 30° increments starting with point A. Label the points on the pivot circle A_1 through A_{11} (pivot point A represents the positions A_0 and A_{12}. Draw the remaining arcs of radius AB with centers on points A_1 through A_{11}. These arcs also start at the prime circle and extend radially outward past the pivot circle.

Step 8 Draw a chord BD equal to the 1 in. follower stroke along arc BC. Because there are six 30° increments in the 180° rise, chord BD is divided into six incremental distances that are equal to the corresponding incremental distances on the follower displacement diagram. Dividers can be used to transfer the distances from the displacement diagram to the layout. For example, the first incremental distance is the displacement during the first 30°; the second incremental distance is the displacement during the second 30°. Thus the sum of all the incremental displacements of the follower displacement diagram equals the 1 in. stroke.

Step 9 Label the distances on chord BD as B_1 through B_6. Using a compass with its pin at point P, rotate each point on chord BD about the camshaft centerline until it meets its corresponding arc. For example, point B_3 is rotated until in reaches its corresponding arc at point B'_3.

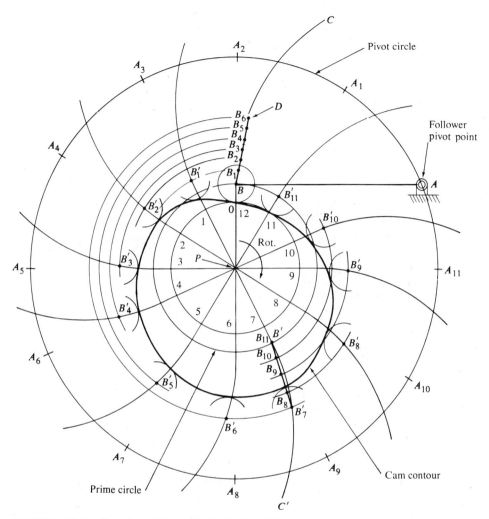

FIGURE 6.24 Cam layout for a pivoted roller follower

Step 10 Point B'_7 is located at the same radial distance as point B'_6 because the follower dwells for the next 30°.

Step 11 At point B'_7 draw a chord similar to BD. Label this chord $B'B'_7$, where B' is located back on the prime circle. Divide it into four incremental distances because the return takes place in 4 × 30°, or 120°. Take the incremental distances once again from the return phase of the displacement diagram. Then rotate the points on $B'B'_7$ as you did in step 9.

Step 12 Draw the bottoms of the rollers whose centers are located at B, B'_1, B'_2 through B'_{11}.

Step 13 The 30° of cam rotation between B'_6 and B'_7 and between B'_{11} and B are perfect arcs of circles to produce the dwells at TDC and BDC.

Step 14 Now draw a smooth curve tangent to the bottoms of all the roller bottoms. This cam is the desired cam contour.

SAMPLE PROBLEM 6.6 _____

Cam Design: SHM with a Flat-Face Follower

PROBLEM: The data are the same as for Sample Problems 6.4 and 6.5, except that we are to lay out a cam profile for a flat-face follower.

Solution

Steps 1 and 2 are the same as the corresponding steps in Sample Problem 6.4.

Step 3   Draw the base circle with center at point P, as shown in Figure 6.25 (there is no prime circle, so it cannot be used).

Step 4   Draw the flat-face follower at BDC. This is the location where the follower is tangent to the base circle (position 0 or 12 in Figure 6.25).

Step 5   Same as step 4 in Sample Problem 6.4.

Step 6   Same as step (5) in Sample Problem 6.4, except the follower displacement must be marked off from the *base* circle on the layout instead of the prime circle. Also, point Q_3 is not the center of the roller but the bottom of the flat-face follower *at its centerline.*

Step 7   At point Q_3 draw a line N_3 normal to radial line 3.

Step 8   Repeat steps 6 and 7 for the other positions. This will result in 12 normal lines (N_0 through N_{11}).

Step 9   Using a french curve, draw a smooth curve tangent to all the normal lines (N_0 through N_{11}). Note that the cam profile during a dwell is the arc of a circle. Therefore, use a compass to draw the cam contour from positions 6 and 7 and from 11 back to 0 or 12.

_____

The minimum width W of a flat-face follower is determined so that the cam will not contact the follower on its edges. It can be found from the cam layout by drawing in the follower at all 12 positions. This has been done for position 3 in Figure 6.25 as an example, and the point of contact is labeled C_3. In this way, the contact point, farthest away from the follower centerline can be found directly from the layout. The result is shown in Figure 6.26a, where dimensions L and M represent the distances from the farthest contact points on both sides of the centerline of the follower. A symmetrical follower can be made even though dimensions L and M are not equal. Increase the short side until it equals the longer side (side M in Figure 6.26). The result is a symmetrical follower of width $2M$, as shown in Figure 6.26b.

The follower should be manufactured with a width slightly larger than the value obtained from a layout to provide a margin of safety as the cam contact point approaches the edge of the follower. An adequate width for a symmetrical follower is $2M + \frac{1}{8}$ in.

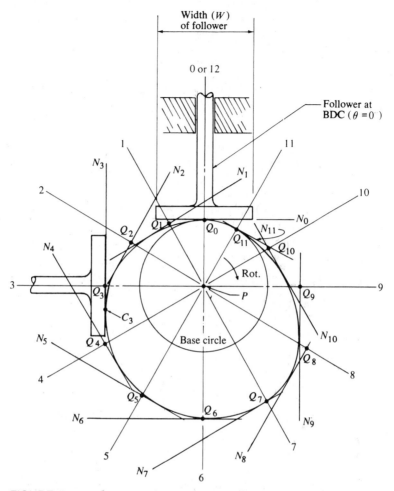

FIGURE 6.25 Cam layout for a flat-face follower

Flat-face followers are commonly used in automobile engines to open and close the intake and exhaust valves. Two distinct advantages of a flat-face follower as compared to a roller-type follower are as follows.

1. The pressure angle for flat-face followers with their faces perpendicular to the follower centerline is always zero. This is so because the normal to the cam contour at the point of contact with the follower is always parallel to the direction of motion of the follower, as shown in Figure 6.27. Thus flat-face followers are less likely to jam or bind in the guide. If the face of the follower is inclined to the centerline at some angle other than 90°, then the pressure angle is some constant value but not zero.

2. When small-size cams are used, a small-size roller is required for a roller follower. This very often results in excessive stresses in the pin of the roller.

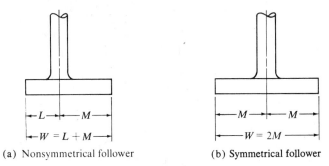

(a) Nonsymmetrical follower   (b) Symmetrical follower

FIGURE 6.26   Dimensions of a flat-face follower

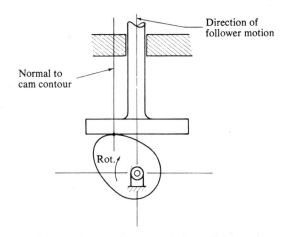

FIGURE 6.27   Zero pressure angle for a flat-face follower

There is, of course, greater wear when using a flat-face follower, as a result of sliding friction. This wear is usually minimized, however, because the area of contact between the follower and cam changes as the cam rotates. Thus, with proper lubrication and design, wear problems can be virtually eliminated for low- to medium-speed applications.

SAMPLE PROBLEM 6.7

Cam Design: SHM with an Offset Roller Follower

PROBLEM: The data are the same as for the previous sample problems, except that an offset roller follower will be used. The offset is in the proportion shown in Figure 6.28.

Solution

Refer to Figure 6.28. Steps 1 and 2 are the same as in Sample Problem 6.4.

Step 3 Draw the base and prime circles, as well as an offset circle, all with centers at the center of the camshaft (point *P*). The radius of the offset circle equals the amount of offset of the follower.

Step 4 Draw radial lines every 30° from the center of the camshaft. This step is similar to step 4 in Sample Problem 6.4, but in this case, do not label the lines and only draw them through the offset circle.

Step 5 Draw lines tangent to the offset circle at the points where the radial lines intersect the offset circle.

Step 6 Label the tangent lines in the same fashion that you labeled the radial lines in the previous problems. Start with the original vertical offset line as number 0 or 12.

Step 7 Using a pair of dividers, mark off the follower displacements on each tangent line. The displacements are measured from the prime circle along each tangent line to the position that the roller center will occupy. For example, the displacement at position 1 on the displacement diagram in Figure 6.23a will be marked off along tangent Line 1, shown on Figure 6.28. The roller center is shown as Q_1.

Step 8 Draw the bottom arcs of the roller positions, and using a french curve, draw a smooth curve tangent to the roller arcs.

Offsetting the follower is one method to reduce the pressure angle.

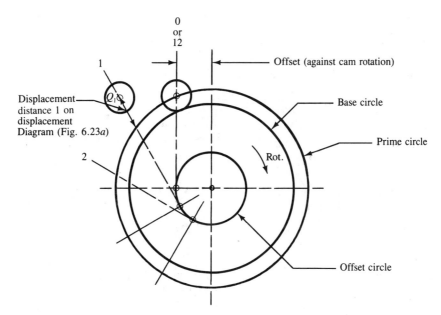

FIGURE 6.28 Offset cam construction

6.11 DETERMINATION OF MAXIMUM PRESSURE ANGLE α_{max} ▬

As was mentioned earlier, a general practice is to design a cam such that the maximum pressure angle does not exceed 30°. Too large a value can cause binding and excessive bending. The maximum pressure angle for a given cam can be found by referring to Figure 6.29. Let us first examine the follower displacement diagram of Figure 6.29a. The maximum pressure angle occurs when the follower is being pushed through its guide at maximum velocity. Therefore, the location of the maximum pressure angle α_{max} can be determined by locating the steepest slope of the rise portion of the displacement diagram. The velocity diagram also may be used to locate α_{max}. The critical cam rotation angle, which corresponds to the maximum pressure angle, will be called θ_{crit}. It is the critical cam rotation angle where binding is most likely to occur. In our specific example $\theta_{crit} = 45°$.

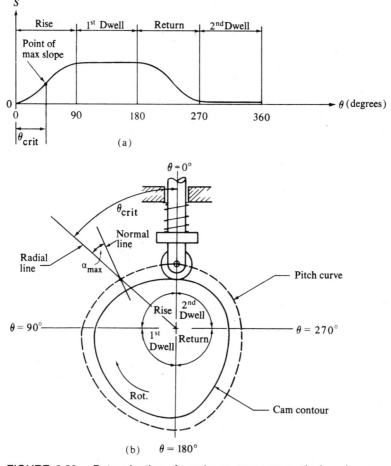

FIGURE 6.29   Determination of maximum pressure angle (α_{max})

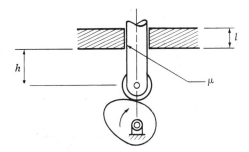

FIGURE 6.30   Parameters for finding $\alpha_{\text{max perm}}$

We can now refer to Figure 6.29*b*, which gives the cam contour for a clockwise cam rotation. The cam cycle is broken into the rise, the first dwell, the return, and the second dwell. The position of the follower is BDC at $\theta = 0°$. We can now measure θ_{crit} (as found from the displacement or velocity diagram) in a counterclockwise direction and draw a radial line. At the point where this radial line intersects the pitch curve, a normal to the pitch curve is drawn. The angle between the radial and normal lines is the maximum pressure angle, which can be measured with a protractor.

6.12 DETERMINATION OF MAXIMUM PERMISSIBLE PRESSURE ANGLE $\alpha_{\text{max perm}}$

The general rule of keeping the maximum pressure angle below 30° is usually satisfactory to prevent binding. However, it is possible to calculate a parameter called the *maximum permissible pressure angle* $\alpha_{\text{max perm}}$, which is defined as the pressure angle below which binding will not occur. Thus, if the actual maximum pressure angle of the cam is less than the maximum permissible value, the follower should not bind.

The value of the maximum permissible pressure angle can be calculated from the following equation, which refers to Figure 6.30:

$$\tan \alpha_{\text{max perm}} = \frac{l}{\mu(2h + l)} \tag{6.11}$$

where l = the length of guide in inches

            h = the maximum follower overhang outside of guide in inches

            μ = coefficient of friction between follower and guide

The value of μ depends on the materials of the follower and guide, as well as on the lubrication condition of the mating surfaces. It is apparent that the maximum permissible pressure angle can be increased by

—   Reducing the overhang h
—   Increasing the guide length l
—   Reducing the value of μ by using lubrication and a good selection of materials

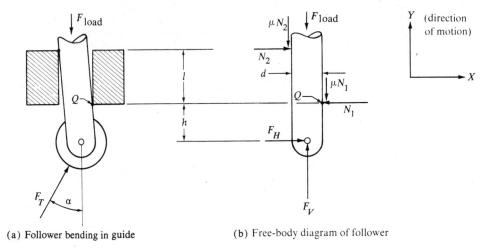

(a) Follower bending in guide   (b) Free-body diagram of follower

FIGURE 6.31   Force analysis of cam follower

Since Equation 6.11 is a theoretical equation, a margin of safety is added to ensure no binding. This is indicated in the following equation, which simply states that if α_{max} perm is greater than or equal to the *actual* maximum pressure angle α_{max} plus 6°, no binding will occur:

$$\alpha_{max\ perm} \geq \alpha_{max} + 6° \qquad\qquad (6.12)$$

A second reason for the 6° margin is to take into account the inherent error in graphically determining the value of α_{max}.

Two additional modifications that will improve the operation of a cam are

— Reduce the clearance between the follower and guide to lower the sideways looseness of the follower
— Increase the follower diameter so that its bending stiffness and thus bending deflection will be reduced in the sideways direction

You will be interested in the derivation of Equation 6.11. It is derived by applying the laws of statics to the follower during the rise. Figure 6.31a shows the follower inside its guide, while Figure 6.31b depicts a free-body diagram of the follower only. For analysis purposes, all forces have units of pounds and all dimensions have units of inches. As shown, the cam pushes on the roller with a total normal force F_T that can be broken into a horizontal component F_H and a vertical component F_V. Notice that the vertical component is the useful one that pushes the follower upward. On the other hand, the horizontal component is undesirable because it produces friction as it attempts to bend the follower about the guide. However, such friction is unavoidable. The clearance between the follower and guide is exaggerated in Figure 6.31a to show this bending effect.

The normal force F_T causes a counterclockwise bending moment that is resisted by the guide reactions N_1 and N_2. Note also that as the follower moves up the guide, frictional forces μN_1 and μN_2 oppose the motion. The following symbols are applicable,

where

F_T = the total applied normal force between cam and roller

α = the pressure angle

F_H = the horizontal component of the normal force ($F_H = F_T \sin \alpha$)

F_V = the vertical component ($F_V = F_T \cos \alpha$)

N_1, N_2 = the horizontal reaction forces between follower and guide

d = the diameter of the follower

μ = the coefficient of friction between follower and guide

l = the length of guide

h = the follower overhang outside of guide

F_{load} = the load the follower must overcome

We will now apply the two laws of statics:

1. *The sum of all the forces acting on the follower, along any direction, equals zero.* For the X axis, we have

$$F_T \sin \alpha + N_2 - N_1 = 0$$

For the Y axis, we have

$$F_T \cos \alpha - \mu(N_1 + N_2) - F_{load} = 0$$

2. *The sum of all the moments about any point equals zero.* If the moments are taken about point Q on the follower, the forces μN_1 and N_1 drop from the equation, and the result is

$$F_T \cos \alpha \frac{d}{2} - (F_T \sin \alpha)\, h - \mu N_2 d + N_2 l - F_{load} \frac{d}{2} = 0$$

where d is the follower diameter.

We therefore have the three preceding equations and three unknowns, F_T, N_1, and N_2. Simultaneous solution of the three equations yield a result for the total applied normal force:

$$F_T = \frac{l}{l \cos \alpha - (2\mu h + \mu l - \mu^2 d) \sin \alpha} F_{load}$$

This equation permits us to solve for the required total applied normal force if the other values are known.

Based on practical applications, F_{load} will always be finite in magnitude. Thus F_T can become infinitely large only when the denominator equals zero, as follows:

$$l \cos \alpha - (2\mu h + \mu l - \mu^2 d) \sin \alpha = 0$$

Since $\mu^2 d$ is negligibly small compared to $2\mu h$ or μl, it is disregarded, and the final result is obtained:

$$\frac{\sin \alpha}{\cos \alpha} = \tan \alpha = \frac{l}{\mu(2h + l)}$$

which is written as

$$\tan \alpha_{max\ perm} = \frac{l}{\mu(2h + l)}$$

Thus, when this equation is solved for α, we will know the maximum permissible pressure angle. Angles above this value will cause binding between the follower and guide.

SAMPLE PROBLEM 6.8

Maximum Permissible Pressure Angle

PROBLEM: A cam has a maximum pressure angle of 34°. The maximum overhang is 2 in., and the guide length equals 1 in. The coefficient of friction between the follower and guide is 0.2. Is the maximum pressure angle of 34° acceptable?

Solution

Based on the general rule of thumb, the 34° pressure angle is not acceptable. However, using Equation 6.11, we have

$$\tan \alpha_{max\ perm} = \frac{1}{0.2(4 + 1)} = 1$$

$$\alpha_{max\ perm} = 45°$$

From Equation 6.12,

$$45° > 34° + 6°$$

Answer: The 34° maximum pressure angle is acceptable when a more refined analysis is made.

6.13 METHODS TO REDUCE THE PRESSURE ANGLE ▬▬▬

If the pressure angle is too large, and binding is predicted, the following steps can be taken to reduce the pressure angle:

1. Increase the cam rotation angle for a given rise. This change provides a lower V_{max}, and thus the cam profile becomes less steep. Note that the timing of the mechanism may preclude this step.

2. Increase the diameter of the base circle, which makes the cam physically larger, and the cam profile becomes less steep. If space is at a premium, this change may not be possible.

3. Change to a different type of follower motion curve, for example, parabolic to SHM.

4. If the cam has an in-line follower, change its position to provide a specified amount of offset in the correct direction. Offsetting the follower in the correct direction will decrease the pressure angle during the rise, at the expense of increasing the pressure during the return. However, since binding is of no concern during the return, the increased pressure angle presents no problem. The following table shows the correct direction of offset to reduce the pressure angle during the critical rise portion of the follower motion:

Direction of Cam Rotation	Direction of Follower Offset
Clockwise	To left of cam centerline
Counterclockwise	To right of cam centerline

Figure 6.32 gives a pictorial representation of the correct direction in which to offset the follower. Figure 6.32a is for clockwise cam rotation, while Figure 6.32b gives the direction for counterclockwise cam rotation. The maximum recommended amount of offset is 30% of the base circle diameter.

5. Decrease the stroke of the follower if permissible.

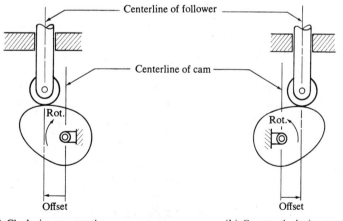

(a) Clockwise cam rotation   (b) Counterclockwise cam rotation

FIGURE 6.32   Offsetting of follower to reduce pressure angle

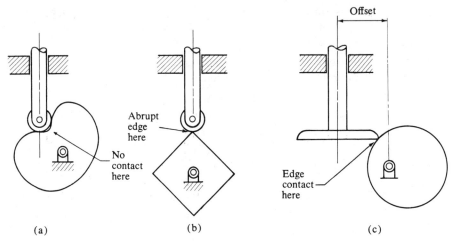

FIGURE 6.33   Unacceptable cam-follower contact for heavily loaded high-speed cams

6. If a pivoted follower is being used, select a different location for the pivot point.

6.14 PRACTICAL DESIGN CONSIDERATIONS FOR CAMS AND FOLLOWERS

When a cam is actuating a low-force mechanism such as an electric switch at low speeds (such applications are in clothes washers and dryers), there is little danger of excessive shock and subsequent cam surface wear. On the other hand, if the forces are significant and the speeds are high, a careful check of the compatibility of the cam-follower contact surfaces should be made. Figure 6.33 shows several unacceptable cam-follower contact relationships for high force-speed applications.

Figure 6.33a reveals a case where the contact path is not continuous because the roller is too large for the abrupt change in cam contour.

In Figure 6.33b we see a situation where the abrupt change in cam contour gives the follower four severe shocks per cam revolution.

Finally, we have an illustration in Figure 6.33c where the offset is too large for the size of the flat-face follower used. Consequently, the cam is engaging at an edge instead of on the flat-face surface.

A machining problem called *undercutting* can sometimes occur if a cam is too small for the amount of follower stroke and roller diameter. In Figure 6.34a we see a cam that has been machined properly, with no undercutting. The cam contour for the desired stroke of the follower is thereby attained during machining. The much smaller cam shown in Figure 6.34b has the same desired stroke as the cam depicted in Figure 6.34a. However, the smaller cam has a more pointed tip as a result of undercutting. As a consequence, the machining operation does not produce a cam contour that will provide the desired stroke.

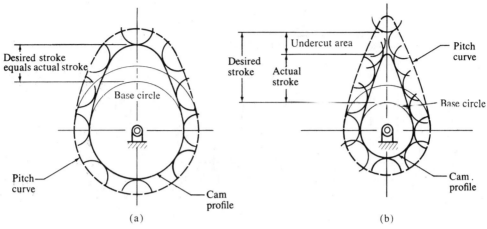

FIGURE 6.34   Undercutting of a small cam

The problem of undercutting can be solved by increasing the size of the base circle (make a larger cam) or by reducing the diameter of the roller (this gives greater contact stresses). As a cam becomes larger, the tendency toward vibration increases, and the cam becomes heavier and exhibits more inertia. Greater contact stress promotes cam-follower contact surface wear. Therefore, a compromise between the two proposed solutions should be made to eliminate the occurrence of undercutting.

6.15 CAM CALCULATIONS IN SI UNITS

Of course, the units for time and angles are the same in the SI system as in the U.S. Customary System. The units for displacement, velocity, and acceleration are meters, meters per second, and meters per second squared. The following sample problem will demonstrate the use of SI units.

SAMPLE PROBLEM 6.9

SI Units: Constant Acceleration (Parabolic) Motion Cam

A camshaft rotates at 400 rpm. The total rise of follower is 70 mm. The rise occurs in 120°, dwell occurs for 60°, return for 120°, and dwell for 60°. Construct the three diagrams for **(a)** displacement, **(b)** velocity, and **(c)** acceleration.

Solution

Refer to Figure 6.35.

(a) Displacement diagram: Table 6.2 is used as an aid. The following data are used to plot points for every 20% of cam rotation to achieve total rise or return.

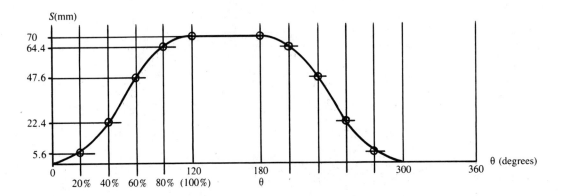

(a) Displacement Diagram

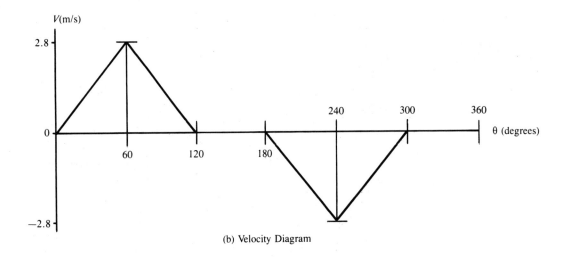

(b) Velocity Diagram

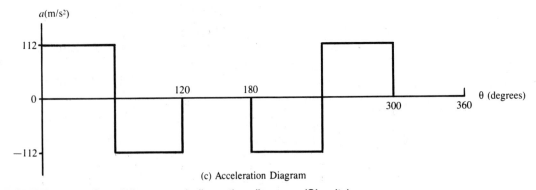

(c) Acceleration Diagram

FIGURE 6.35   Cam follower parabolic motion diagrams (SI units)

TABLE 6.3   Data for Plotting Points of Cam Rotation

Percent of γ	Cam Angle (degrees)	Percent of h	Stroke (mm)
20	24	8	5.6
40	48	32	22.4
60	72	68	47.6
80	96	92	64.4

(b) Velocity diagram: Note that velocity increases smoothly for $\theta = 1/2 \times 120°$. The time for the cam to rotate 60° is

$$t_{60} = \frac{60 \text{ s/m}}{400 \text{ r/m}} \left(\tfrac{1}{6} \text{ rev.}\right) = 0.025 \text{ s}$$

$$V_{avg} = \frac{S}{t} = \frac{0.035 \text{ m}}{0.025 \text{ s}} = 1.4 \text{ m/s}$$

$$V_{max} = 2 \times V_{avg} = 2 \times 1.4 = 2.8 \text{ m/s}$$

This information plus the aid of Figure 6.17b will allow us to construct Figure 6.35b.

(c) Acceleration diagram:

$$a = \frac{\Delta V}{\Delta t} = \frac{2.8}{0.025} = 112 \text{ m/s}^2$$

This plus the information in Figure 6.17c will allow us to construct Figure 6.35c.

6.16   SUMMARY

There are two general classifications of cams. The first class requires some external force (such as a spring) to maintain contact between the follower and cam on the return stroke. The second class, called *positive-motion cams,* uses the cam design to force the follower through the return stroke.

The most common types of cams in the first class are plate cams and translation cams. The second class includes grooved-cylinder cams, grooved-plate cams (also called *face cams*), and matched-plate cams.

The followers can have sliding contact with the cams, such as the knife-edge, flat-face, and Scotch yoke types, or there can be rolling contact by attaching a roller to the follower. The followers also can be the type that slide in their guides or rotate about pivots.

The main types of follower motions are described as constant velocity, constant acceleration (parabolic), simple harmonic, and cycloidal. The following table details their areas of application.

Feature	Constant Velocity (Uniform Motion)	Constant Acceleration (Parabolic)	SHM	Cycloidal
Smooth acceleration curve			x	x
Lowest theoretical maximum acceleration		x		
Smallest maximum pressure angle			x	
Least power required			x	
Zero acceleration at BDC and TDC				x
Operational at low speeds	x	x		
Operational at medium speeds		x	x	
Operational at high speeds				x

Pressure angles must be held to a value below approximately 30°; otherwise, the follower may jam in its guide as a result of bending and increased friction. A number of ways to reduce the pressure angle are discussed.

6.17 QUESTIONS AND PROBLEMS

Questions

1. Name three types of cam configurations.
2. What action can be taken to make sure that a follower will have continuous contact with its cam profile?
3. Name one advantage and one disadvantage that a knife-edge follower has relative to a roller follower.
4. When a roller follower is at BDC, it is resting on the cam surface represented by the (a) base circle, (b) prime circle, (c) pivot circle, or (d) pitch curve.
5. What advantage does parabolic motion have over constant velocity motion relative to cam design?
6. What advantage does cycloidal motion have over simple harmonic motion relative to cam design?
7. Name three steps that can be taken to increase the maximum permissible pressure angle for a given cam system.
8. For an in-line follower, what is the actual pressure angle during dwell?
9. What steps can be taken to reduce the pressure and angle of a cam?
10. Describe what is meant by the phrase *undercutting of a cam*.

Problems

For the displacement diagrams, plot points for every 15° of cam rotation for the rise and return (or at 15% intervals if you use Table 6.2). Use a cam angle scale of 60° per inch and graph paper with 10 squares per inch. For the velocity and acceleration curves, you need plot only those points indicating a maximum or minimum.

1. A cycloidal cam raises a follower 1½ in. during a cam rotation angle of 120°. Find the follower displacement at 40° using Table 6.2.

2. Construct the displacement, velocity, and acceleration diagrams using the following data:
    (a) Stroke = 2 in. (make full scale)
    (b) Constant velocity rise for θ = 0° to 120°
    (c) Constant velocity return from 150° to 330°
    (d) Dwells from 120° to 150° and 330° to 360°
    (e) Cam speed = 500 rpm

3. Refer to Problem 2. Superimpose the displacement curve for parabolic motion.

4. Refer to Problem 2. Superimpose the displacement curve for simple harmonic motion.

5. Find the maximum velocity during the rise for Problems 3 and 4.

6. Construct the follower displacement curve for cycloidal motion using the following data. Also, find V_{max} during the rise.
    (a) Stroke = 2 in. (make full scale)
    (b) Cycloidal rise for θ = 0° to 120°
    (c) Cycloidal return from 150° to 300°
    (d) Dwells from 120° to 150° and 300° to 360°
    (e) Cam turns clockwise at 1000 rpm.

7. Construct the displacement diagram and lay out the plate cam profile (full scale) from the data supplied below. Also, specify the maximum pressure angle.
    (a) Base circle diameter = 2.5 in
    (b) Stroke = 1.75 in.
    (c) SHM rise for θ = 0° to 120°
    (d) A dwell from 120° to 240°
    (e) Parabolic return from 240° to 360°
    (f) One-half inch diameter in-line roller
    (g) Cam turns clockwise

8. Using the data from Problem 7 and Figure 6.36, construct the plate cam profile for the pivoted roller follower. Specify the maximum pressure angle during the rise.

9. The roller follower shown in Figure 6.37a is driven by a plate cam turning clockwise at constant speed about the camshaft centerline A. The follower rise takes place for θ = 0° to 180°. The follower velocity during the rise is shown in Figure 6.37b. The follower dwells from 180° to 240° and then has a SHM return to 360°.

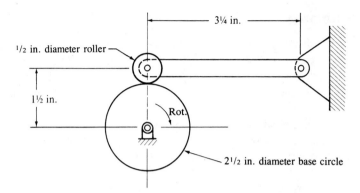

FIGURE 6.36   Sketch for Problem 8

(a) Construct the follower displacement diagram.
(b) Lay out the cam profile.
(c) Find the angular velocity of the cam.
(d) Specify the maximum pressure angle during the rise.
(e) Find the maximum acceleration of the follower during the rise.

10.   Refer to Problem 9 and Figure 6.37. Use $l = 2$ in., $h = 4$ in., and $\mu = 0.25$.
(a) Find the maximum permissible pressure angle.
(b) Is the actual maximum pressure angle acceptable?

11.   Design a positive-motion cam for the follower illustrated in Figure 6.38. The follower is to go through 10 rises and 10 returns per minute of SHM. There are no dwell times.
(a) Specify the cam speed in rpm.
(b) Specify the cam contour.

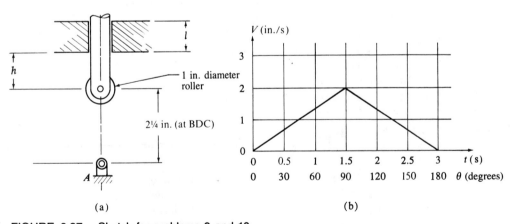

FIGURE 6.37   Sketch for problems 9 and 10

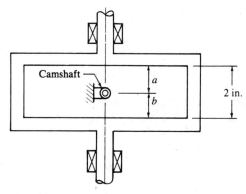

FIGURE 6.38   Sketch for Problem 11

(c) If the follower stroke is 1 in., specify the dimensions *a* and *b* when the follower is at TDC.

SI Unit Problems

12. Solve Sample Problem 6.9, except change the rise and return to 100° each and the two dwells to 80° each.

13. Construct a constant velocity displacement diagram given the following information:
    (a) Follower stroke = 80 mm.
    (b) There are no dwells, and the rise and return each require 180° to complete.

14. Make a layout of the cam profile at 20% intervals of a plate cam having an in-line roller follower. The specifications are
    (a) Base circle diameter = 50 mm.
    (b) Follower stroke = 30 mm.
    (c) Roller diameter = 15 mm.
    (d) The follower motion is SHM rise for $\theta = 0°$ to 180°, SHM return for $\theta = 210°$ to 330°, and dwells for $\theta = 180°$ to 210° and 330° to 360°.

15. If cam speed in Problem 14 is 800 rpm, find the maximum velocity and the maximum acceleration during the rise.

7

Fundamentals of Gears

Objectives

On completing this chapter, you will

- Understand gear terminology.
- Be able to analyze the design of spur gears by using the appropriate equations.
- Be able to design spur gears for the more common applications.

7.1 INTRODUCTION

Throughout the evolution of machinery, gears have proven to be the most widely used method of transmitting power from one shaft to another. In most applications, the shafts are required to rotate at varying speeds, but the speed ratio (output shaft speed divided by input shaft speed) must remain a constant.

The kinematics of gear systems has been developed to the point where the following orientations of any two shaft centerlines can be adequately handled:

— Parallel
— Collinear
— Perpendicular and intersecting
— Perpendicular and nonintersecting
— Inclined at any arbitrary angle

Modern gears are made to high precision standards. As a result, they normally are purchased from gear manufacturers rather than designed and machined at the user's plant. However, a person cannot arbitrarily order any gear from a manufacturer's catalog for a particular application. One must have a working knowledge of gear design, including design limitations, in order to produce a satisfactory gear drive system.

If the loads are low, with resulting low torque transmission, friction-drive roller systems can satisfactorily provide a constant speed ratio. Phonograph turntables generally employ a roller drive. The following equation provides a mathematical definition of speed ratio in relation to the friction roller system (refer to Figure 7.1).

$$\text{Speed ratio} = \frac{\text{output shaft speed}}{\text{input shaft speed}} = \frac{\omega_2}{\omega_1} = \frac{r_1}{r_2}$$

Notice that speed is inversely proportional to the roller radius.

In many applications, however, large torques are involved. Of course, the frictional resistance could be increased by roughing the surfaces of the two rollers. Historically, this was practiced, leading to the addition of cogs on one of the rollers and pockets on the other roller. Records show that these wooden cogwheels, or so-called toothed wheels, have been in existence since 2600 B.C. This produced a positive drive, but the cogs and pockets were very crudely made and did not properly mesh with each other. It is not enough merely to have positive drive using arbitrarily shaped toothed wheels—the teeth must have a proper form and be precisely positioned so that smooth rolling contact is assured at all speeds.

7.2 GEAR TERMINOLOGY

Before we can properly learn how to design gears, it is necessary to learn some of the important terms. The following definitions should be studied in conjunction with Figure 7.2, which depicts a portion of a very common type of gear, the *spur gear*. Spur

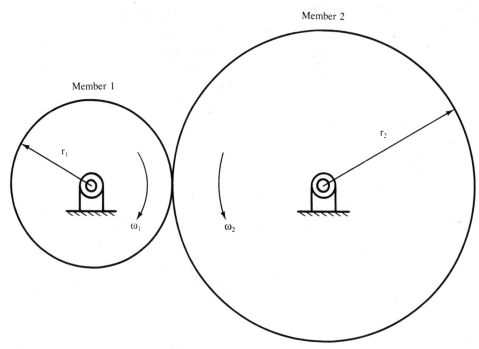

FIGURE 7.1   Friction drive roller system

gears are gears that have their teeth projecting radially from the axle, like spokes on a wheel. Spur gears will be considered first because they are the simplest and, in general, the definitions used for them apply to other types of gears. One distinguishable feature of spur gears is that the teeth are formed on a cylindrical surface and run parallel to the axis of the gear. As a result, spur gears are used for transmitting power between parallel shafts. Figure 7.3 shows a large and a small spur gear mounted on the same shaft.

Pitch Circle.   The *pitch circle* is the imaginary circle on which most gear calculations are based. When two gears mesh, their pitch circles are tangent to each other. If two friction rollers (with circles equal to the pitch circles of a pair of meshing spur gears) were to roll together without sliding, they would have the same speed ratio as the gears.

Pitch Diameter (D) and Pitch Radius (r).   *Pitch diameter* and *pitch radius* are the diameter and radius of the pitch circle.

Pitch Point.   The *pitch point* is the point on the imaginary line joining the centers of two meshing gears where the pitch circles touch.

Addendum Circle.   The *addendum circle* is the circle that bounds the outer ends of the teeth and whose center is at the center of the gear.

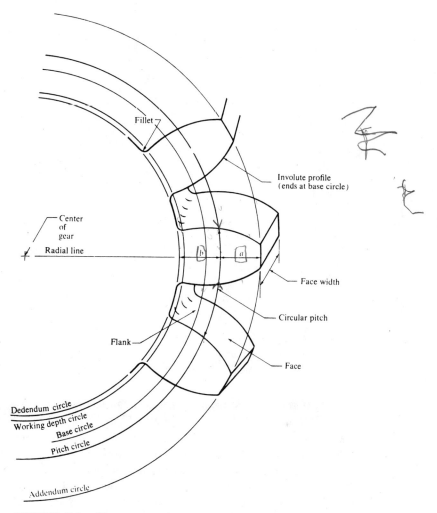

FIGURE 7.2   Gear nomenclature

Dedendum Circle.   The *dedendum circle* is the circle that bounds the bottoms of the teeth and whose center is at the center of the gear.

Addendum (a).   The *addendum* is the radial distance from the pitch circle to the outer end of the tooth.

Dedendum (b).   The *dedendum* is the radial distance from the pitch circle to the bottom of the tooth.

Circular Pitch (p).   The *circular pitch* is the distance between corresponding points on adjacent teeth, measured along the pitch circle. The circular pitches of meshing gears must be equal if they are to operate properly. This is because our method of gear

FIGURE 7.3   Two spur gears mounted on a single shaft
*(Courtesy of the Tool Steel Gear and Pinion
Company, Cincinnati, Ohio)*

manufacture requires that the teeth on each mating gear be the same size. Thus, a tooth takes up half of the circular pitch and the space for the mating tooth takes up the other half. This is theoretical, since in practice, clearance (to be discussed later) must be allowed.

Diametral Pitch (P).   *Diametral pitch* specifies the number of teeth **per inch** of pitch diameter. For example, a gear wheel with 20 teeth and a pitch diameter of 4 inches has a diametral pitch of 5. A small diametral pitch implies a large tooth size. (See Figure 7.4 for the physical sizes of teeth having various values of diametral pitches.) Diametral pitches have been standardized. Figure 7.4 shows most of the standard values. Additional standard sizes not shown are 2, 1 3/4, 1 1/2, 1 1/4, and 1. When the word *pitch* is used by itself, it implies the diametral pitch (not circular pitch). Gear catalogs usually list gears according to their diametral pitches, listing all the sizes available for a diametral pitch of, say, 2 1/2 together. The reason for arranging catalogs this way is that mating gears must have the same diametral pitch (as will be proved later).

Tooth Space.   *Tooth space* is the space between adjacent teeth, measured along the pitch circle.

Tooth Thickness.   *Tooth thickness* is the thickness of the tooth, measured along the pitch circle. The pitch circle cuts each tooth at such a location that the tooth thickness equals the tooth space, assuming there is no backlash (see Definition below).

Face Width (W).   *Face width* is the length of the tooth, measured parallel to the axis of the gear.

Face.   The *face* is the surface between the pitch circle and the top of the tooth.

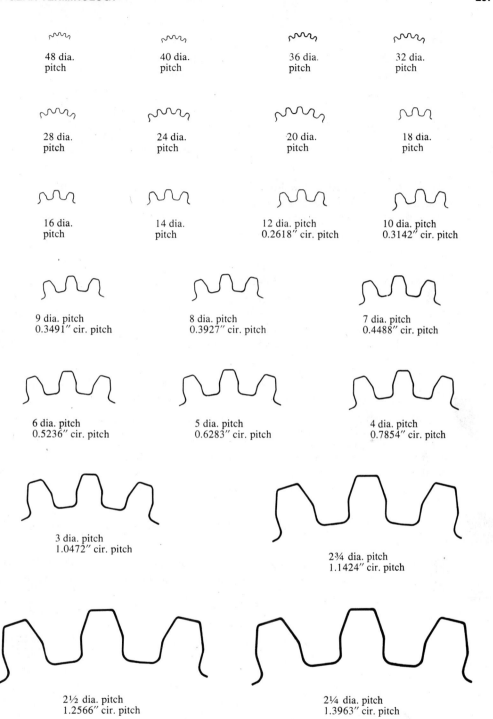

48 dia.
pitch

40 dia.
pitch

36 dia.
pitch

32 dia.
pitch

28 dia.
pitch

24 dia.
pitch

20 dia.
pitch

18 dia.
pitch

16 dia.
pitch

14 dia.
pitch

12 dia. pitch
0.2618″ cir. pitch

10 dia. pitch
0.3142″ cir. pitch

9 dia. pitch
0.3491″ cir. pitch

8 dia. pitch
0.3927″ cir. pitch

7 dia. pitch
0.4488″ cir. pitch

6 dia. pitch
0.5236″ cir. pitch

5 dia. pitch
0.6283″ cir. pitch

4 dia. pitch
0.7854″ cir. pitch

3 dia. pitch
1.0472″ cir. pitch

2¾ dia. pitch
1.1424″ cir. pitch

2½ dia. pitch
1.2566″ cir. pitch

2¼ dia. pitch
1.3963″ cir. pitch

FIGURE 7.4 Size of gear teeth of various diametral pitches *(Courtesy of the Tool Steel Gear and Pinion Company, Cincinnati, Ohio)*

Flank.   The *flank* is the surface between the pitch circle and the bottom of the tooth.

Pressure Angle (φ).   The *pressure angle* is the angle between the line of action and a line tangent to the two pitch circles at the pitch point.

Line of Action.   The *line of action* is the locus of all the points of contact between two meshing teeth from the time the teeth go into contact until they lose contact. Thus the load is transmitted from one gear to another along the line of action.

Pinion.   The *pinion* is the smaller of two meshing gears. The larger is called the gear.

Backlash.   *Backlash* is the difference (clearance) between the tooth thickness of one gear and the tooth space of the meshing gear, measured along the pitch circle. With backlash, there is looseness between meshing teeth, which becomes apparent when the direction of rotation of the driver is reversed by hand.

To be precise, the tooth thickness along the pitch circle equals one-half the circular pitch minus the backlash. The values of backlash are somewhat standardized in the following amounts: 0.030in./*P*, 0.040in./*P*, and 0.050in./*P*.

Hence, a typical range of values for backlash would be 0.005 in. to 0.020 in., depending on the teeth size. The value selected should be the largest one acceptable, since the manufacture of gears with small values of backlash is very expensive. Also, too small a value of backlash can cause binding (see Figure 7.5 for an illustration of backlash).

Clearance (c).   *Clearance* is the addendum minus the dedendum (see Figure 7.5). Clearance and backlash are both required to prevent binding.

Working Depth.   *Working depth* is the distance that one tooth of a meshing gear penetrates into the tooth space.

Base Circle.   The *base circle* is an imaginary circle about which the tooth involute profile is developed. Most spur gear teeth have an involute shape that runs from the base circle to the top of the tooth. The base circle is always tangent to the line of action.

Fillet.   The *fillet* is the radius that occurs where the flank of the tooth meets the dedendum circle.

Module.   The term *module* replaces diametral pitch in the metric system. It is found by dividing the pitch diameter by the number of teeth (the reciprocal of diametral pitch).

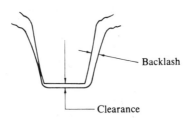

FIGURE 7.5   Clearance and backlash in gears

7.3 BASIC FORMULAS FOR SPUR GEARS

The sizes and proportions of spur gears are governed by the following formulas. All dimensions are in inches.

$$\text{Diametrical pitch} = P = \frac{N}{D} \quad \overset{\text{\# of teeth}}{\underset{\text{pitch Diameter}}{}} \quad Need\ D \tag{7.1}$$

$$\text{Circular pitch} = p = (\pi)\frac{D}{N} \tag{7.2}$$

$$\text{Addendum} = a = \frac{1}{P} \tag{7.3}$$

$$\text{Dedendum} = b = \frac{1.250}{P}$$

(for the more common diametral pitches.

For details see Table 7.1 in Section 7.12.) $\tag{7.4}$

$$\text{Clearance} = c = b - a = \frac{0.250}{P} \tag{7.5}$$

where $\quad D = $ pitch diameter

$\quad N = $ number of teeth on gear wheel

A very interesting result can be obtained by manipulating the first two formulas for diametral pitch and circular pitch:

$$p = \frac{\pi D}{N} = \frac{\pi}{N} \times \frac{N}{P} = \frac{\pi}{P}$$

rearranging:

$$p \times P = \pi \tag{7.6}$$

Since the circular pitches of meshing gears are equal, the diametral pitches must also be equal. Otherwise, the product of the circular pitch and diametral pitch for a given gear would not equal the constant π as shown above.

The use of these formulas is illustrated in Sample Problem 7.1 below.

SAMPLE PROBLEM 7.1

Gear Dimensions

PROBLEM: A 20-tooth gear has a diametral pitch of 4. Find the (a) pitch diameter; (b) circular pitch; (c) addendum; (d) dedendum; (e) outside diameter (D_o—same as addendum circle diameter); (f) clearance.

Solution

(a) $D = \dfrac{N}{P} = \dfrac{20}{4} = 5$ in.

(b) $p = (\pi)\dfrac{D}{N} = \pi \times \dfrac{5}{20} = 0.785$ in.

(c) $a = \dfrac{1}{P} = \dfrac{1}{4} = 0.250$ in.

(d) $b = \dfrac{1.250}{P} = \dfrac{1.250}{4} = = .312$ in.

(e) $D_0 = D + 2a = 5 + 2(0.250) = 5.500$ in.

(f) $c = b - a = 0.312 - 0.250 = 0.062$ in.

7.4 SPEED RATIOS

Normally, the smaller of two mating spur gears is called the *pinion* and the larger is referred to as the *gear*, regardless of which is the driver and which is the follower. Therefore, note in Figure 7.6 that the larger gear has a pitch circle radius designated as r_g and the smaller gear (the pinion) radius is designated r_p. The speed ratio of two mating spur gears can now be established.

If we arbitrarily say that the pinion has a circumference of 10 in. and the gear has a circumference of 20 in., it is obvious that the pinion must make two revolutions for every one revolution of the gear. Thus the gear and pinion speeds have an inverse relationship to their dimensions:

$$\frac{\omega_p}{\omega_g} = \frac{2\pi r_g}{2\pi r_p} = \frac{D_g}{D_p} \tag{7.7}$$

where ω = speed in revolutions per minute

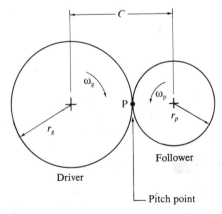

FIGURE 7.6   Speed ratio of mating spur gears

and the subscripts (p) and (g) refer to pinion and gear. Noting that $D = N/P$ and substituting, we get:

$$\frac{\omega_p}{\omega_g} = \frac{(N_g/P)}{(N_p/P)} = \frac{N_g}{N_p} \tag{7.8}$$

(since the diametral pitch P is the same for both gears.)

Thus the speed ratio is not only inversely proportional to the gear size ratio, but is also inversely proportional to the ratio of the number of teeth. Equating the right-hand sides of Equations 7.7 and 7.8 provides still another intuitive relationship:

$$\frac{D_g}{D_p} = \frac{N_g}{N_p} \tag{7.9}$$

Thus, if a gear is twice as large as its mating pinion, it will possess twice as many teeth.

A useful relationship for the center distance (dimension C of Figure 7.6) can also be developed:

$$C = r_p + r_g = \frac{D_p}{2} + \frac{D_g}{2}$$

$$C = \frac{N_p}{2P} + \frac{N_g}{2P} = \frac{N_p + N_g}{2P} \tag{7.10}$$

SAMPLE PROBLEM 7.2

Center Distance and Gear Size

PROBLEM: Two spur gears have a speed ratio of 3 to 1 and a diametral pitch of 4. The center distance should be close to 11.600 in. (It usually won't be exact because we must use a whole number of teeth.) Find the number of teeth on each gear.

Solution

$$C = \frac{N_p + N_g}{2P}$$

$$11.6 = \frac{N_p + 3N_p}{2 \times 4}$$

$$4N_p = 92.8$$

$$N_p = 23.2$$

$$N_g = 69.6.$$

To maintain the speed ratio, the number of pinion teeth must be 23 and the number of gear teeth must be 69. This gives us a center distance of

$$C = \frac{(23 + 69)}{8} = 11.5 \text{ in.}$$

7.5   INVOLUTE TOOTH PROFILES

The gear tooth profile must be designed to provide smooth transmission of motion and to not interfere with the mating tooth. It is possible to generate any number of different profiles that satisfy these requirements. It was not until the seventeenth century, however, that a geometric curve was proposed. This was the cycloidal curve which had one unique advantage—there was no sliding between the mating teeth. However, this advantage of pure rolling action prevailed only if the actual center distance matched precisely with the design value. Since this rarely occurred, and since the cost of producing cycloidal teeth is high, the involute profile has become the most widely used tooth form. It has the following practical advantages:

1.   It is easily manufactured.
2.   It is interchangeable with gears having the same diametral pitch. Thus matched sets are not needed.
3.   It provides efficient transmission of power. Sliding action is minimized.
4.   It provides linear path of contact, thus producing a constant pressure angle.
5.   It can accommodate center distance values which are not as critical as with other types of profiles (such as cycloidal).
6.   One tool can be used to generate gears of the same diametral pitch having anywhere from ten to hundreds of teeth.

An involute is the trace generated by the end of a taut string as it is unwrapped from a fixed cylinder, as shown in Figure 7.7. The size of the cylinder to use for the gear tooth profile is one having a diameter equal to the diameter of the base circle. Thus the circle representing the cylinder from which the string unwinds is called the base circle.

Notice the one very distinct characteristic of an involute curve: the involute, at any point, has a direction that is perpendicular to the tangent emanating from the point on the base circle. Thus the normal to an involute curve is always tangent to the base circle.

The size of the base circle can be determined if the gear pitch diameter and pressure angle are known. The procedure for construction can be found in most engineering graphics texts. Figure 7.8 shows the mathematical relationship. Triangle *ABP* in the figure is a right triangle, where angle *BAP* equals the pressure angle ϕ. Using trigonometry, we have

$$\cos \phi = \frac{r_b}{r}$$

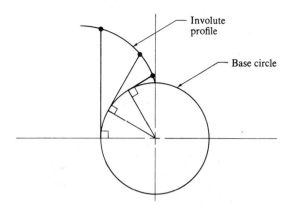

FIGURE 7.7   Definition of involute profile

and therefore

$$r_b = r \times \cos \phi \qquad\qquad (7.11)$$

where r = radius of the pitch circle.

r_b = radius of the base circle.

Figure 7.9 helps explain why involute teeth of mating gears do not interfere with each other. The figure shows two pulleys connected by a crossed belt. Now we glue a large piece of paper to pulley A (so that the paper rotates when the pulley does) and fasten a pencil to the belt at point P. Pulley A is rotated clockwise and the pencil moves

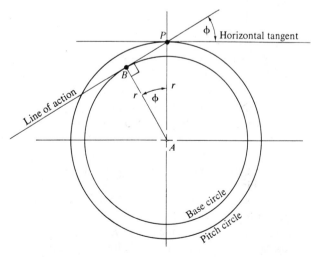

FIGURE 7.8   Determination of the base circle

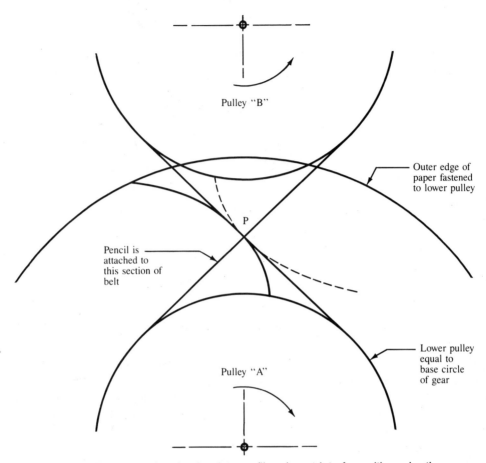

FIGURE 7.9   Demonstration that involute profiles do not interfere with each other

with the belt toward pulley *B*. As the pencil moves, it traces an involute curve on the paper. Leaving the pencil in place, we remove the paper from pulley *A* and attach a paper to pulley *B*. On rotating the pulleys we see that an involute is also traced on this sheet of paper (represented by the dotted line). Since the same point on the belt (where the pencil is attached) drew both curves, they cannot interfere with each other.

7.6   INTERFERENCE AND UNDERCUTTING OF GEAR TEETH ▬▬

Mating teeth will interfere with each other if contact is made where either of the two surfaces are not involute in shape. In Figure 7.10*a*, the contact points lie on the line of action which goes through the pitch point *P*. Point *C* is a point on the tooth of the lower gear. For the position shown in Figure 7.10*a*, point *C* is also a point on the base circle of the upper gear. Thus, between points *P* and *C*, the mating action is between two

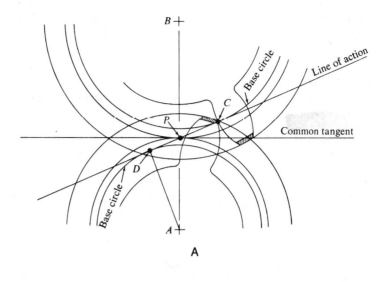

A

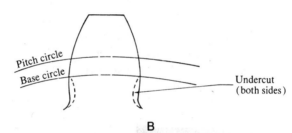

B

FIGURE 7.10   Interference and undercutting: (A) teeth interference; (B) undercut tooth

involute profiles. However, if mating action continues along the line of action beyond point *C*, which is the case in our figure, interference will occur. This is because the flank of the upper gear is a radial surface (not involute).

As the action goes beyond point *C*, the tip of the lower tooth will dig into the flank of the upper tooth because the point of contact will deviate from the line of action. Any contact that is not on the line of action will produce interference, resulting in gouging of the flank area of the interfering tooth. The tips of the upper and lower teeth that do the gouging are shown by crosshatching in Figure 7.10*a*.

Interference can be eliminated by machining a concave surface into the blank areas sufficient to avoid contact in these areas. This process, called *undercutting*, weakens the tooth at areas close to the point of maximum bending moment, as shown in Figure 7.10*b*. Undercutting is therefore undesirable, and the amount of undercutting should not be allowed to become excessive to the point of causing tooth failure. One way of eliminating the need to undercut is to machine the crosshatched areas from the tips of the teeth. This process is called *tip relieving* and is an alternate to undercutting. A sufficient number of teeth will make undercutting unnecessary. For pressure angles of 14 1/2°, 20° and 25°, the minimum number of teeth recommended is 34, 18, and 12, respectively.

7.7 CONTACT RATIO

The *contact ratio* can be defined as the number of pairs of teeth that are in contact at any instant of time. It is necessary that continuous action take place between mating teeth. Hence, it is desirable to have more than one pair of teeth in contact at all times during operation. For smooth, continuous motion, the contact ratio should not be less than 1.4.

Mathematically, the contact ratio can be defined as the length of the path of contact along the line of action divided by the circular pitch along the base circle. Referring to Figure 7.11, we see that the length of the path of contact Z is measured from point B to point C. We also know that

$$r_b = r \times \cos \phi$$

From our knowledge of geometry we know the circular pitch along the pitch circle subtends the same angle as the circular pitch along the base circle. This leads us to the relationship that the circular pitches are in the same proportion to each other as their radii. Thus the circular pitch along the base circle equals the circular pitch along the pitch circle multiplied by the cosine of the pressure angle. We can now write a convenient formula for contact ratio:

$$\text{Contact Ratio} = \frac{Z}{\rho \cos \phi}$$

and

$$Z = \sqrt{(R + a)^2 - R_b^2} + \sqrt{(r + a)^2 - r_b^2} - C \sin \phi$$

where
  Z = the length of the path of contact
  $R + a$ = the addendum circle radius for the gear
  $r + a$ = the addendum circle radius for the pinion
  R_b = the base circle radius of the gear
  r_b = the base circle radius of the pinion
  C = the center distance
  p = circular pitch measured along pitch circle

The derivation of the formula for Z will be left as a student problem. Use the procedure outlined below.

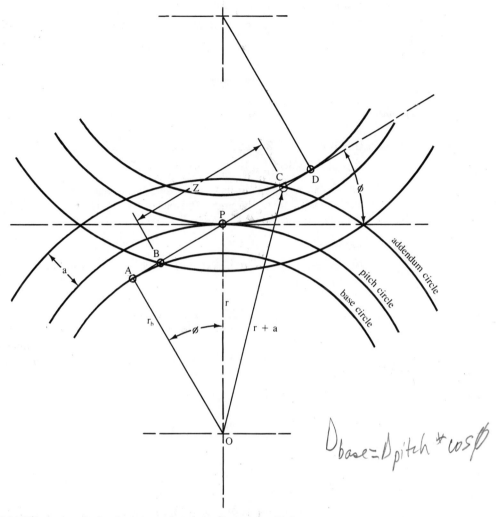

FIGURE 7.11 Path of contact: determination of the Z distance

PROCEDURE

Deriving the formula for Z

Step 1 Refer to Figure 7.11.

Step 2 You will know, or can find easily, the pressure angle ϕ, r, r_b and $(r + a)$ for the pinion.

Step 3 Find AC using the Pythagorean theorem.

Step 4 Find AP by using the formula $r\sin \phi$.

Step 5 You can now find distance *PC* by subtraction.

Step 6 Use the other gear dimensions and repeat this process to find distance *BP*.

Step 7 Of course, $Z = BP + PC$.

The following sample problem will illustrate the use of the contact ratio formula.

Review

SAMPLE PROBLEM 7.3

Contact Ratio

PROBLEM: A pair of meshing spur gears has a diametral pitch of 4 and a pressure angle of 14 1/2 degrees. The pinion has 20 teeth and the gear 40 teeth. Find the contact ratio.

Solution

We need to find the addendum circle radii and the base circle radii.
For the pinion

$$r = \frac{D_p}{2} = \frac{N_p/P}{2} = \frac{20/4}{2} = 2.5 \text{ in.}$$

$$(r + a) = 2.5 + \frac{1}{P} = 2.5 + 0.25 = 2.75 \text{ in.}$$

$$\sqrt{(r+a)^2 - r^2b}$$
using

$$r_b = r \cos 14\frac{1}{2}° = 2.5 \times 0.968 = 2.42 \text{ in.}$$

$$p = \pi \frac{D}{N} \text{ pitch dia} \atop \text{teeth}$$

$$p = \pi \frac{}{} \atop 40$$

For the gear

$$R = \frac{40/4}{2} = 5 \text{ in.}$$

$$(R + a) = 5 + 0.25 = 5.25 \text{ in.}$$

$$R_b = R \cos 14\frac{1}{2}° = 5 \times 0.968 = 4.84 \text{ in.}$$

$$C = \frac{N_p + N_g}{2P} = \frac{20 + 40}{(2 \times 4)} = 7.5 \text{ in.}$$

$$Z = \sqrt{(5.25^2 - 4.84^2)} + \sqrt{(2.75^2 - 2.42^2)} - 7.5 \sin 14\frac{1}{2}°$$

$$Z = 1.46$$

$$p = \frac{N}{D}$$

$$= \pi \frac{D}{N}$$

p

$$CR = \frac{Z}{p \cos \phi} = \frac{1.46}{(\pi/4) \times \cos 14\frac{1}{2}°} = 1.92$$

7.8 RACK AND PINION

A *rack* is a gear whose pitch diameter has become infinite in size, thereby resulting in a straight line for the pitch circle, which is called the *pitch line*. An involute of a very large base circle approaches a straight line. Therefore, surfaces of the rack teeth are flat, but mesh properly with the involute surfaces of the mating pinion, as shown in Figure 7.12. When a pinion meshes with a rack, the rotary motion of the pinion is transformed into translation of the rack, or vice versa. The linear velocity of the rack equals the tangential velocity of the pitch circle of the pinion. The teeth of the rack have the same values of addendum and dedendum as the mating pinion. Note that the tooth thickness and space between teeth of the rack are measured along the pitch line. The flat surfaces of the teeth of a rack are perpendicular to the line of action which goes through the pitch point, *P*. The angle included between the line of action and the pitch line is the pressure angle. The mating action for a rack and pinion begins where the addendum line of the rack touches the line of action. The mating action stops down along the line of action where the addendum circle of the pinion touches the line of action.

Applications of rack-and-pinion drives are found in virtually all types of machine tools, where the rack is usually mounted on some sort of slide. For example, a hydraulic motor can drive a rotating pinion, which in turn translates a rack mounted on the table of a milling machine. Also, drill presses use rack and pinions to change the rotary motion of a handwheel into linear motion of the spindle towards the workpiece. Figure 7.13 is a picture of a segment of a very large radius gear on which is positioned a pinion. On the far side of the gear segment and pinion is a large gear on top of which is positioned a small pinion. Notice that the very large radius gear segment comes very close to approximating a rack.

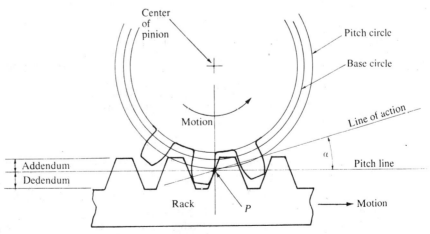

FIGURE 7.12   Action of rack and pinion

FIGURE 7.13   Very large radius gear segment and pinion
next to a gear and pinion *(Courtesy of the
Tool Steel Gear and Pinion Company, Cincinnati, Ohio)*

7.9 THE INTERNAL SPUR GEAR

All the gears we have discussed so far are external spur gears because their teeth are
located on their outer surfaces. *Internal spur gears* (also called *annular gears*) are circular rings with teeth cut into the inner surfaces (see Figure 7.14). Internal spur gears
with mating pinions provide much more compact drive systems than external spur
gears, because the pinions lie within the internal gear. Since the pinion and internal
gear rotate in the same direction, there is less sliding action and wearing of teeth as
compared to external spur gears. Also, internal gears provide a larger contact ratio and
thus can transmit more power. Some formulas may need adjusting. For example, the
center distance = $r_g - r_p$, not $r_g + r_p$.

7.10 METHODS OF MANUFACTURING GEARS

Gears are manufactured by most of the common manufacturing processes such as machining, die casting, stamping, powdered metal manufacturing, cold drawing, extru-
sion, and rolling. Gears for most power applications are commonly formed by the following four machining processes.

Form Milling.   In this method, the spur gear is machined from a blank using a cutter
that has a form to match the shape of the tooth space of the gear to be cut. At least
eight different form cutters are required to handle fairly accurately each diametral pitch
required, because the number of teeth affects the profile shapes. The reason is simple:
since $P = N/D$, different numbers of teeth require different pitch diameters for a given

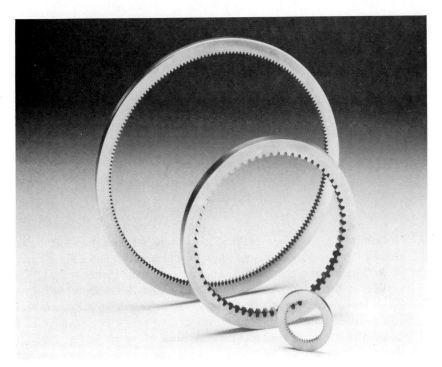

FIGURE 7.14   Photograph of internal gears *(Courtesy of the Boston Gear Company)*

diametral pitch. This in turn produces different sizes of base circles, and subsequently, different involute profiles, even when the pressure angle remains constant. This method is limited to the production of low-speed gears because a high degree of precision can rarely be achieved. Also, form milling is expensive because of the lengthy cutting times required and the large cutter inventories needed. Cutters with less than 12 teeth are not used because of the resulting undercutting. A picture of a rotating form milling cutter is shown in Figure 7.15.

Rack Generation.   A *rack* is considered to be a gear of infinite radius (as stated before). Thus a tool shaped like a rack can be made of hardened steel with cutting edges all around the ends of the teeth. The tool is given a reciprocating motion parallel to the gear blank axis. Simultaneously, the gear blank is rotated very slowly while the rack cutter continues to reciprocate axially. The rack cutter is given a slow transverse velocity to match the pitch line velocity of the gear blank. Thus the material between the gear teeth is removed, leaving the involute teeth. Only one tool for each diametral pitch is required to cut gears of any number of teeth. Because the rack cutter has a transverse velocity, the process is not continuous, and indexing is required each time the rack cutter length has been exhausted. See Figure 7.16 for a picture of a rack cutter.

FIGURE 7.15   Rotary form cutter *(Courtesy of the Tool Steel Gear and Pinion Company, Cincinnati, Ohio)*

Hobbing.   A *hob* is a cylindrical tool around which a thread of the same form as a rack tooth has been helically cut. In appearance, a hob looks like a large-lead threaded screw, except that the thread is given a series of axial gashes around its periphery. This is done so that cutting edges can be ground on the hardened steel hob (See Figure 7.17). Initially, the hob is placed to give the correct depth of cut into the gear blank. It is then rotated and, kinematically, the cutting action is equivalent to that of a rack cutter. The transverse motion of the rack cutter of the preceding method is simulated by the lead of the helix of the thread of the hob as it rotates. The hob is fed axially along the blank until the teeth are cut along the entire face width of the gear blank. A hob is thus similar to a tap used for threading holes. Hobbing is a continuous process, and therefore no indexing is required. It is presently the most popular method of generating gear teeth and is recommended for gears whose pitch line speeds are in the 2000 fpm to 4000 fpm range (for higher speeds the teeth manufactured by this method should be ground or lapped).

FIGURE 7.16   Rack cutter *(Courtesy of the Tool Steel Gear and Pinion Company, Cincinnati, Ohio)*

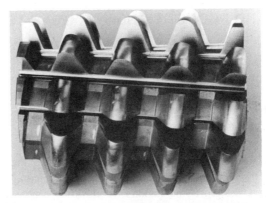

FIGURE 7.17   Hob *(Courtesy of the Tool Steel Gear and Pinion Company, Cincinnati, Ohio)*

Fellows Gear Shaper Method.   This process uses a hardened steel cutter that is similar in appearance to the gear to be machined. Figure 7.18 shows a pinion shaped cutter. The cutter and gear blank are mounted on parallel axes. The cutting process is begun with the gear blank held stationary and the cutter fed radially while it reciprocates axially until the proper depth for the tooth is obtained. Then the gear blank and cutter are rotated slowly as the cutter continues to reciprocate axially. The big advantage of this method is that only one cutter is needed for all gears of the same diametral pitch and pressure angle, regardless of the number of teeth required. As in the hobbing process, the Fellows gear shaper method is considered a precision machining operation.

FIGURE 7.18   Pinion shape cutter *(Courtesy of the Tool Steel Gear and Pinion Company, Cincinnati, Ohio)*

7.11   STANDARD PROPORTIONS FOR GEAR TEETH ■■■■■■

At one time, each gear manufacturing company developed its own standards for gear cutting sizes. This resulted in

— Loss of interchangeability from one company to another
— A large inventory of replacement gears
— Problems in making repairs

Because of the preceding undesirable factors, the American Standards Association (ASA) and the American Gear Manufacturers Association (AGMA) have adopted standards for gear design. Some standard tooth forms that have become obsolete, but may require a mating gear replacement (so data for them is still available), are the Brown and Sharp System, the AGMA 14 1/2° Composite System, the AGMA 14 1/2° Full-depth System, and the Fellows 20° Stub-tooth System (the degree values refer to the pressure angle). The present standards, which embody the most modern technology, have pressure angles of 20° and 25°. A few specifications are shown in Table 7.1.

7.12   GEAR TOOTH FORCES AND BENDING STRENGTH OF SPUR GEARS ■■■■■■

Let us examine the forces acting on the teeth of mating spur gears as they transmit power. The total tooth force (F_{total}) is normal to the involute profiles at the point of contact. Therefore, the total tooth force is transmitted from the driver to the driven along the line of action. The total force can be broken into two components—the *tangential force* F_t and the *separating force* F_s, as shown in Figure 7.19. F_t is used in the horsepower-torque formula mentioned in Chapter 2. Thus

TABLE 7.1   Specifications for Standard Systems of Gear Teeth

Item	Full Depth & Pitches Coarser Than 20		Full Depth & Pitches 20 & Finer	14½° Full Depth
Pressure Angle	20°	25°	20°	14½°
Addendum (in.)	1.000/P	1.000/P	1.000/P	1/P
Dedendum (in.) (minimum)	1.250/P	1.250/P	$\frac{1.200}{P} + 0.002$	1.157/P

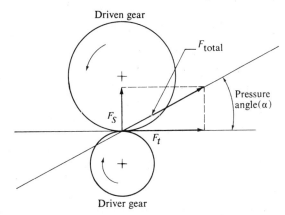

FIGURE 7.19   Force transmission through spur gears

$$hp = \frac{T \times \omega}{63,000} = \frac{(F_t \times r) \times \omega}{63,000}$$

where   hp = the horsepower transmitted.

ω = the speed of the gear or pinion in revolutions per minute

r = the pitch radius of gear or pinion in inches

F_t = the tangential tooth force in pounds   It is assumed to be acting at the pitch radius.

Obviously the same tooth force exists on the teeth of the pinion and gear. Hence, the product of the radius and the speed must be either the pinion or the gear and not have mixed values. Once the tangential force has been found, trigonometry can be used to find the separating force (F_s) and the total force (F_{total}).

A gear tooth failure can be very costly, since an entire machine must be shut down. In addition, the replacement of a gear is usually a complex job because it requires proper installation, as well as removal of other machine parts to get to the damaged gear.

A gear tooth can fail by fracturing at its base. Essentially, a gear tooth behaves as a cantilever beam and hence, the maximum bending stress occurs at the base. As shown in Figure 7.20a, tiny fatigue cracks initially occur because of repeated applications of load as a tooth goes into and out of mesh. The cracks begin on the side where the force is applied because the stress is tensile in this area. The initial crack increases the stress concentration above the value caused by the fillet, which in turn accelerates the crack propagation process until a dangerous stage is reached (see Figure 7.20b). Because of the greatly reduced cross section, the tooth fractures suddenly, as shown in Figure 7.20c.

The tangential force in the horsepower-torque formula above is the **transmitted force,** but is not necessarily the maximum force that the gear tooth is subjected to. F_t is the maximum force on a gear tooth only at low speeds, such as 10 rpm or less.

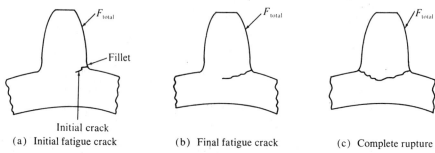

(a) Initial fatigue crack   (b) Final fatigue crack   (c) Complete rupture

FIGURE 7.20   Fatigue crack propagation of a gear tooth

Above this speed, impact loads (or dynamic loads) occur and these can exceed the value of F_t many times. Under such conditions, F_t is considered the average value. The causes of dynamic loading are

— Slight errors in spacing of mating gears during mounting
— Slight discrepancies in teeth profile during machining
— Deflection of teeth under load

The bending strength of a gear tooth acting as a cantilever beam can be found by using a modified version of the Lewis equation. Wilfred Lewis graduated from the Massachusetts Institute of Technology (MIT) in 1875 as a mechanical engineer, and during his lifetime received about 60 patents for his inventions. In the early 1890s he developed an equation for the bending strength of a gear tooth based on a geometrical factor that depends on the number of teeth on the gear and the system of gearing used. This equation has been modified to account for dynamic loading and is still used in gear design. The modified version is:

$$F_{db} = \frac{SWYk}{P} \qquad\qquad (7.14)$$

where F_{db} = allowable dynamic bending strength in pounds

S = allowable stress in pounds per inch^2

W = face width in inches

Y = the Lewis form factor (See Table 7.2.)

k = the factor accounting for speed

P = the diametral pitch

Once F_{db} has been found, it is substituted into the horsepower-torque formula as F_t so that the transmitted power may be obtained.

The value of k in Equation 7.14 can be obtained from the following equations. These equations make use of the pitch line velocity "V" in units of feet per minute (fpm).

$$V = \frac{\pi D \omega}{12\text{in./ft}}$$

TABLE 7.2   Form Factors *(Y)* of Gear Teeth

Number of Teeth	14½° Full-depth Involute	20° Full-depth Involute	20° Stub Involute
12	0.210	0.245	0.311
13	0.223	0.261	0.324
14	0.236	0.277	0.339
15	0.245	0.290	0.346
16	0.254	0.296	0.361
17	0.264	0.302	0.368
18	0.270	0.308	0.377
19	0.276	0.314	0.386
20	0.283	0.321	0.393
21	0.289	0.327	0.399
22	0.292	0.330	0.405
24	0.298	0.337	0.415
26	0.308	0.346	0.424
28	0.314	0.352	0.430
30	0.317	0.359	0.437
34	0.327	0.371	0.447
38	0.333	0.384	0.455
43	0.339	0.397	0.462
50	0.346	0.410	0.474
60	0.355	0.421	0.484
75	0.361	0.434	0.496
100	0.368	0.447	0.505
150	0.374	0.460	0.518
300	0.383	0.472	0.534
rack	0.390	0.484	0.550

where V = pitch line velocity in feet per minute

D = pitch diameter in inches

ω = the speed of the gear in revolutions per minute

The value k for ordinary quality commercial gears where V < 2000 fpm is derived as follows:

$$k = \frac{600}{600 + V}$$

The value k for precision machined gears generated by hobbing or by the Fellows gear shaper method where V < 4000 fpm is derived as follows:

$$k = \frac{1200}{1200 + V}$$

TABLE 7.3   Allowable Stresses for
Various Gear Materials

Gear Material	Allowable Stress (lb/in.^{2})
Ordinary cast iron	8,000
Medium-grade cast iron	10,000
Highest-grade cast iron	15,000
Untreated steel	20,000
Heat-treated alloy steel	60,000
Case-hardened steel	50,000
SAE 62 bronze	10,000
SAE 65 phosphor bronze	12,000
Bakelite	7,000
Nylon	4,000

The value k for nonmetallic gears such as nylon and bakelite is derived as follows:

$$k = \frac{150}{200 + V} + 0.25$$

When using Equation 7.14, the allowable stress S must be based on the endurance limit of the material and must include an appropriate stress concentration factor because of the fillet. Appropriate allowable stress values for various materials are given in Table 7.3.

Plastic gears are used in those applications where a reduced noise level is desired. A common arrangement is to have the driving pinion made of metal and the driven gear made of plastic.

From Equation 7.14, we conclude that the dynamic bending strength is proportional to the face width of the gear tooth. However, if the face width is made too large, any misalignment can cause the load to be concentrated at one end of the tooth, a situation that can actually hasten tooth failure. As a general rule of thumb, the face width should not exceed the pinion pitch diameter. To increase the dynamic beam strength of a gear, the following steps can be taken

— Increase tooth size by decreasing the diametral pitch
— Increase face width up to the pitch diameter of the pinion
— Select material of greater endurance limit
— Machine tooth profiles more precisely
— Mount gears more precisely
— Use proper lubricant and reduce contamination

PROCEDURE

Applying the Lewis Equation

Step 1   Determine the pitch line velocity of the gear being analyzed and then find the appropriate k value.

Step 2   Substitute the appropriate values into Equation 6.14 and solve for F_{db}.

Step 3   Substitute F_{db} for F_t in the horsepower-torque formula and find the horsepower that the gear can transmit.

SAMPLE PROBLEM 7.4

Application of the Lewis Equation

PROBLEM: A 30-tooth spur gear has a diametral pitch of 5 and a face width of 1.5 in. The gear teeth are 14 1/2° full-depth involute and are made of case-hardened steel. If the rotational speed is 1000 rpm, how much horsepower can be transmitted?

Solution

Step 1
$$D = \frac{N}{P} = \frac{30}{5} = 6 \text{ in.}$$

$$V = \frac{\pi D \omega}{12} = \frac{3.14 \times 6 \times 1000}{12} = 1570 \text{ fpm}$$

$$k = \frac{600}{600 + 1570} = 0.276.$$

Step 2 S (from Table 7.3) = 50,000 psi
Y (from Table 7.2) = 0.317

$$F_{db} = \frac{50,000 \times 1.5 \times 0.317 \times 0.276}{5}$$

$$= 1310 \text{ lb}$$

Step 3 $\text{hp} = \dfrac{F r \omega}{63,000} = \dfrac{1310 \times 3 \times 1000}{63,000}$

Power transmitted = 62.4 hp

7.13 WEAR STRENGTH OF SPUR GEARS

Excessive wear of the contact surfaces of gear teeth is an equally common cause of gear failure. This wear usually appears as scoring, scratching, pitting or normal wearing away of metal due to friction. General wear has the same effect on dynamic

TABLE 7.4   Values of Load-Stress Factor (K)
for Steel Pinion Mating with
Steel Gear of Equal Hardness

Hardness (Bhn)	K (lb/in.^{2})	
	$\phi = 14\frac{1}{2}°$	$\phi = 20°$
150	30	41
200	58	79
250	96	131
300	144	196
350	201	275
400	268	366
450	344	470
500	430	588
600	630	861

loads as do manufacturing discrepancies in tooth profiles. Hence, for long life requirements, gears undergoing continuous operation must also be designed on the basis of
anticipated wear.

Lubrication is required to reduce wear and prevent excessive temperature buildup.
Grit must not be allowed to get into the lubricant since it would produce scoring, which
in turn would hasten tooth failure.

The wear strength of a spur gear can be determined by using the Buckingham wear
equation:

$$F_w = D_p WK \left(\frac{2D_g}{D_p + D_g} \right) \tag{7.15}$$

where F_w = tooth wear strength in pounds

D_p = pitch diameter of the pinion in inches

D_g = pitch diameter of the gear in inches

W = face width in inches

K = a load-stress factor that depends on the modulus of elasticity and
hardness of the gear material as well as on the gear pressure angle
(ϕ). Values of K are given in Table 7.4 for a steel pinion mating
with a steel gear of equal hardness.

To increase the wear strength of a gear pair, the following steps can be taken:

—   Increase tooth size by decreasing diametral pitch
—   Increase face width up to a maximum of the pinion diameter
—   Select materials of greater hardness

— Perform more precise machining of tooth profile and proper mounting of gears to produce rolling action rather than sliding action
— Use proper lubricant and reduce contamination

SAMPLE PROBLEM 7.5

Application of the Wear Equation

PROBLEM: Solve Sample Problem 7.4, taking into account failure from wear. The mating pinion has a 3 in. diameter. The gear and pinion are made of steel having equal hardness values of 400 Brinell (Bhn).

Solution

From Table 7.4, $K = 268$

$$F_w = D_p W K \left(\frac{2 D_g}{D_p + D_g} \right) = 3 \times 1.5 \times 268 \, \frac{2 \times 6}{3 + 6}$$

$$F_w = 1610 \text{ lb}$$

$$\text{Power transmitted} = \frac{F \times r \times \omega}{63,000} = \frac{1610 \times 3 \times 1000}{63,000}$$

Note: The radius and speed of either the gear or pinion may be used.

Power = 76.7 hp

Comparing the two Sample Problems 7.4 and 7.5, we see that the bending strength is the limiting factor. However, note that the bending strength applies just to the gear, whereas the wear strength applies to the pair of gears. Therefore when applying the bending strength equation, both pinion and gear should be checked, since the answers are not necessarily the same. Solving for the bending strength of the pinion in Sample Problem 7.5 will be left as a student problem.

7.14 SPUR GEAR DESIGN

Gear design is a very complicated process and depends on many factors obtained from experience and experiment. Most manufacturers do not design their own gears, but rely on gear design specialists. We intend to provide you with sufficient data so that you may determine approximate sizes and specifications needed for the most common types of spur gear applications. You can then proceed with the preliminary overall design of your drive system while the specialist proceeds with a more refined design. Or, you may find that in many instances your design is satisfactory. Actually, the specialist may make similar preliminary calculations, but he or she must take into consideration such variables as gear cutting accuracy, mounting errors, elastic deflections of gear teeth, type of material, tooth stiffness and strength, type of machinery involved, surface durability of the teeth, expected life of the gear set, operating temperatures, and various efficiencies available (good spur gear design can have efficiencies varying from 80% to 98%).

TABLE 7.5 Load Rating of Gears—X Factor (Courtesy of McGraw-Hill Book Company's Gear Handbook, edited by D. W. Dudley)

| Application | Service characteristics | | BHN of material | | Pitch-line speed, fpm | Accuracy | X factor |
	Driver	Driven	Pinion	Gear			
Turbine driving generator			225	180	Over 4,000	High precision	110
Internal-combustion engine driving compressor			225	180	Over 4,000	High precision	60
Motor driving compressor			225	180	Over 4,000	High precision	55
General-purpose industrial drive	Uniform	Uniform	575 350 210	575 300 180	1,000 1,000 1,000	Commercial Commercial Commercial	500–1,000 350–450 170–250
	Uniform	Uniform	575 300 210	575 300 180	3,000 3,000 3,000	Commercial Commercial Commercial	475–750 275–375 125–200
Large industrial gears— hoists, kilns, mills	Uniform	Moderate shcok	225	180	1,000 max.	Generated	80–100
	Uniform	Moderate shock	200	210		Generated	130–170
Aircraft (single pair)	Engine	Auxiliary drive	58 Rockwell C min.	58 Rockwell C min.	10,000	High ground	1,000 (at take-off)
Aircraft planetary	Engine	Propeller	58 Rockwell C min.	58 Rockwell C min.	3,000–10,000	Ground	600 (at take-off)

Automotive transmission		In low gear	58 Rockwell C min.	58 Rockwell C min.		High	1,500
Small commercial	Uniform	Uniform	Steel 350 BHN	Phenolic laminate	Under 1,000	Commercial	75
Small commercial	Uniform	Uniform	Steel 350 BHN	Nylon	Under 1,000	Commercial	50
Small gadget	Uniform	Uniform	Steel 350 BHN	Zinc-alloy die casting	Under 1,000	Commercial	25
Small gadget	Uniform	Uniform	Steel 200 BHN	Brass or aluminum	Under 500	Commercial	25
Small gadget	Uniform	Uniform	Brass or aluminum	Brass or aluminum	Under 500	Commercial	15

NOTES: These values for 10 hr. per day.

Because there are so many variables to consider and so many equations, text space requires that we limit our design discussion to the most common applications and to

— Spur gears
— Uniform loads and uniform or light shock power sources
— Gear materials shown in Table 7.5
— Standard 20° involute teeth

Before discussing the procedure to use in gear design, another equation will be presented to aid your design process. This equation provides you with a preliminary center distance so that approximate gear sizes may be determined. Equations 7.14 and 7.15 may then be used. The equation for center distance and face width is:

$$C^2 W = \frac{31,500}{X} \times \frac{hp}{\omega_p} \times \frac{(m+1)^3}{m} \qquad (7.16)$$

where C = the center distance in inches

W = the face width in inches

X = a factor that involves force on the tooth, face width, pinion diameter and speed ratio   Obtain this from Table 7.5.

hp = the horsepower to be transmitted

ω_p = the rotational speed of the pinion in rpm

m = the tooth ratio of the gear set (N_g/N_p)

N_g = the number of teeth on the gear

N_p = the number of teeth on the pinion

The following procedure for gear design is general because gear design does have a somewhat educated trial-and-error approach. You will have to use your judgment in a number of instances. For example, after going through most of the steps, you may find the teeth are not strong enough, or too strong, so you may want to repeat the steps with new data.

PROCEDURE

Spur Gear Design

Step 1   You will normally be given the power requirements, input and output speeds, power source and driven equipment. Go to Table 7.5 and decide on the approximate pitch-line speed and the X factor. Remember that for a given rpm, the pitch-line speed varies directly as the gear diameter. It is suggested that you use the largest appropriate pitch-line speed, the lowest X factor and the lowest hardness for your application (the lower the hardness, the lower the cost).

Step 2   Solve Equation 7.16 for the center distance C. Calculate the pitch diameters of the gears (they can be determined from the center distance and the speed ratio information). For convenience, choose a face width W of 1 inch if the power requirements are 50 horsepower or greater. Choose 1/2 inch or 1/4 inch as a face width for the lower horsepowers.

Step 3 Knowing the gear diameters, choose a standard diametral pitch (see Figure 7.4) and have at least 19 teeth on the pinion and a pressure angle of 20°. Doing so eliminates undercutting and other special manufacturing procedures (a specialist may go as low as 7 teeth—the lowest number recognized by the American Gear Manufacturers Association—AGMA). Determine the number of teeth on each gear, the pitch diameters, the addendum and the new center distance.

Step 4 Check to see that the face width does not exceed the pinion pitch diameter.

Step 5 Check to see if the gear set will provide the required horsepower, based on wear. Use Equation 7.15 and the horsepower-torque formula.

Step 6 Check to see that the bending stress will provide the desired horsepower. Use Equation 7.14 and the horsepower-torque formula. Normally the pinion is the weaker, but both gear and pinion should be checked as a matter of policy, especially if the gear material differs from the pinion.

Step 7 Check to see that the contact ratio is satisfactory.

Step 8 If any of the steps 4 through 7 give an unsatisfactory answer, then either: change the face width, choose a harder material, choose a different diametral pitch, or change the gear sizes. Make any or all of these changes and go back to Step 4. It is unnecessary to start over at Step 3, as it merely provided an estimate to build on.

SAMPLE PROBLEM 7.6

Spur Gear Design

PROBLEM: Design a gear drive for a 100 HP motor driving a blower. Motor speed = 1800 rpm. Blower speed = 600 rpm. This system is a general purpose industrial gear drive with uniform service characteristics. Use a pressure angle of 20°.

Solution

Step 1 A pitch line speed of 3000 fpm is chosen. The set with hardnesses of 210 and 180 Bhn and an **X** value of 125 are chosen.

Step 2 Let the face width **W** = 1 inch.

$$C^2 W = \frac{31,500}{X} \times \frac{hp}{\omega_p} \times \frac{(m + 1)^3}{m}$$

$$C^2 \times 1 = \frac{31{,}500 \times 100\ hp \times (3 + 1)^3}{125 \times 1800\ rpm \times 3}$$

$$C = \sqrt{298.67} = 17.28\ in.$$

$$C = \frac{(D_g + D_p)}{2}$$

and $D_g = 3D_p$

$$17.28 = \frac{4D_p}{2}$$

$$D_p = 8.64\ in.$$

$$D_g = 25.92\ in.$$

Step 3 $P = \dfrac{N}{D} = \dfrac{19}{8.64} = 2.20$

Choose $P = 3$.

Therefore: $D_p = \dfrac{19}{3} = 6.333\ in.$

and $D_g = \dfrac{57}{3} = 19.00\ in.$

Step 4 Face width is satisfactory.

Step 5 $F_w = D_p WK \left(\dfrac{2D_g}{D_p + D_g}\right)$

K is found by referring to Table 7.4. Assume both pinion and gear have a Bhn hardness of 200. $K = 79$.

$$F_w = 6.333 \times 1 \times 79 \times \left(\frac{2 \times 19}{25.333}\right) = 750\ lb$$

100 HP requires a tangential force of:

$$F = \frac{hp \times 63{,}000}{\omega_p r_p}$$

$$= \frac{100 \times 63{,}000}{1800 \times 3.167} = 1105\ lb$$

F_w needs to be increased by about 30%. The face width will be increased to 1.5 in. (a standard size), which gives an increase of 50%.
Now:

$$F_w = 1125\ lb \quad \text{This is satisfactory.}$$

Step 6 Pinion: $F_{db} = \dfrac{SWYk}{P}$

$$V = \frac{\pi D_p \omega}{12}$$

$$V = \frac{3.14 \times 6.333 \times 1800}{12} = 2983 \text{ fpm}$$

$$k = \frac{1200}{(1200 + 2983)} = 0.287$$

$S = 20,000$ psi

(From Table 7.3. Hardness below about 250 Bhn may be considered to be untreated steel.)

$$Y = 0.314 \text{ (From Table 7.2)}$$

$$F_{db} = 20,000 \times 1.5 \times 0.314 \times 0.287/3 = 901 \text{ lb}.$$

This is too low, so the face width must be increased to 2 in. With $W = 2$ in., $F_{db} = 1201$ lb, which is satisfactory.
Gear: V, S, and k are the same.

$$Y = 0.418 \quad \text{(Interpolated from Table 7.2)}$$

$$F_{db} = \frac{20,000 \times 2 \times 0.418 \times 0.287}{3} = 1600 \text{ lb} \quad \text{satisfactory}$$

Step 7 $CR = \dfrac{Z}{p \cos 20°} = \dfrac{Z}{(\pi/P) \cos 20°}$

$p \cos 20° = 0.984$

$$Z = \sqrt{((R + a)^2 - R_b^2)} + \sqrt{((r + a)^2 - r_b^2)} - C \sin 20°$$

$a = \dfrac{1}{P} = 0.333$ in.

$r + a = 3.167 + 0.333 = 3.5$ in.

$r_b = r \cos 20° = 2.976$ in.

$R + a = 9.5$ in. $+ 0.333$ in. $= 9.833$ in.

$R_b = R \cos 20° = 8.927$ in.

$C = R + r = 12.667$ in.

$$Z = \sqrt{(9.833^2 - 8.927^2)} + \sqrt{(3.5^2 - 2.976^2)} - 12.667 \sin 20°$$
$$= 1.632$$

$$CR = \frac{1.632}{0.984} = 1.66.$$

This is greater than 1.4, therefore satisfactory.

Answer: For the selected gear set, Diametral pitch = 3, face width = 2 in., center distance = 12.667 in.

	Pinion	Gear
Pitch diameter	6.333 in.	19.000 in.
Number of teeth	19	57
Bhn hardness	200	200

7.15 SI UNITS

In the SI system the term *module* is used in gear calculations instead of diametral pitch. The module is a number which gives the millimeters of pitch circle diameter per tooth. It is obtained by dividing the pitch circle diameter by the number of teeth on the gear wheel.

$$M_o = \frac{D}{N} \tag{7.17}$$

where M_o = the module number in millimeters

D = the pitch diameter in millimeters

N = the number of teeth on the gear

Thus, the module is the reciprocal of diametral pitch ($P = N/D$). Therefore, whenever a spur gear formula has the diametral pitch listed, simply place the module in the reciprocal position. Table 7.6 lists a number of standardized module numbers and the equivalent diametral pitches. Note that when converting from module to diametral pitch, the reciprocal of the module must be multiplied by the conversion factor, 25.4 mm/in. For the present, SI metric standards cover only the 20° pressure angle gear form.

SAMPLE PROBLEM 7.7

SI Units

PROBLEM: A 22 tooth gear has a module of 3. Find **(a)** the pitch diameter, **(b)** circular pitch, **(c)** addendum, **(d)** dedendum, **(e)** outside diameter (D_o), and **(f)** clearance.

Solution

(a) $M_0 = \dfrac{D}{N}$. $D = 3 \times 22 = 66$ mm

(b) $p = M_0 \times \pi = 3 \times \pi = 9.42$ mm

(c) $a = M_0 = 3$ mm

(d) $b = M_0 \times 1.250 = 3.75$ mm

(e) $Do = D + 2a = 66 + 2(3) = 72$ mm

(f) $c = b - a = 3.75 - 3 = 0.75$ mm

7.16 SUMMARY

In order to understand a technical discussion on gears, one must be familiar with gear terminology, including such terms as pitch diameter, diametral pitch, circular pitch, addendum, pressure angle, line of action, and pinion.

The speed ratio of two mating gears is inversely proportional to

— The circumferences
— The diameters
— The number of teeth on each gear

The involute profile is the most widely used tooth form. A pair of mating gears must have the same diametral pitch for proper operation.

Good gear design requires that more than one pair of teeth be in contact at all times to help distribute the load and provide smooth, continuous motion. The contact ratio should not be less than 1.4.

Variations of the spur gear are the rack and pinion and the internal spur gear. There are numerous methods of manufacturing gears but the four most common machining methods are

TABLE 7.6 · Table of Preferred Modules *(Courtesy of the Boston Gear Co.)*

Module	Diametral Pitch	Circular Pitch (Inches)
.4	63.500	.0495
.5	50.800	.0618
.6	42.333	.0742
.8	31.750	.0989
1	25.400	.1237
1.25	20.320	.1546
1.5	16.933	.1855
2	12.700	.2474
2.5	10.160	.3092
3	8.467	.3711
4	6.350	.4947
5	5.080	.6184
6	4.233	.7422
8	3.175	.9895

Modules 1 and larger listed above are preferred modules listed in ISO 54. The metric module of a gear is a number which designates the millimeters of pitch diameter per tooth. Metric standards (ISO R53) cover only the 20° pressure angle, involute gear form.

-–   Form milling
—   Rack generation
—   Hobbing
—   Fellows gear shaper method

Good gear design requires the designer to consider the type of power source and driven load; the expected life and speeds; the precision with which the gear is manufactured; the gear material properties, including elastic deflection, and temperature, lubrication, and mounting and alignment data. The gear design procedures in this text should give you good preliminary design information for your particular gear set.

7.17 QUESTIONS AND PROBLEMS ▬▬▬▬▬▬▬▬▬

Questions

1. What is the difference between circular pitch and diametral pitch.
2. What is meant by the term *contact ratio*.
3. Explain the difference between interference and undercutting of gear teeth.
4. What is the difference between backlash and clearance in gearing? Why are they both needed?
5. Give three reasons why the involute profile was selected for gear teeth.
6. Name one advantage that an annular gear drive has over an external spur gear drive.
7. Define *pinion*.
8. Is the following statement correct? "To find the outside diameter, add two to the number of teeth and divide by the diametral pitch."
9. Name 3 ways to increase the horsepower capacity of gears.
10. What is the reason for limiting the face width of a gear?

Problems

1. A spur gear of the 14 1/2° involute system has 32 teeth of diametral pitch 8. Find (a) the pitch diameter; (b) the circular pitch; (c) the outside diameter.
2. Gear A drives gear B with a speed ratio of 5 to 1. If gear A has 30 teeth and a diametral pitch of 10, what is the center distance between the shafts?
3. A pair of mating spur gears (20° involute system) has teeth of diametral pitch 10. The pinion has 30 teeth and the gear has 60. Find: (a) the pitch diameters; (b) the center distance between the shafts; (c) the outside diameters.
4. A pinion has 20 teeth of 5 diametral pitch. It drives a rack while it is rotating at 100 rpm. Find the linear speed of the rack.
5. A gear has a base circle diameter of 3.850 in. and a pitch circle diameter of 4.000 in. Find the pressure angle.
6. Derive the equation for the length of contact path, Z.

7. Each of a pair of mating spur gears (20° involute) has teeth of 6 diametral pitch. The driving pinion is rotating clockwise and has 24 teeth. The driven gear has 36 teeth. Find the contact ratio.

8. A pair of meshing spur gears has the following specifications: number of pinion teeth = 10; number of gear teeth = 12; pressure angle = 14 1/2°; diametral pitch = 2. Find the contact ratio. *1,55*

9. In a pair of mating spur gears, the pinion has 30 teeth and the gear has 50 teeth. If either gear can rotate at 1000 rpm, find: (a) the fastest output speed; (b) the slowest output speed.

10. Gear A turns at 100 rpm, driving gear B at 400 rpm. The center distance is 8 in. and the diametral pitch is 5. Find: (a) the number of teeth on each gear if they rotate in opposite directions; (b) the number of teeth on each gear if they rotate in the same direction.

11. A 20° involute coarse pitch spur gear has a total tooth height of 1.125 in. and contains 24 teeth. Find the diametral pitch and the pitch diameter.

12. Solve for the dynamic bending strength (F_{db}) of the pinion in Sample Problem 7.5.

13. Solve Sample Problem 7.5, except use 20° full depth involute gears.

14. Calculate the horsepower that can be supplied by the pair of 14 1/2° full-depth involute gears described as follows: (a) material = ordinary cast iron with a Bhn hardness of 150; (b) diametral pitch = 2.5; (c) pinion speed = 500 rpm; (d) teeth of pinion = 20; (e) teeth of gear = 35; (f) face width = 0.75 in.

15. Solve Sample Problem 7.6 except change the blower speed to 300 rpm.

16. Design a pair of large industrial steel gears for a limestone kiln. The system is rated at 500 hp, the driving motor rotates at 210 rpm, and the kiln rotates at 30 rpm.

17. Design a pair of gears for commercial application with the following specifications: system to be rated for 8 hp. The pinion speed = 900 rpm and is made of case hardened (350 Bhn) steel. The nylon gear rotates at 180 rpm. Start with a face width of 2 in. Consider wear strength adequate and investigate only for bending strength.

SI Unit Problems

18. Solve Sample Problem 7.7, except with these values: Module = 8 and the gear has 35 teeth.

19. Two spur gears have a speed ratio of 2.5 to 1 and a module of 6. The center distance should be close to 275 mm. Find the number of teeth on each gear.

20. A pair of meshing spur gears has a module of 4. The pinion has 20 teeth and the gear has 34 teeth. Find the contact ratio.

21. Two spur gears have a speed ratio of 3 to 2 and a desired center distance of approximately 200 mm. The pinion should have at least 18 teeth to avoid undercutting. Based on these criteria only, choose a suitable module.

8
Gear Drive Systems

Objectives

On completing this chapter, you will

- Be able to describe various types of gears, gear systems, their characteristics, and their applications.
- Be able to use the equations to determine the speed ratios of various gear trains, including the epicyclic (or planetary) gear train.

8.1 INTRODUCTION

In addition to spur gears, many other types of gears have been developed for various applications. Some of the types that we will consider are helical, herringbone, bevel, and worm-and-wheel. It is important to know their significant features as well as their limitations so as to select the appropriate gears for a particular application. One of the first considerations in gear selection is the geometrical relationship between the center-lines of the shafts. The size limitations, speed ratios, and horsepower requirements should then be introduced.

Quite often, a gear train of more than two meshing gears is required. For example, a large speed ratio may be needed in addition to a compact size arrangement. If only two gears are used, the size of the large gear may exceed the space limitations. Cost considerations are also involved, since a very large gear is often very expensive.

The following types of gear trains (or systems) will be analyzed:

— Simple
— Compound
— Reverted
— Epicyclic (also referred to as planetary)

A gear train is a system that contains two or more meshing gears, and is designed to provide a desired speed ratio within specified space limitations, shaft centerline orientations, and horsepower requirements. Figure 8.1 shows a huge gear train consisting

FIGURE 8.1 Large gear train *(Courtesy of the Tool Steel Gear and Pinion Company, Cincinnati, Ohio)*

solely of herringbone gears. In the forefront is one of the herringbone gears, which can be compared in size to the man standing in front of the gear train.

When contemplating the design of a gear train, the following should be taken into account.

Repetition of Tooth Contact.   A gear pair may have 40 teeth for the gear and 20 teeth for pinion, to produce a speed ratio of 2 to 1. However, with this ratio, the same pairs of teeth will make contact once every other revolution of the pinion. It would be preferable to use 41 and 21 teeth even though the speed ration of 2 to 1 will not be maintained exactly, because the same pair of teeth will make contact only once in every 41×21, or 861 revolutions. Having contact that varies from tooth to tooth produces a more uniform pattern of wear. This varying contact actually slows down wear and helps compensate for minor discrepancies in manufacturing tolerances.

Number of Gears in Train.   It is often possible to obtain a specified speed ratio with any number of pairs of gears. For the case where the speed ratio is accomplished in one step, the gear sizes and, hence, space requirements can be quite large. If many steps (or speed reductions) are used, a large number of shafts and smaller gears are needed, which usually means added cost. Thus a proper compromise between space requirements, economy, and efficiency of operation must be made to arrive at the best overall system.

Strength of Tooth.   When a large amount of power is transmitted at low speeds, a huge value of torque occurs. This torque exerts a high load on the gear teeth, necessitating large size teeth. The high torque value also means that at the low-speed end of a gear train the teeth need to be larger than at the high-speed end. The amount of load determines the tooth size, which in turn affects the pitch diameter and number of teeth.

Integral Number of Teeth.   A gear must have a whole number of teeth. This requirement frequently makes it difficult to arrive an an exact speed ratio or center distance between shafts. In many cases, an exact speed ratio or center distance is not mandatory and slight deviations from design values are acceptable.

8.2 HELICAL GEARS

Let us take a number of spur gears formed from a thin plate and assemble them with a small angular displacement between each adjacent gear. The result is a *stepped spur gear* as shown in Figure 8.2. We can see the advantage of such a gear. There are more

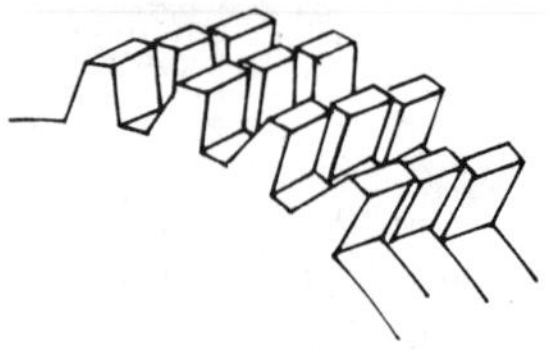

FIGURE 8.2   Stepped spur gear

FIGURE 8.3   Pair of helical gears *(Courtesy of the Tool Steel Gear and Pinion Company, Cincinnati, Ohio)*

teeth in contact at the same time (higher contact ratio). Also, the load is applied in smaller steps as successive teeth come into mesh. *Helical gears,* as shown in Figure 8.3, are basically the limiting case of stepped spur gears where the thickness of the plates and the angular displacement become smaller and smaller as the number of plates becomes larger and larger. This is accomplished physically by cutting the teeth at an angle to the gear axis. This angle is called the helix angle and normally ranges from 7° to 23°. The helix angle must be of the same magnitude for both mating gears, but also must be of the opposite hand. This means that if the pinion is left-handed, the gear must be right-handed, as shown in Figure 8.4.

Helical gears can transmit more power and operate at higher speeds than spur gears because they run more smoothly and quietly. Transmissions of automobiles are one of the very common places where helical gears are used. The main disadvantage of helical gears is the resulting axial thrust forces produced due to the helix angle (angle alpha (α) in Figure 8.5). Bearings which support the geared shafts must be designed to

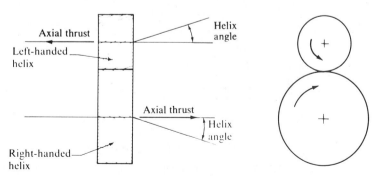

FIGURE 8.4   Helix angles of mating helical gears

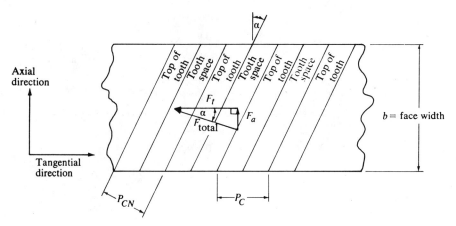

FIGURE 8.5 Partial view of helical gear

absorb these axial thrust loads. The magnitude of the axial thrust force can be determined by referring to Figure 8.5. Observe that, for a helical gear, we have the usual diametral and circular pitches. In addition, we have these pitches in a plane normal to the line of contact along the tooth surface.

Helical gears can also be designed to transmit power between nonparallel shafts. Such gears are usually used under relatively light load situations because the teeth have only point contact. One example is the drive between the camshaft and the distributor shaft of the automobile engine. Figure 8.6 shows a schematic of two meshing helical gears mounted on two nonparallel shafts. For helical gears mounted on nonparallel shafts, the gears can be either of the same hand or of the opposite hand. Furthermore the helix angles do not have to be equal. For proper operation, the following two geometric requirements must be met where theta (θ) is the angle between shaft centerlines:

(1) For gears of the same hand,

$$\theta = \alpha_1 + \alpha_2$$

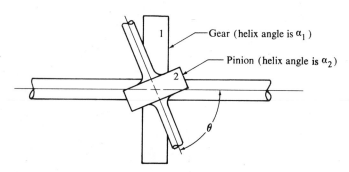

FIGURE 8.6 Helical gears mounted on nonparallel shafts

(2) For gears of the opposite hand,

$$\theta = \alpha_1 - \alpha_2$$

8.3 HERRINGBONE GEARS

A *herringbone gear* consists basically of two rows of helical teeth cut into one gear. Figure 8.7 shows a pair of meshing herringbone gears. One of the rows of each gear is right-handed and the other is left-handed to cancel out the axial thrust force. In Figure 8.8, we can see that the axial thrust forces are canceled by direct subtraction. Since the

FIGURE 8.7   Pair of meshing herringbone gears *(Courtesy of the Tool Steel Gear and Pinion Company, Cincinnati, Ohio)*

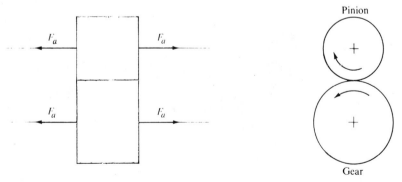

FIGURE 8.8   Axial force cancellation on herringbone gears

axial thrust force is eliminated, herringbone gears are used for heavy and continuous load applications. They have all the advantages of helical gears without the undesirable feature of axial thrust forces. One excellent application is for driving rolling mills in modern steel plants. Also observe the size of the herringbone gear train of Figure 8.1. A large amount of power is obviously transmitted by this train.

8.4 BEVEL GEARS

INTERSECTING SHAFTS

Bevel gears are characterized by their conical shape, as shown in Figures 8.9 and 8.10. Three basic types of bevel gears are the straight-tooth, the spiral-tooth, and the hypoid gears, briefly described below.

Straight-Tooth Bevel Gear.   The *straight-tooth bevel gear* is used for nonparallel shafts that would intersect if extended. The teeth are straight and are inclined to the gear axis by a conical angle called the pitch angle (See Figures 8.9 and 8.10). The speed ratio of two meshing bevel gears is the ratio of the number of teeth in each gear. The speed ratio is also the ratio of their pitch diameters and the ratio of the sines of their pitch angles. If the shafts are perpendicular, the bevel gears are called miter gears. Miter gears have the same number of teeth on each gear, giving a 45° cone (pitch) angle. Figure 8.11 shows a gear train consisting of two parallel shafts connected by spur gears. The left shaft contains a straight bevel gear that drives a rear shaft at right angles through a second bevel gear. Since miter gears have the same number of teeth on each gear, the speed ratio equals one. The sum of the two pitch angles equals the angle between the centerlines of the two shafts.

Spiral Bevel Gear.   The *spiral bevel gear* is also used for nonparallel shafts that would intersect if extended. However, the teeth are not straight but are curved in the

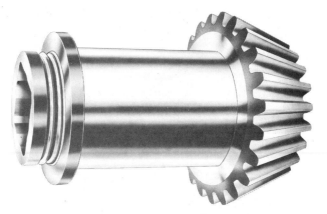

FIGURE 8.9   Straight-tooth bevel gear (*Courtesy of the Tool Steel Gear and Pinion Company, Cincinnati, Ohio*)

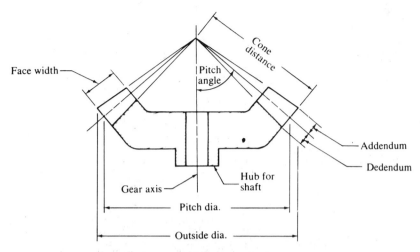

FIGURE 8.10   Cross section of a bevel gear

FIGURE 8.11   Gear train containing miter gears (*Courtesy of the Tool Steel Gear and Pinion Company, Cincinnati, Ohio*)

form of spirals around the conical surface. This is depicted in Figure 8.12, where the teeth are represented by spiral lines for simplicity.

Spiral bevel gears enjoy the same advantages over straight-tooth bevel gears that helical gears have over spur gears. (See Section 8.3) Thus spiral bevel gears are used for high-speed, high-load applications. They also produce large axial thrust loads and thus complicate the design of the shaft bearings.

Hypoid Gears.   The *hypoid gear* is similar in appearance to the spiral bevel gear, except for meshing hypoid gears the shaft centerlines are perpendicular and offset from

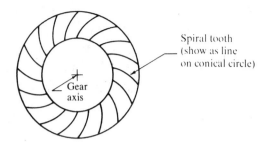

FIGURE 8.12   Spiral bevel gear

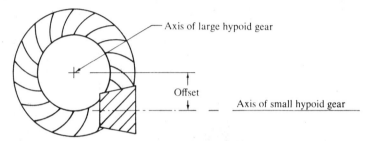

FIGURE 8.13   Pair of meshing hypoid gears

each other. The hypoid gear was developed for the rear axles of automobiles, and enables the drive shaft to pass below the level of the floor. Since the shaft centerlines do not intersect for hypoid gears, even if extended, bearings can be mounted on both sides of either gear, promoting greater system stiffness and subsequent smoother operation. This arrangement is shown schematically in Figure 8.13, where the teeth are represented by lines.

8.5 THE WORM AND WHEEL (WORMGEAR SET) ▬▬▬▬

A *worm and wheel* are used for large speed reductions between two perpendicular but nonintersecting shafts (see Figure 8.14). The driver, which is called the *worm*, has a small diameter and a low helix angle. In appearance it is similar to a threaded screw, and thus the helical teeth are commonly called *threads*. The follower is called the *worm gear* or *worm wheel* and has a face which is made concave to match the curvature of the worm. Hence, the wheel is said to *envelop* the worm, and their matching curvatures provide a large area of contact and reduce wear.

Worms can be made with either right- or left-hand threads similar to those used for threaded fasteners such as screws and bolts. Also, the worm can have single, double, triple, or quadruple threads. Note the following two definitions used in connection with threaded fasteners and worms.

Pitch.   The *pitch* is the axial distance from one point of a thread to the corresponding point on an adjacent thread;

FIGURE 8.14   Worm and wheel *(Courtesy of the Eaton Corporation, Cleveland, Ohio)*

Lead.   The *lead* of the worm is the axial distance that a point on a thread advances in one revolution of the worm.

By geometric necessity, the pitch of a worm equals the circular pitch, *p*, of the gear. We can now write an equation to relate the lead of the worm to its pitch or to the circular pitch of the gear:

$$L = kp$$

where L = lead of the worm

p = circular pitch of the gear or pitch of the worm

k = 1 for a single-threaded worm (lead = pitch)

k = 2 for a double-threaded worm (lead = 2 × pitch)

k = 3 for a triple-threaded worm (lead = 3 × pitch)

k = 4 for a quadruple-threaded worm (lead = 4 × pitch)

If the worm gear has 20 teeth, it would require 20 turns of a single-threaded worm to rotate the worm gear one revolution. This relationship exists because one revolution of a single-threaded worm indexes the worm gear exactly one tooth. A double-threaded worm indexed its worm gear two teeth with each revolution, and so forth. As a result, the speed ratio of a worm and wheel can be found from

$$\text{Speed ratio} = \frac{\omega_{wg}}{\omega_w} = \frac{k}{N_{wg}} \tag{8.1}$$

where ω = the rotational speed in revolutions per minute

N = the number of teeth

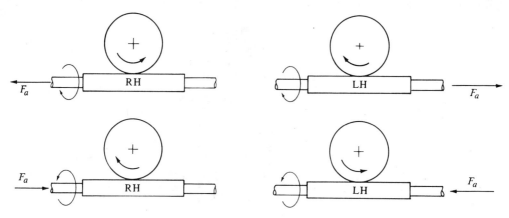

FIGURE 8.15   Axial thrust force on worm shaft

$_{\text{wg}}$ = the subscript for the worm gear

$_{\text{w}}$ = the subscript for the worm

Obviously, the greatest speed reduction occurs with a single-threaded worm. Also, note that the speed ratio does not depend upon diameter ratios or tooth ratios, as was the case for spur gears. Because of the low helix angle of the worm, the worm gear cannot normally drive the worm. This is a self-locking feature that is usually desirable. However, if for example a quadruple-threaded worm having a large helix angle is used, it would be possible to have the worm gear drive the worm. In practice, this usage is rare, since large speed reductions are normally desired when using a worm and wheel.

Note that the use of small helix angles (less than 20°) results in poor efficiencies, which can be as low as 25%. If the helix angle is optimized (30–45°), the efficiency can be as high as 95%.

Figure 8.15 shows the direction of the axial thrust force on the worm shaft. The direction depends on the direction of rotation of the worm and whether it is left- or right-handed.

We can find the output torque if the efficiency and input torque are known. This formula is derived from the horsepower-torque formula:

$$\text{hp} = \frac{T \times \omega}{63,000}$$

and from the power relationship formula:

Power out $= \eta \times$ (power in)

where η is the efficiency of the gearset

Therefore

$$T_{\text{wg}}\omega_{\text{wg}} = \eta T_{\text{w}}\omega_{\text{w}}$$

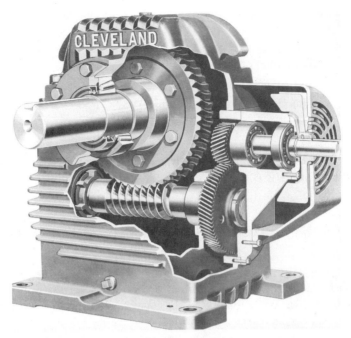

FIGURE 8.16   Gear train with a worm and wheel *(Courtesy of the Eaton Corporation, Cleveland, Ohio)*

$$\text{Output torque} = T_{wg} = \eta \, \frac{T_w \times \omega_w}{\omega_{wg}}$$

Substituting the worm gear set speed ratio equation, we

$$T_{wg} = \eta T_w \left(\frac{N_{wg}}{k} \right) \tag{8.2}$$

As expected, there is a torque multiplication whenever a speed reduction occurs. For an efficiency of 100 percent, the torque multiplication ratio equals the inverse of the speed ratio.

Figure 8.16 shows a gear train which contains a worm and wheel. Also mounted on the worm shaft is a helical gear that meshes with a smaller helical gear mounted on a parallel shaft. The input shaft contains the small helical gear and hence, the output shaft contains the worm gear. This gear train reduces the speed in two steps: first by the helical gear pair and second by the worm and wheel. Notice the large size of the output shaft as compared to the input shaft. This shows the significant torque multiplication that occurs as a direct result of the high prevailing speed reduction. The total speed reduction equals the product of the two individual speed reductions. For example, let us assume that the large helical gear is three times as large as its meshing pinion.

Let us further assume a single-threaded worm driving the worm gear having 50 teeth. Therefore

Speed ratio of the helical gear pair = 1/3
Speed ratio of worm and wheel = 1/50
Total speed ratio of gear train = 1/3 × 1/50 = 1/150
Torque multiplication ratio = 150/1

Table 8.1 lists the principal advantages, disadvantages, and shaft angular relationships for the various types of gear drives just discussed.

8.6 THE SIMPLE GEAR TRAIN

A *simple gear train* is one in which each shaft contains only one meshing gear. Figure 8.17 shows a simple gear train consisting of three gears. The middle gear is called an *idler* because it does not affect the speed ratio. Its purpose is to give the output shaft the same direction of rotation as the input shaft. It also can be used to assist in providing the required center distance between the input and output shafts.

The total speed ratio of any gear train equals the product of the individual speed ratios of each meshing pair as a series path is followed from the input gear to the output gear.

$$Speed\ ratio = \frac{\omega_o}{\omega_i} = \frac{D_i}{D_o} = \frac{N_i}{D_o}$$

where   subscript, o, refers to the output gear

subscript, i, refers to the input gear

D = the pitch diameter of the gear

N = the number of teeth on the gear

The speed ratio for the simple gear train of Figure 8.17 is

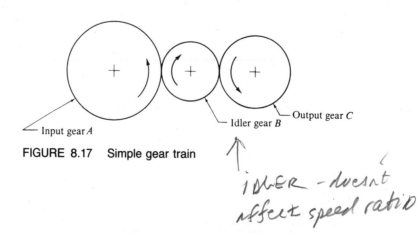

— Input gear A
— Idler gear B
— Output gear C

FIGURE 8.17   Simple gear train

idLER – duesn't affect speed ratio

TABLE 8.1   Characteristics of Various Types of Gear Drives

Gear Drive	Advantages	Disadvantages	Shaft Relationships
External spur	no axial thrust force	low contact ratio	parallel shafts
Rack and pinion	compact	low contact ratio; low speeds	rotary to linear motion and vice versa
Internal spur	no axial thrust force; large contact ratio; compact	costly	parallel shafts
Helical	quiet and smooth operation; high speeds	axial thrust force	parallel shafts and nonparallel shafts
Herringbone	no end thrust; large contact ratio; large load capacity	costly	parallel shafts
Bevel (a) straight-tooth	less expensive than spiral and hypoid	low to medium speeds	nonparallel shafts (would intersect if extended)
(b) spiral	large load capacity; large contact ratio	costly	nonparallel shafts (would intersect if extended)
(c) hypoid	large load capacity; very rigid support	costly	perpendicular shafts (would not intersect even if extended)
Worm-and-wheel	high-speed reduction; can be self-locking	axial thrust force on worm shaft	perpendicular, nonintersecting shafts

$$\text{Speed ratio} = \frac{\omega_C}{\omega_A} = \frac{N_A}{N_B} \times \frac{N_B}{N_C} = \frac{N_A}{N_C} = \frac{D_A}{D_C}$$

Having just proved that the idler does not affect the speed ratio, we can now establish the following rule: **The speed ratio of a simple gear train equals the number of teeth on the input gear divided by the number of teeth on the output gear, regardless of the number of intermediate idler gears.**

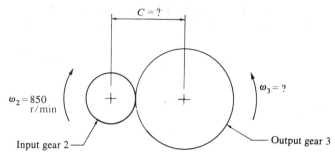

FIGURE 8.18   Simple gear train

Directions for Solving Simple Gear Train Problems

(1) The usual gear train problem involves adjusting gear diameters, diametral pitch, and/or the number of teeth on each gear to meet the required specifications. Several trials are usually involved because each gear must have a whole number of teeth.

(2) Use the appropriate equations in Chapter 7 and in this chapter.

SAMPLE PROBLEM 8.1

Speed Ratio

PROBLEM: Design a simple gear train consisting of two gears which will provide a speed ratio of 2/7. The minimum number of teeth is 12.

Solution

Multiply the numerator and denominator of the speed ratio by a whole number (starting with 6) to keep the minimum number of teeth at 12.

$$\text{Speed ratio} = \left(\frac{2}{7}\right)\left(\frac{6}{6}\right) = \frac{12}{42}$$

$$= \left(\frac{2}{7}\right)\left(\frac{7}{7}\right) = \frac{14}{49}$$

$$= \left(\frac{2}{7}\right)\left(\frac{8}{8}\right) = \frac{16}{56}$$

$$= \left(\frac{2}{7}\right)\left(\frac{9}{9}\right) = \frac{18}{63} \quad (etc)$$

The preceding results show that, if the input gear has 12, 14, 16, or 18 teeth, the output gear will have 42, 49, 56, or 63 teeth, respectively.

SAMPLE PROBLEM 8.2

Speed Ratio and Center Distance

PROBLEM: For the simple gear train of Figure 8.18, the center distance, *C*, should be as close as possible to 4.870 in.; the diametral pitch is 8. The input speed is 850 rpm and the output speed is to be as close to 300 rpm as possible (closeness of center distance has top priority). Find: **(a)** the number of teeth for each gear; **(b)** the exact output speed; **(c)** the exact center distance.

Solution

The approach will be to find the total number of teeth from the center distance formula in Chapter 7 and, combining this with the speed ratio, find the theoretical number of teeth on each gear. Various tooth numbers will be tried around the theoretical figures to get the closest center distance required.

(a) $C = \dfrac{(N_A + N_B)}{2P}$

$(N_A + N_B) = 4.870 \times 2 \times 8 = 77.92.$

Speed ratio $= \dfrac{\omega_B}{\omega_A} = \dfrac{N_A}{N_B} = \dfrac{300}{850}$

Rearranging

$$N_A = \dfrac{N_B \times 300}{850} = 0.353 N_B$$

And substituting in the first formula

$(0.353 N_B + N_B) = 77.92$

$N_B = 57.59$ and $N_A = 20.33$

Since the number of teeth has to be a whole number, the final results are $N_A = 20$ teeth and $N_B = 58$ teeth.

(b) From the speed ratio formula we have

$$\omega_B = \omega_A \left(\frac{N_A}{N_B}\right) = 850 \left(\frac{20}{58}\right) = 293 \text{ rpm}$$

(c) $C = \dfrac{N_A + N_B}{2P} = \dfrac{20 + 58}{2 \times 8} = 4.875 \text{ in.}$

You should try combinations of 20 and 57 teeth and 21 and 57 teeth to verify that the solution provided gives the best overall results.

8.7 THE COMPOUND GEAR TRAIN ▬▬▬▬▬▬▬

A compound gear train is one in which there are two or more different gear pairs connected in series. As a result, some of the shafts will contain more than one gear (see Figure 8.19). (Note that the symbol E is used to represent the output gear instead of D because D is used to represent the pitch diameter.)

Notice that gears B and C are mounted on the same shaft and thus have the same speed. The total speed ratio for the subject gear train will now be found.

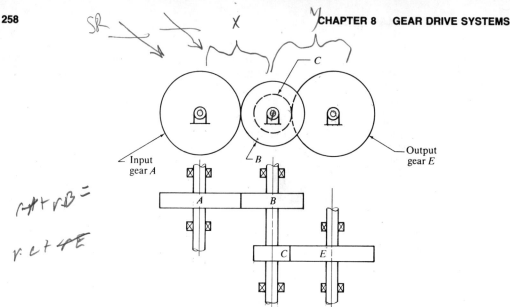

FIGURE 8.19   Compound gear train

$$\text{Total speed ratio} = (SR)_{AB} \times (SR)_{CE}$$

$$SR = \left(\frac{N_A}{N_B}\right)\left(\frac{N_C}{N_E}\right)$$

where SR = the speed ratio   (Subscripts identify the gears.)

Based on the preceding result, the following rule is established: **The speed ratio of a compound gear train equals the product of all the tooth numbers of the driving gears divided by the product of all the tooth numbers of all the driven gears.**

In Figure 8.19, gear A drives gear B and gear C drives gear E. Compound gear trains with two pairs of mating gears are referred to as *double reduction* or *two-stage* gear trains. Compound gears may also have three stages. Problems in this area require you to (1) use the speed ratio formula to find the total speed ratio given the gear sizes; or (2) design a gear train given the total speed ratio and the number of stages.

PROCEDURE

Solving Compound Gear Train Problems

Step 1 For case one above, simply insert data into the speed ratio formula.

Step 2 For case two, where you have to design the gear train—that is, choose the number of teeth on each gear or find the gear sizes, follow this procedure to find the speed ratio of each stage:

(a) For a two-stage gear train, take the square root of the total speed ratio and use it for the speed ratio of each stage.

(b) For a three-stage gear train, take the cube root of the total speed ratio.

Finding the square root or cube root is recommended because it is desirable to have all the stages equal in size.

SAMPLE PROBLEM 8.3

Speed Ratio and Torque of a Compound Gear Train

PROBLEM: For the compound gear train of Figure 8.19, the following data are given:

$$\omega_A = 600 \text{ rpm}$$

$$T_A = 100 \text{ lb-in.}$$

$$D_A = 10 \text{ in.}$$

$$D_B = 12 \text{ in.}$$

$$D_C = 8 \text{ in.}$$

$$D_E = 14 \text{ in.}$$

Find **(a)** the output speed and **(b)** the output torque.

Solution:

(a)
$$SR = \frac{\omega_E}{\omega_A} = \frac{D_A}{D_B} \times \frac{D_C}{D_E} = \frac{10 \times 8}{12 \times 14}$$

$$SR = \frac{\omega_E}{\omega_A} = \frac{1}{2.1}$$

$$\omega_E = \frac{600 \times 1}{2.1} = 286 \text{ rpm}$$

(b) Assume Power in = Power out

$$T_A \omega_A = T_E \omega_E$$

$$T_E = \frac{100 \times 2.1}{1} = 210 \text{ lb-in.}$$

SAMPLE PROBLEM 8.4

Compound Gear Design

PROBLEM: Design a compound gear train with a speed ratio of 1/150. There can be no more than 98 and no fewer than 14 teeth on any gear. Find the number of teeth for each gear.

Solution

Since the maximum speed reduction for a single gear pair is 98/14 = 7, three pairs of gears are needed, as shown in Figure 8.20. Two pairs can give a

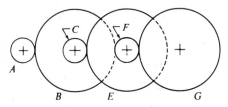

FIGURE 8.20   Compound gear train (speed ratio = 1/150)

maximum speed reduction of $7 \times 7 = 49$, while three pairs give $7^3 = 343$. If all three pairs were made identical, each pair would give a speed reduction of

$$\sqrt[3]{150} = 5.3.$$

This means that the speed ratio equation would look like this:

$$SR = \frac{\omega_{out}}{\omega_{in}} = \frac{1}{5.3} \times \frac{1}{5.3} \times \frac{1}{5.3} = \frac{1}{150}$$

To be as close as possible to the ideal, let us use ratios of $1/6 \times 1/5 \times 1/5 = 1/150$. (Note that 1/6 is below the limit of 7 established above.) Also, since the minimum number of teeth is 14, then all the ratios are multiplied by this number.

$$SR = \frac{N_A}{N_B} \times \frac{N_C}{N_E} \times \frac{N_F}{N_G} = \frac{1}{6} \times \frac{1}{5} \times \frac{1}{5}$$

$$SR = \frac{14}{84} \times \frac{14}{70} \times \frac{14}{70}$$

Therefore

$$N_A, N_C, N_F = 14 \text{ teeth each}$$

$$N_B = 84 \text{ teeth}$$

$$N_E, N_G = 70 \text{ teeth each}$$

The reason for the limitation of a maximum of 98 teeth is to prevent the largest gear size from becoming excessive (based on the diametral pitch required).

8.8   THE REVERTED GEAR TRAIN

A reverted gear train is actually a compound gear train specifically designed so that the input and output shafts are collinear (see Figure 8.21). Notice that the center

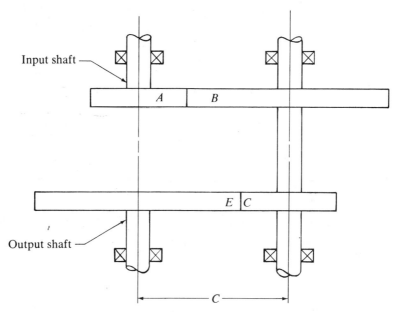

Input shaft

Output shaft

FIGURE 8.21   Reverted gear train

distance C is the same for each pair of gears. Thus, for Figure 7.21, we have the following relationship:

$$\text{Distance } \boxed{C} = \frac{N_A + N_B}{2 \times P_{AB}} = \frac{N_C + N_E}{2 \times P_{CE}} \qquad (8.3)$$

The subscripts for the diametral pitches refer to the pair of gears for that particular diametral pitch. That is, the diametral pitches need not be the same for both pairs. Normally, the high-speed gears have the larger diametral pitch number because they transmit a lower value of torque. One of the principle advantages of a reverted gear train is its compactness. Also, in many applications it is necessary that the input and output shafts be collinear. One example is the automotive transmission described in Section 8.10.

PROCEDURE

Designing Reverted Gear Trains

The problems in this section will provide you with the total speed ratio and the diametral pitch for each pair of mating gears. You will be required to find the number of teeth on each gear.

Step 1   Find the speed ratio for each pair of gears.

Step 2   Solve Equation 8.3 as far as possible with the known information. Doing so will provide you with the tooth ratios for each pair of gears.

Substitute different numbers of teeth for the pinion and try to find a value that will give you a whole number for the mating gear. Normally, you won't get an exact speed ratio, but try about half a dozen numbers and pick the set that comes closest to the specified speed ratio. (If more time were available we could try more numbers or vary the speed ratios of each pair of gears so they wouldn't be identical.) See Sample Problem 8.5.

Step 3　If the diametral pitches are different for each pair of gears, then your trials will have to include a check for the second set of gears, as demonstrated by Sample Problem 8.6.

SAMPLE PROBLEM 8.5

Reverted Gear Train—All Gears With the Same Diametral Pitch

PROBLEM: The reverted gear train of Figure 8.21 is to have a speed ratio of 1/20. No gear can have less than 12 teeth. Assume that all gears have the same diametral pitch. Find the number of teeth for each gear.

Solution

Factor the total speed ratio.

$$\sqrt{\frac{1}{20}} = \frac{1}{4.47}$$

$$SR = \frac{1}{4.47} \times \frac{1}{4.47}$$

Thus

$$\frac{N_A}{N_B} = \frac{1}{4.47} \quad \text{and} \quad \frac{N_C}{N_E} = \frac{1}{4.47}$$

From Equation 8.3 we see that the diametral pitches drop out since they are the same and we have

$$N_A + N_B = N_C + N_E$$

and

$$N_A + 4.47N_A = N_C + 4.47N_C$$

From this expression we can see that both pairs have the same number of total teeth, and that

$$N_A = N_C \quad \text{and} \quad N_B = N_E$$

Starting with the minimum number of teeth allowed, 12, we find that if $N_A = 12$, then

$$N_B = 4.47 \times 12 = 53.6 \text{ teeth.}$$

We try 13, 14, 15, 16, 17, and 18 and find that 17 teeth for N_A gives us 75.99 teeth for N_B—the closest to a whole number. So we have

$$N_A = N_C = 17 \text{ teeth}$$

And

$$N_B = N_E = 76 \text{ teeth}$$

Check:

$$SR = \frac{17}{76} \times \frac{17}{76} = \frac{1}{19.99}$$

SAMPLE PROBLEM 8.6

Reverted Gear Train With Different Diametral Pitches for the Two Pairs of Gears

PROBLEM: The data are the same as Sample Problem 8.5, except the the diametral pitch for gears A and B is $P_{AB} = 5$, and for gears C and E $P_{CE} = 2\frac{1}{2}$. The coarser diametral pitch (which means larger teeth) is needed at the low-speed end of the train because of the greater torque load.

Solution

As before, we will use the same speed ratio for each pair of gears. Equation 8.3 now has this arrangement:

$$\frac{N_A + 4.47N_A}{2 \times 5} = \frac{N_C + 4.47N_C}{2 \times 2\frac{1}{2}}$$
$$5.47N_A = 2(5.47N_C)$$

This means that $N_C = (\frac{1}{2})N_A$. Therefore N_C must have one half the number of teeth of N_A. Therefore, since 17 teeth is the smallest number, this value must be for gear C. Maintaining our ratios, we have

$$N_A = 34 \text{ teeth}$$

$$N_B = 152 \text{ teeth}$$

$$N_C = 17 \text{ teeth}$$

$$N_E = 76 \text{ teeth.}$$

Check:

$$SR = \frac{34}{152} \times \frac{17}{76} = \frac{1}{19.99}.$$

8.9 THE EPICYCLIC (PLANETARY) GEAR TRAIN ▰▰▰▰

Up to now, all the gears discussed have been mounted on shafts supported by bearings in stationary housings. Thus the axis of rotation was fixed. An *epicyclic, or planetary,* gear train is one that has one or more gears rotating about a moving axis. It is normally used for speed ratios ranging from 1.2:1 to 12:1. The planetary gear train is usually more compact than other types of trains for the same speed ratio and also a singe planetary gear train can provide several speed ratios. Usually, automatic transmissions in automobiles employ the epicyclic gear train. In Figure 8.22, we see one example of an epicyclic gear train containing the following gears:

S = sun gear.

P = planet gear (there are 2 planet gears in Fig. 8.22).

R = ring gear (an internal, or annular, gear).

C = planet carrier which connects the planet gears to a single shaft.

The following four modes of operation are possible.

1. The sun gear shaft is the input and the planet carrier shaft the output. The ring gear is fixed. As the sun gear rotates, it causes the two planet gears to turn about their axes as their axes revolve about the sun gear. This arrangement is similar to the planet earth revolving about the sun as it spins on its own axis. It is possible to have one, two, three, or more planet gears equally spaced (see Figure 8.23, which shows four planet gears). The number of planet gears does not affect the speed ratio, although more torque can be transmitted as the number of planet gears increases.

2. The planet carrier shaft is fixed and the ring gear is free to rotate. As the sun gear rotates, each planet gear rotates about its own, fixed, axis. In this way, the plan-

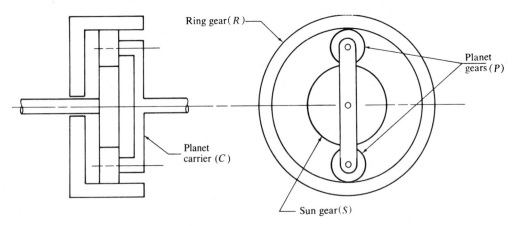

FIGURE 8.22   Epicyclic gear train

Takes up less space

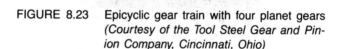

FIGURE 8.23   Epicyclic gear train with four planet gears
*(Courtesy of the Tool Steel Gear and Pin-
ion Company, Cincinnati, Ohio)*

etary, or epicyclic, phenomenon does not take place and the system behaves as a simple
gear train with an internal gear.

3. The sun gear is fixed and the ring gear and planet carrier are connected directly
to the input and output shafts.

4. All the gears are permitted to rotate. In this case, the internal gear has external
teeth as well as internal teeth. A pinion is used to drive the ring gear through its
external teeth, as shown in Figure 8.24. In calculations for this case, the pinion drive
of the ring gear is treated as a simple gear train of two external gears. Thus the speed
of the ring gear becomes the input to the epicyclic gear train. This is an unusual mode
of operation, so we will be concerned with problems only in the first three modes.

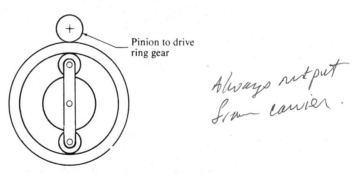

Pinion to drive
ring gear

*Always output
from carrier.*

FIGURE 8.24   Pinion drive of epicyclic gear train

The gear speeds in the epicyclic gear train of Figure 8.22 can be solved by the following equation:

$$\omega_s = \left(1 + \frac{N_R}{N_S}\right)\omega_C - \left(\frac{N_R}{N_S}\right)\omega_R \tag{8.4}$$

where
ω_s = angular velocity of sun gear.

ω_C = angular velocity of planet carrier.

ω_R = angular velocity of ring gear.

N_R = number of teeth on the ring gear.

N_S = number of teeth on the sun gear.

Note that, since the planet gears mesh with the sun and ring gears, they must all have the same diametral pitch. Also, from geometry, the pitch diameter of the ring gear is equal to the diameter of the sun gear plus twice the diameter of the planet gear. Since the pitch diameter of a gear equals the number of teeth divided by the diametral pitch, we have

$$D_R = D_S + 2D_P$$

Also

$$D = \frac{N}{P}$$

Substituting:

$$\frac{N_R}{P} = \frac{N_S}{P} + 2 \times \frac{N_\rho}{P}$$

This leaves us with $N_R = \left(N_S + 2N_P\right)$

Thus the number of teeth on the ring gear equals the number of teeth on the sun gear plus twice the number of teeth on the planet gear.

PROCEDURE

Solving Planetary Gear Train Problems

Normally you will need to find some unknown value using Equation 8.4 and be asked to determine the direction of rotation. If you are asked to find the speed and direction of rotation of a gear, a minus sign indicates a direction opposite to that of the given input. This can be checked graphically.

SAMPLE PROBLEM 8.7

Epicyclic Gear Train

PROBLEM: For the epicyclic gear train of Figure 8.22, the following data are given: The sun gear is the input, the ring gear is fixed, and the planet carrier is the output. The speed of the sun gear is 1000 rpm CW; $N_S = 80$ teeth, $N_P = 20$ teeth, and $N_R = 120$ teeth. Find the direction and magnitude of the planet carrier shaft speed.

Solution

Substitute directly into Formula 8.4.

fixed

$$1000 = \left(1 + \frac{120}{80}\right)\omega_C - \left(\frac{120}{80}\right)(0)$$

$$\omega_C = +400 \text{ rpm CW}$$

same as input

The direction is clockwise because of the plus sign.

SAMPLE PROBLEM 8.8

Epicyclic Gear Train

PROBLEM: For the epicyclic gear train of Figure 8.22, the following data are given: The ring gear is fixed and has 120 teeth. $\omega_s = $ **80 rpm and** $\omega_c = $ **20 rpm.** Find the number of teeth on the sun and planet gears.

Solution

Substitute directly into Formula 8.4.

①

$$80 \text{ rpm} = \left(1 + \frac{N_R}{N_S}\right)20 - \frac{N_R}{N_S}(0)$$

$$\frac{N_R}{N_S} = 3$$

6

$$80 = \left(1 + \frac{120}{NS}\right)20 - \frac{120}{NS}(6)$$

$$Ns = 40$$

Therefore

$$N_S = \frac{120}{3} = 40 \text{ teeth}$$

From the formula

② $$N_S + 2N_P = N_R$$

$$40 + 2N_P = 120$$

$$N_P = 40 \text{ teeth}$$

8.10 THE AUTOMOTIVE TRANSMISSION

A *gear box* is an assembly that houses gear trains that produce different speeds at the output shaft for a given input shaft speed. The automotive gear box (transmission) is an excellent example of where reverted gear trains are used. By manually shifting the position of a particular gear along its shaft, which is splined, a different output speed is obtained.

Figure 8.25*a* shows a sketch of a three speed, manual-shift automobile transmission. This type of transmission is rather out of date for automobiles, but a similar system is still used on lathes.

The operation is described below.

1. The input shaft is connected to the engine through a clutch. Gear 1 is keyed to the input shaft and also is constantly meshing with gear 2, which is keyed to the lower shaft (called a countershaft). Gears 3, 4, and 5 are also keyed to the countershaft. As a result, when the input shaft rotates, it forces the countershaft with its gears to rotate.

2. Gears 6 and 7 are mounted on the splined output (drive shaft), which extends to the rear axle through a universal joint. Gears 6 and 7 are free to slide along the output shaft (when shifting gears) because it is splined as shown in Figure 8.25*b*. The sliding action of gears 6 and 7 is accomplished by the actuation of a shifting mechanism not

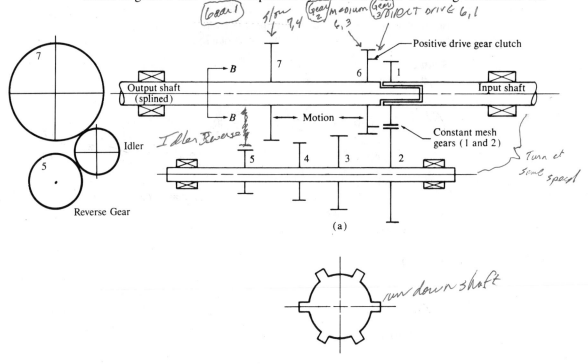

(a)

(b) Section *B-B* (enlarged)

FIGURE 8.25   Three-speed automotive transmission

shown. Note that there is no positive connection between the input and output shafts except where the input shaft supports the right end of the output shaft.

3. Gear 6 also contains a positive drive gear clutch. When gear 6 is shifted to the right, the positive drive gear clutch engages with gear 1. Under this condition, the output shaft is directly connected to the input shaft. This is called direct drive, and it provides a high-speed output (speed 3, which we call third gear) that equals the input shaft speed.

4. When gear 6 is shifted to the left, it meshes with gear number 3. We then have a reverted gear train providing an intermediate output speed (speed 2).

5. If gear 6 is returned to its neutral position, then gear 7 can be shifted to mesh with gear 4. This is accomplished by the design of the shifting mechanism. In this arrangement, we have the lowest forward speed (speed 1).

6. If gear 7 is shifted to the left, it clears gear 5 and meshes with an idler, which constantly meshes with gear 5. The idler causes the output shaft to reverse, providing reverse speed.

To permit gears 6 and 7 to slide into mesh, the ends of the teeth are rounded. Of course, shifting must be done with the clutch disengaged on the input shaft so that the input shaft is free-wheeling to avoid stripping the gears. However, the shifting gears will still tend to clash because one of them is usually rotating faster than the other during the meshing operation. This clashing is eliminated by making the speeds of the gears nearly equal just prior to meshing. A friction clutch (not shown), which is mounted on the driving gear, engages a friction clutch on the driven gear just prior to meshing. By the time the gears begin to mesh, the speeds are about equal. A transmission having this feature where the teeth of two gears mesh quietly is called a *synchromesh transmission*.

Figure 8.26 illustrates a fully synchronized three-speed automotive manual transmission. This design used constant mesh helical gears in all speeds and thus the gears are not physically shifted as was done for the design of Figure 8.25. During the shifting into a particular speed range, a device called a synchronizer is moved on its splined main shaft. The inner cone-shaped hub portion of the synchronizer then makes contact with the side of a gear which is free to turn on the main shaft (see Figure 8.27 which is a simplified schematic of a fully synchronized transmission). This gear is in constant mesh with a driver gear on the lower countershaft. The initial contact of the synchronizer with the side of the free rotating gear is a frictional contact of flat, cone-shaped surfaces, which causes the synchronizer and gear to start rotating as a unit. The synchronizer also contains an outer sleeve (called a *dog sleeve*), which is splined to the inner hub. The outer sleeve and inner hub can move axially relative to each other and are held axially together by a spring-loaded ball and socket. As shifting continues, the outer dog sleeve slides over its inner hub, which contacts the side of the constant mesh gear, and engages splines (called dogs) attached to the gear. Shifting is completed when the splines are fully meshed and positive drive in the desired speed range occurs. Shifting, of course, is done only with the clutch disengaged so that the transmission is disconnected from the engine.

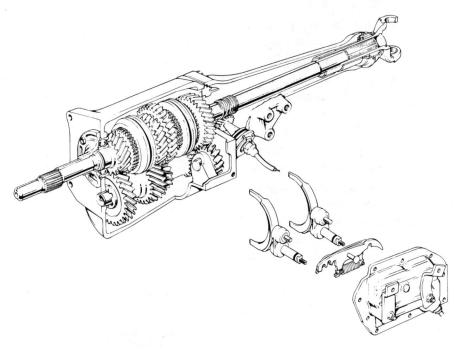

FIGURE 8.26   Fully synchronized three-speed manual transmission (constant
mesh gears) *(Courtesy of Chrysler Corp.)*

Figure 8.28 is a photograph of a sophisticated automatic transmission. Two planetary gear sets, operating individually or in combination, provide the various gear ratios.

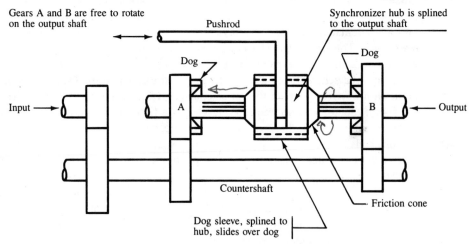

FIGURE 8.27   Schematic of a fully synchronized transmission

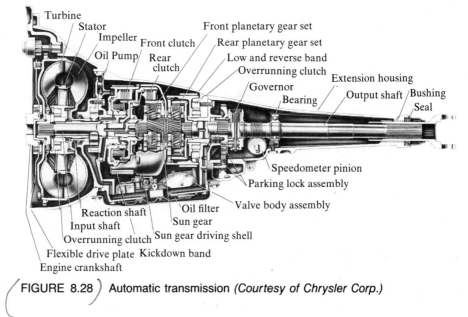

Turbine
Stator
Impeller   Front clutch
Oil Pump   Rear
clutch
Front planetary gear set
Rear planetary gear set
Low and reverse band
Overrunning clutch
Governor
Bearing
Extension housing
Output shaft   Bushing
Seal
Speedometer pinion
Parking lock assembly
Valve body assembly
Reaction shaft   Oil filter
Input shaft   Sun gear
Overrunning clutch   Sun gear driving shell
Flexible drive plate   Kickdown band
Engine crankshaft

FIGURE 8.28   Automatic transmission *(Courtesy of Chrysler Corp.)*

8.11   SUMMARY

In addition to the spur gear covered in Chapter 6, this chapter explained other types of gears such as the helical, herringbone, bevel, and worm-and-wheel gears. The teeth of the helical gear follow a helix around the shaft centerline. This design is quieter and runs more smoothly than the spur gear, but a disadvantage is that it develops axial thrust. The herringbone gear corrects this thrust problem by having two rows of helical teeth where the helix for each row opposes the other. Thus the axial thrust of one row is canceled by the axial thrust in the opposite direction of the other row. Bevel gears have their teeth mounted on a conical surface, as opposed to spur and helical gears, which have their teeth mounted on a cylindrical surface. Bevel gears are useful when their respective shafts are mounted at an angle to each other. The worm-and-wheel gear set provides large speed reductions in a small compact space.

The following types of gear trains were analyzed: simple, compound, reverted, and epicyclic (planetary). A simple gear train is one in which each shaft contains only one meshing gear. If there are more than two gears, all of the intermediate gears between the input and output gears are called *idler gears*. Idler gears do not affect the speed ratio but are used to change the direction of rotation of the output shaft. The speed ratio can be obtained by using data only from the input and output gears. A compound gear train is one in which there are two or more different gear pairs connected in series, with some of the shafts containing more than one gear. Large speed reductions can result. The reverted gear train is a compound gear train in which the input and output shafts are collinear. Epicyclic, or planetary, gear trains offer speed ratios from 1.2:1 to about 12:1. They are compact, their input and output shafts are collinear, and one gear train set can offer several speed ratios, depending on which gear is fixed. This chapter also discussed the use of gears in automotive transmissions. In manual transmissions, the gear train is essentially a reverted gear train, while automatic transmissions use epicyclic gear trains.

8.12 QUESTIONS AND PROBLEMS ▬▬▬▬▬▬▬▬▬▬▬▬

Questions

1. For a pair of meshing spur gears, the speed ratio should be only approximately equal to two. Give one advantage for using gears with 41 and 20 teeth instead of 40 and 20 teeth.
2. Name three factors that should be taken into account when designing gear trains.
3. Name one advantage and one disadvantage of helical gears relative to spur gears.
4. What is a herringbone gear? What advantage does it have over a helical gear?
5. List the three types of bevel gears and indicate a significant feature of each.
6. What are miter gears?
7. What is meant by the self-locking feature of a worm-and-wheel?
8. How does a compound gear train differ from a simple gear train?
9. Name one desirable feature of a reverted gear train.
10. What is an epicyclic gear train? How does it differ from a simple or compound gear train?
11. What type of gear train is commonly used in the automatic transmission of an automobile?

Problems

1. A speed reducer has a 52-tooth worm gear and a single threaded worm. The worm is connected to a motor shaft rotating at 1800 rpm. Find the output speed in rpm.
2. A triple-threaded worm rotating at 1800 rpm drives a worm gear at 50 rpm. How many teeth does the worm gear have? If the input torque is 1000 lb-in. and the efficiency is 75%, find the output torque.
3. For a simple gear train of Figure 8.17, $N_A = 32$, $N_B = 18$ and $N_C = 26$. Find the speed ratio.
4. For the compound gear train of Figure 8.19, $N_A = 50$, $N_B = 30$, $N_C = 15$, and $N_E = 40$. If $\omega_A = 1000$ rpm, find ω_E.
5. Design a gear pair with a speed ratio of 5/8. The diametral pitch is 6 and the minimum number of teeth on any gear is 14. Find the number of teeth on each gear and the center distance.
6. The input gear of Figure 8.18 rotates at 1800 rpm and the output speed should be as close as possible to 700 rpm. The diametral pitch is 10 and the center distance should be as close as possible to 4.500 in. The center distance value is very critical and thus has priority over the speed ratio from a precision point of view. Find: **(a)** the number of teeth on each gear; **(b)** the exact output speed; and **(c)** the exact center distance.

7. Design the compound gear train of Figure 8.19 to provide a speed ratio as close as possible to 1/24. The minimum number of teeth is 14. Specify the number of teeth on each gear.

8. Solve Sample Problem 8.4 except make the total speed ratio of the gear train 1/200.

9. A speed reducer is to accept an 1800 rpm input shaft and provide a 90 rpm output. The reduction is to take place in two steps. If the minimum number of teeth is eighteen, find the number of teeth in each gear.

10. For the reverted gear train of Figure 8.21, the input shaft speed is 1800 rpm and the output shaft speed is to be 100 rpm. The diametral pitches of all gears are equal and the minimum number of teeth on any gear is 14. Find the number of teeth on each gear.

11. Repeat Problem 10 except that gears A and B have a diametral pitch of 6 and gears C and E have a diametral pitch of 4.

12. For the epicyclic gear train of Figure 8.22, the following data are given: $N_S = 85$ and $N_P = 25$. The ring gear is fixed and the sun gear rotates at 1800 rpm clockwise. Find the output speed of the planet carrier and its direction.

13. For the automotive transmission of Figure 8.25, specify the number of teeth on gears 1, 2, 4, and 7 needed to provide a speed ratio of 1:3 for low speed (speed 1). Assume that all gears have the same diametral pitch and that the minimum number of teeth on any gear is 14. Remember, this is a reverted gear train.

9

Belts and Chains

Objectives

This chapter explains that there are a number of factors to consider, other than size, when selecting belts or chains for various applications and diagnosing problems with equipment.

On completing this chapter, you will be able to

- Decide whether a chain or a belt is more appropriate for a given application. Calculate the length of a belt or chain, which depends on center distances and sizes of pulleys or sprockets.
- Calculate the power a belt or chain can transfer.
- Describe the advantages and disadvantages a V-belt has over flat belts.

9.1 INTRODUCTION TO BELTS

Those of us who enjoy cowboy movies can easily picture the desperado stopping in front of the local saloon and tying his horse by casually wrapping the reins a few times around the hitching rail. The horse is not strong enough to pull away! This illustrates an important concept in the application of belting—the *angle of wrap*. This angle of wrap (or *angle of contact*) must be great enough to transmit the power required. (Angle of contact is further discussed in Section 9.3.)

Before we analyze the angle of contact, let us discuss a number of other factors important in the selection of belting.

Consequence of Failure.   In this consideration, belts have a distinct advantage over chains and gears since a sudden overload will not break a belt. Instead, the belt slips until the overload is ended. However, even momentary overloads can break gear teeth or chain links.

Versatility in Shaft Connection.   Since belts are more flexible than chains and gears, they are the more versatile in connecting two shafts with unusual geometrical arrangements or large center distances. Gears are the least versatile from a practical point of view, especially if the center distances are large.

A little bad

Effect on Shaft Bearing Life.   Belts are driven by friction, and therefore require initial tensioning resulting in tension on the slack side of the belt and increased bearing loads. Gears and chains are positive drives and do not place this increased load on the bearings.   →○ ℞ *belts in tension, bearing*

little bad

Speed Ratio.   Generally speaking, belt drives do not provide an exact speed ratio as do gear systems. The slippage that protects belt drives from damage by sudden overloads ironically prevents an exact timing between the driving and driven shafts. However, special timing belts that produce positive drive are available.

least expensive

Cost.   Belts are the least expensive of either gears or chains, while chains are less expensive than gears. The required precision of machining and mounting of gears is the principal reason for their higher cost. When using chains, the alignment of the shafts must be more precise than for belts.

Noise and Vibration.   Belt drives produce the least amount of noise and vibration, and thus are used where vibration levels must be low.

Speed and Power.   Gears can operate at higher speeds and transmit more power than chains or belts.

Maintenance.   Chains and belts require periodic adjustment resulting from wear and stretch, respectively. Chains and gears require lubrication. Neglecting unexpected overloads, properly designed gear systems require the least amount of maintenance.

Common applications of belts and chains are given in Table 9.1.

TABLE 9.1   Common Applications
of Belts and Chains

Common Applications	
Belts	Chains
Compressors	Bicycles
Sewing machines	Motorcycles
Textile machines	Power lawn mowers
Automotive devices	Chain saws
water pumps	Cranes
alternators	Hoists
fans	Paper-mill machinery
Mixing machines	Conveyers
Washing machines	Textile machinery
Printing machinery	All terrain vehicles
Pumps	
Machine tools	
Crushing machinery	

9.2   FLAT BELTS   *most simple*

Flat belts are rectangular in cross section and are mounted on pulleys that are usually crowned at the periphery as shown in Figure 9.1. The purpose of the crowning is to prevent the belt from running off the pulley.

In addition to the open-belt design of Figure 9.2a, other typical drives are the cross-belt, serpentine, and quarter-turn drives. Figure 9.2b shows the cross-belt drive, which unlike the open-belt design, has the driver and driven pulleys rotating in opposite directions. The serpentine configuration is illustrated in Figure 9.2c. It is essentially an open-belt drive with an idler pulley used to increase the angle of contact and provide an adjustment for the belt tensions. Generally speaking, the angle of contact for flat belts should not be less than 160°.

*cross section
is retangle.
most used.
simplest.*

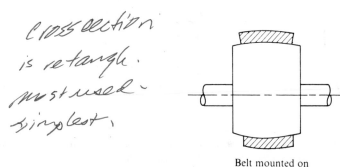

Belt mounted on
crowned pulley

FIGURE 9.1   Flat belt and crowned pulley

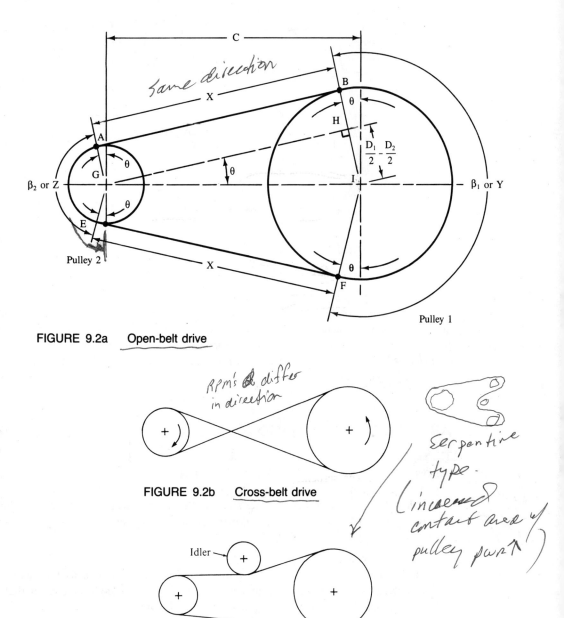

FIGURE 9.2a   Open-belt drive

FIGURE 9.2b   Cross-belt drive

FIGURE 9.2c   Serpentine drive

Flat belts are frequently used for drives between nonparallel shafts. The quarter-turn drive is one example in which the shafts are perpendicular, as shown in Figure 9.2*d*. Guide pulleys are usually required so that belts approaching a pulley will be oriented in the correct plane. Flat belting is usually made of leather, rubber, or canvas.

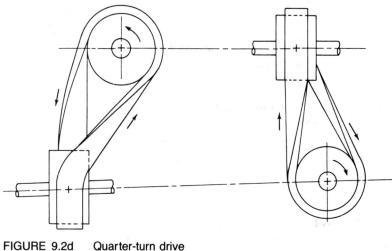

FIGURE 9.2d   Quarter-turn drive

Rubber belts are sometimes reinforced with tensile cords. A flat belt can be made endless at the manufacturer's plant or come in any lengths with a supply of standard fasteners to join the ends. Flat belts are specified by the number of layers, such as, single-ply, double-ply, or triple-ply. The maximum allowable tensile force can be obtained from manufacturers' catalogs.

9.3 ANGLE OF CONTACT AND BELT LENGTH ████████████

The angle of contact is the angle over which a belt makes contact with a pulley. If two pulleys are equal in size, each will have a 180° angle of contact. Normally the pulleys will not be the same size because a speed change is usually required. Figure 9.2a shows that the angle of contact for the small pulley is less than 180° for an open-belt drive. For cross-belt drives, the values of the angles of contact of the two pulleys are equal and are always greater than 180°.

The amount of horsepower that a pulley can transmit decreases as the angle of contact decreases. Hence, we need to evaluate the amount of angle of contact that a pulley possesses. The following definitions apply to Figure 9.2a

where: D_1 = diameter of large pulley in inches

D_2 = diameter of small pulley in inches

$X = AB = EF$ = belt length not contacting pulleys in inches

Y = belt length in contact with large pulley in inches

Z = belt length in contact with small pulley in inches

β_1 = angle of contact for large pulley in degrees

β_2 = angle of contact for small pulley in degrees

C = center distance between pulleys in inches

2θ = angular deviation from 180° angle of contact in degrees

To derive equations for the angle of contact and belt length of an open-belt drive, line GH is drawn parallel to belt segment AB. This produces right triangle GHI from which we obtain the following trigonometric equation:

$$\sin\theta = \frac{D_1/2 - D_2/2}{C} \tag{9.1}$$

Test stuff (handwritten)

find first (handwritten)

For a cross-belt drive

$$\sin\theta = \frac{D_1/2 + D_2/2}{C} \tag{9.2}$$

From this equation we can find angle θ in degrees. We will need to find θ in radians, so use this conversion:

$$\theta_{radians} = \frac{\theta°}{57.3^{deg/rad}} \tag{9.3}$$

Thus the angle of contact can be expressed as follows:
For open-belt drives

$$\beta_1 = 180° + 2\theta° \tag{9.4}$$

and

$$\beta_2 = 180° - 2\theta° \tag{9.5}$$

For cross-belt drives

$$\text{Both } \beta_1 \text{ and } \beta_2 = 180° + 2\theta \tag{9.6}$$

We can now conclude that, to find the angle of contact, all that is needed are the diameters of the two pulleys and the center distance. Also, note that the sum of the two angles of contact for an open-belt configuration equals 360°:

$$\beta_1 + \beta_2 = 360°$$

The total belt length, L, equals the sum of the lengths of the individual segments:

$$L = 2X + Y + Z \tag{9.7}$$

Using the Pythagorean Theorem on triangle GHI yields

$$X = \sqrt{C^2 - \left(\frac{D_1}{2} - \frac{D_2}{2}\right)^2} \tag{9.8}$$

For a cross-belt drive

$$X = \sqrt{C^2 - \left(\frac{D_1}{2} + \frac{D_2}{2}\right)^2} \tag{9.9}$$

Note that the length of any arc of a circle equals the radius of the circle multiplied by the arc central angle measured in radians. This relationship allows for a relationship for Y and Z:

For open or crossed-belt drives

$$Y = \frac{D_1}{2}(\pi + 2\theta) \tag{9.10}$$

And for open-belt drives

$$Z = \left(\frac{D_2}{2}\right)(\pi - 2\theta) \tag{9.11}$$

and for cross-belt drives

$$Z = \left(\frac{D_2}{2}\right)(\pi + 2\theta) \tag{9.12}$$

Note that for Equations 9.10, 9.11, and 9.12, θ must be measured in radians. Another useful relationship can be obtained by referring to Figure 9.2a. Theoretically, the speed ratio of the pulleys is inversely proportional to their diameters, just as for gears. This relationship can be demonstrated as follows: In Figure 9.2a, the belt speed at point A (the linear velocity of point A) equals the velocity of point B, since the belt has one linear velocity:

$$V_A = V_B$$

In terms of pulley diameter D and rpm ω, we have

$$\pi D_2 \omega_2 = \pi D_1 \omega_1$$

or

$$\frac{\omega_1}{\omega_2} = \frac{D_2}{D_1}$$

Again, as with gears, the larger pulley rotates at the slower speed. This equation, as stated above, is theoretical because it does not take into account *slip* and *creep*, both of which reduce the output speed of the driven pulley. Slip is the actual sliding of the belt relative to the pulley and occurs when the net belt pull exceeds the available frictional force. Creep is the stretching and contracting of the belt as it alternately moves into and out of the high and low tension sides of the pulleys. The speed loss resulting from slip and creep is normally about 1% each, giving a total speed loss of 2%. This value, or course, assumes proper initial belt tensions and normal loading conditions.

SAMPLE PROBLEM 9.1

Angle of Contact and Belt Length

PROBLEM: An open-belt drive has two pulleys with 4 in. and 8 in. diameters and a 24 in. center distance. The small pulley is the driver and rotates at 500 rpm. Find: **(a)** the angle of contact of the small pulley; **(b)** the total belt length; and **(c)** the speed of the large pulley, assuming combined slip and creep causes a 3% speed loss.

Solution

(a)
$$\text{Sin } \theta = \frac{D_1/2 - D_2/2}{C} = \frac{4 - 2}{24} = 0.0833$$

$$\theta = 4.778° = 0.08339 \text{ radians}$$

$$\beta_2 = 180° - 2 \times 4.8° = 170.4°$$

(b)
$$L = 2X + Y + Z$$

$$X = \sqrt{\left(C^2 - \left(\frac{D_1}{2} - \frac{D_2}{2}\right)^2\right)} = \sqrt{(24^2 - (4 - 2)^2)}$$

$$X = 23.92 \text{ in.}$$

$$Y = \left(\frac{D_1}{2}\right)(\pi + 2\theta) = 4(3.142 + (2 \times 0.08339))$$

$$Y = 13.23 \text{ in.}$$

$$Z = \left(\frac{D_2}{2}\right)(\pi - 2\theta) = 2(3.142 - (2 \times 0.08339))$$

$$Z = 5.950 \text{ in.}$$

$$L = 2 \times 23.92 + 13.23 + 5.95 = 67.02 \text{ in.}$$

(c)
$$\omega_1 = \frac{D_2}{D_1} \omega_2 \times \text{efficiency}$$

$$\omega_1 = \left(\frac{4}{8}\right)(500)(.97) = 243 \text{ rpm}$$

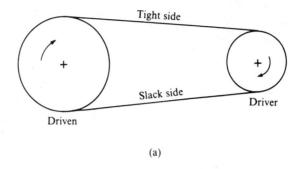

(a)

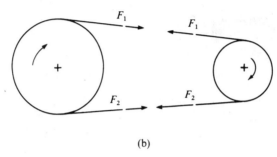

(b)

FIGURE 9.3   Belt tensions F_1 and F_2

9.4   FLAT BELT DESIGN ▬▬▬▬▬▬▬▬▬▬▬▬▬▬▬▬▬▬

Figure 9.3a shows a flat belt drive with the driver and driven pulleys specified. A cutting plane is passed through the belt and the resulting two sections are illustrated in Figure 9.3b. F_1 and F_2 are the individual belt tensions on each side of the pulleys. When there is no power input and thus no rotation, F_1 and F_2 are equal and represent the initial adjusted belt tension. When the driver transmits power, F_1 is commonly called the *tight tension* and F_2, the *slack tension*. Note that the tight and slack sides can be determined by examining the direction of rotation of the driver.

For a flat belt, the relationship between F_1 and F_2 is expressed by

$$\frac{F_1 - F_C}{F_2 - F_C} = e^{\mu\beta} \tag{9.13}$$

where F_1, F_2 = belt tension in pounds

μ = coefficient of friction

β = angle of contact in radians

e = 2.718 (the base for natural logs)

F_c = belt centrifugal force in pounds

TABLE 9.2   Belt Coefficient of
Friction When Using
a Steel Pulley

Belt Material	Coefficient of Friction
Leather	0.40
Rubber	0.35
Canvas	0.30

The belt centrifugal force reduces the horsepower capacity for a given allowable belt tensile force. The belt centrifugal force must be considered when the belt speed exceeds 2100 fpm. The value of F_c can be found from

$$F_C = 0.000,104\gamma btV^2 \tag{9.14}$$

where γ = weight density of belt in pounds per inch^3 For leather, γ = 0.035 in pounds per inch^3.

b = belt width in inches

t = belt thickness in inches

V = belt speed in feet per minute

F_c = belt centrifugal force in pounds

The input torque and output torque are described by Equations 9.15 and 9.16 respectively:

$$T_i = (F_1 - F_2)\left(\frac{D_{driver}}{2}\right) \tag{9.15}$$

$$T_o = (F_1 - F_2)\left(\frac{D_{driven}}{2}\right) \tag{9.16}$$

where D_{driver}, D_{driven} = diameters of the driver and driven pulleys, respectively in inches

T_i, T_o = torques of the driver and driven pulleys, respectively in pound-inches

$F_1 - F_2$ = net belt pull in pounds

It should be clear that F_2 has a negative effect on power transmission and ideally should be small compared to F_1. Thus a large ratio of F_1/F_2 is desired. V-belts can ofer a larger ratio than flat belts and are discussed in Section 9.5.

From Equations 9.15 and 9.16, we see that there is a direct relationship between torque and diameter:

$$\frac{T_o}{T_i} = \frac{D_{driven}}{D_{driver}} \tag{9.17}$$

Thus, if the output pulley is the larger pulley, we have a torque multiplication at the expense of a speed reduction. Since hp = $T \omega/63\ 000$, the horsepower capacity of a flat belt can be determined by substituting the right sides of Equations 9.15 or 9.16 for torque. Just be sure you use the same pulley for both speed and diameter.

$$\text{hp} = \frac{(F_1 - F_2)\ (D/2)\omega}{63,000} \tag{9.18}$$

where D is the pitch diameter of the pulley in inches, either the driver or driven pulley

ω is the speed of the same pulley in rpm.

PROCEDURE

Finding Belt Horsepower

Step 1   Obtain the maximum allowable belt tension (F_1) from manufacturers' catalogs. This maximum is usually based on the belt material allowable stress.

Step 2   Calculate the angle of contact (radians) and the belt velocity (fpm) using equations in the previous section.

Step 3   Calculate the centrifugal force, using Equation 9.14, if the belt speed exceeds 2100 fpm.

Step 4   Find F_2 and $F_1 - F_2$ using Equation 9.13. Assuming equal values of μ and an open-belt drive, **the small pulley limits the horsepower capacity** because it has the smaller angle of contact. For other belt drive designs or for pulleys having different coefficients of friction, check both pulleys. The pulley that produces the smallest ratio of F_1/F_2 cannot transmit as much power.

Step 5   Calculate the horsepower capacity using Equation 9.18.

SAMPLE PROBLEM 9.2

Horsepower of Open-Belt Drive

PROBLEM: A ¼ in. thick flat leather belt is 5 in. wide. The driving pulley has an 8 in. diameter and the driven pulley a 16 in. diameter. The center distance is 48 in. The coefficient of friction is 0.4 for both pulleys and the maximum belt tension is 400 lb. The driving pulley rotates at 1800 rpm. Find the belt horsepower capacity.

Solution

Since both pulleys have the same value of μ, the small pulley limits the horse-power capacity:

Steps 1 and 2 $\sin \theta = \dfrac{(D_1/2) - (D_2/2)}{C} = \dfrac{8 - 4}{48} = 0.0833$

$\theta = 0.0834$ rad

$\beta_2 = \pi - (2 \times 0.0834) = 2.97$ rad

$V = \pi \dfrac{D}{12} \omega = 3.14 \dfrac{8 \text{ in.}}{12 \text{ in./ft.}} \times 1800 \text{ rpm}$

$V = 3770$ fpm

Step 3   Since the belt speed exceeds 2100 fpm, the centrifugal force must be considered:

$F_c = (0.000,104) (0.035) (5) \left(\dfrac{1}{4}\right) (3770)^2 = 64.7$ lb

Step 4   Using Equation 9.13

$\dfrac{400 - 64.7}{F_2 - 64.7} = e^{(0.4)(2.97)}$

$F_2 = 167$ lb

Step 5   Using Equation 9.18

$hp = \dfrac{(400 - 167) \times \dfrac{8}{2} \times 1800}{63,000} = 26.6$ hp

9.5 V-BELTS

The V-belt is the most popular type of belt design. It has a trapezoidal cross section which has been standardized into two different classes. The light-duty belts are designated as 2L, 3L, 4L, or 5L. These are mostly for fractional horsepower applications, although the 5L designation may be used for as high as about 2¾ horsepower. Larger horsepower ratings are handled by another class of belts, having 5 different sizes designated by the letters A, B, C, D, and E. Their descriptions and horsepower ranges are shown in Figure 9.4. The maximum rating for a single size E belt is about 400 hp. For higher horsepowers, multiple belts are used. The 2000 hp to 3000 hp diesel engines in locomotives are connected to generators by multiple V-belts (the generators supply power to the motors on the drive wheels).

V-belts are usually made with synthetic or steel tensile cords molded in rubber and encased in an outer jacket, as illustrated in Figure 9.5. The cords provide great tensile strength and permit flexing of the belt. The lower, rubberized area is capable of withstanding compression. The outer fabric jacket protects the belt from damage from moisture, heat, and dust.

A V-belt is designed to ride inside the groove of the pulley, or *sheave,* as it is usually called (see Figure 9.6). Proper mating of the sheave and V-belt size is required so that the V-belt **does not** ride on the bottom of the groove and the top of the belt rides approximately flush with the top of the groove. The sheave pitch diameter is measured to the centroids of the belt cross-sectional areas, as shown in Figure 9.6. A photograph of an actual V-belt sheave is shown in Figure 9.7.

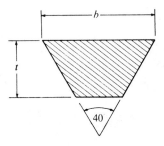

Belt Section	Width, b (in.)	Thickness, t (in.)	Minimum Sheave Diameter (in.)	HP Range (1 or more belts)
A	½	5/16	3.0	¼-10
B	21/32	13/32	5.4	1-25
C	⅞	17/32	9.0	15-100
D	1¼	¾	13.0	50-250
E	1½	23/32	21.6	100 and higher

FIGURE 9.4   Standard V-belt cross sections

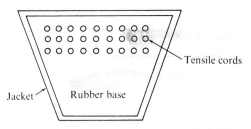

FIGURE 9.5   Construction of a typical V-belt

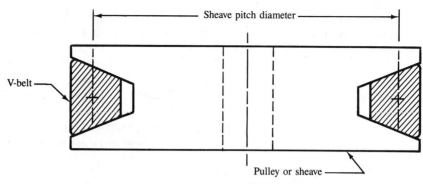

FIGURE 9.6   V-belt in groove of sheave

FIGURE 9.7   V-belt sheave *(Courtesy of Maurey Manufacturing Corp., Chicago, Illinois)*

Several advantages of V-belts over flat belts are the following:

1. V-belts operate more smoothly and quietly.
2. V-belts are more compact because shorter center distances are permitted. The center distance can be as small as the two pulleys physically allow.
3. V-belts operate with lower belt tensions which result in smaller bearing loads.
4. V-belts can absorb greater shock loads.

The groove angle of a sheave is slightly smaller than the 40° included angle of the V-belt cross section. The sheave groove angle usually varies from 32° to 38°, depending on the belt section size, the sheave diameter, and the belt angle of contact. This slightly smaller sheave groove angle forces the belt to wedge into the groove, resulting in increased friction. Figure 9.8 shows the forces acting on the cross section of a V-belt. The normal unit force P between the belt and sheave is perpendicular to the sides, and hence is inclined to the horizontal by an angle $\phi/2$, where ϕ is the included angle of the sheave groove. The frictional unit force is proportional to P. If this were a flat belt, the frictional force would be proportional to the component $P \sin(\phi/2)$, which, as you can see from Figure 9.8, is much smaller than P. Another way of stating this is to say that the coefficient of friction has been increased for the V-belt. The new value of the coefficient of friction is obtained by dividing "μ" in Equation 9.13 by $\sin(\phi/2)$. Thus the relationship between F_1 and F_2 for V-belts is expressed by

$$\frac{F_1 - F_c}{F_2 - F_c} = e^{\mu\beta/\sin(\phi/2)} \tag{9.19}$$

The smaller the groove angle ϕ, the greater the value of P compared to its vertical component $P \sin(\phi/2)$. This characteristic allows V-belts to operate with lower belt tensions than flat belts. On the other hand, decreasing the angle ϕ increases the force required to pull the belt out of the groove. An angle of about 40° produces a practical compromise. Also, observe that a flat belt is actually a special case of a V-belt with a groove angle of 180°. Since $\sin(180°/2)$ equals unity, Equation 9.19 reduces to Equation 9.13.

We mentioned in Section 9.4 that the ratio F_1/F_2 should be ideally as large as possible. For a flat belt, a typical value is about 2.0 assuming $\mu = 0.4$, $\beta = 170°$, and the belt speed $V = 4000$ fpm. A V-belt using a 40° groove angle produces a corresponding value of about 4.5 because of the greatly increased frictional force. These ratios become substantially larger for both flat and V-belts as the belt speed decreases because of the smaller centrifugal force. A demonstration of this will be left as a student problem.

In terms of horsepower per square inch of belt cross-sectional area, V-belts operate most efficiently at a speed of about 4000 fpm. Difficulty may be encountered if the belt speed exceeds 5000 fpm because the centrifugal force greatly reduces the belt effectiveness and the life of the belt is significantly shortened.

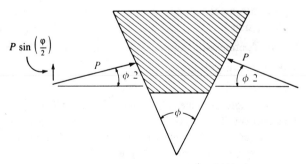

FIGURE 9.8   Forces acting on sides of V-belt

Equations 9.14 through 9.19 permit the calculation of the horsepower capacity for a given V-belt size if the belt's allowable stress is known. For normal operating conditions, an allowable stress of 350 lb/in.^{2} will give a reasonable belt life. For heavy-duty operation, the allowable stress should be reduced to 250 lb/in.^{2}

Manufacturers' catalogs are available that provide tables to simplify selecting V-belts and determining their horsepower capacity. These catalogs include correction factors for belt speed, angle of contact, and severity of application. Sample Problem 9.3 illustrates the mathematical technique. The procedure is the same as that for flat belts, with the exception that Equation 9.19 should be used instead of Equation 9.13. Also note that the cross-sectional area of the V-belt (which may be needed to find centrifugal force and belt tension) is given by the equation:

$$\text{Area} = bt - t^2 \tan (\phi/2) \tag{9.20}$$

where b = width of belt shown in Figure 9.4.

t = thickness of the belt

SAMPLE PROBLEM 9.3 _____

V-Belt Selection

PROBLEM: A 9 in. pitch diameter sheave is used with a 12 in. diameter sheave, both of which have a 36° groove angle. The center distance is 18 in. and the coefficient of friction, μ, is 0.4. The belt has a weight density of 0.04 lb/in.^{3} The belt speed is 4000 fpm and the allowable belt stress is 300 lb/in.^{2} because of the expected operating conditions. If 35 hp is to be transmitted, how many belts are required?

Solution

Step 1   From Figure 9.4, select a C-type belt section.

$$S = 300 \text{ lb/in.}^2$$

$$A = bt - t^2 \tan (\phi/2) = 0.875 \times 0.531 - 0.531^2 \tan 18°$$

$$A = 0.373 \text{ in.}^2$$

$$F_1 = S \times A = 300 \times 0.373 = 112 \text{ lb}$$

Step 2 $\sin \theta = \dfrac{(6 - 4.5)}{18} = 0.0833$

$$\theta = 0.0834 \text{ rad.}$$

$$\beta = 3.1416 - 2(.0834) = 2.97 \text{ rad.}$$

$$V = 4000 \text{ fpm}$$

Step 3 $F_c = (0.000,104)\,(\gamma)\,(A)\,(V^2)$

$\qquad\quad = (0.000,104)\,(0.04)\,(0.373)\,(4000^2)$

$\qquad\quad = 24.8\ \text{lb}$

Step 4 $\dfrac{112 - 24.8}{F_2 - 24.8} = e^{(0.4)\,(2.97)/\sin 18°}$

$\qquad\quad F_2 = 26.7\ \text{lb}$

$\qquad\quad F_1 - F_2 = 85.3\ \text{lb}$

Step 5   Speed of small pulley $= \omega = V \times \dfrac{12}{(\pi)D}$

$\qquad\qquad\quad = 4000 \times \dfrac{12}{3.1416 \times 9} = 1700\ \text{rpm}$

hp per belt $= 85.3\,(9/2)\,(1700)\,/\,63\,000 = 10.4\ \text{hp}$

Number of belts required $= 35\ \text{hp}/10.4\ \text{hp/belt} = 3.37$

Therefore, use 4 belts.

Most V-belts are made endless. However, V-belts can also come in any desired lengths and be spliced at the ends. Standard rolls of 100, 200, 300, 400, and 500 ft lengths are available. Some applications make it impractical to install endless V-belts because, if a belt has to be replaced, and endless belt may require major dismantling of a complex machine.

Figure 9.9 illustrates a V-belt designed specifically to be spliced. The ends of the belt are connected together by special fasteners, as shown in Figure 9.10. To overcome the tendency of the fastener cleats to tear away from the belt, the tension member is made up of layers of square-cut fabric. The cross weave results from the square-cut design and grips the cleats securely, providing a tough, durable anchor for the fasteners. Figure 9.11 illustrates an application where a multiple-groove sheave contains four individual spliced V-belts.

9.6   VARIABLE-SPEED V-BELT DRIVES

Frequently it is desired that the speed ratio of a V-belt drive system be variable. Two typical systems for accomplishing variable speed are as follows.

Stepped Pulley Drive.   To provide a finite number of speed ratios, the two pulleys are stepped as shown in Figure 9.12. Typical applications are the motor-to-spindle drives of drill presses, vertical milling machines, and grinding machines.

FIGURE 9.9   V-belt designed to be spliced *(Courtesy of the Gates Rubber Co. Denver, Colorado)*

FIGURE 9.10   Special fastener to connect ends of a V-belt *(Courtesy of the Gates Rubber Co. Denver, Colorado)*

FIGURE 9.11   Multiple-groove sheave with four spliced belts *(Courtesy of the Gates Rubber Co. Denver, Colorado)*

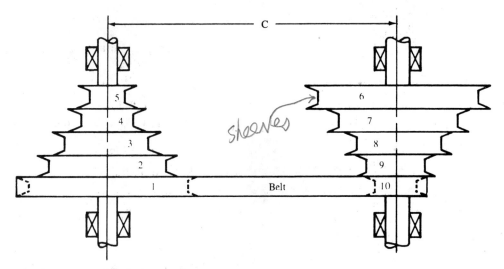

FIGURE 9.12   Stepped pulleys with V-belt drive

Variable-Pitch Pulley Drive.   An infinitely variable speed ratio can be obtained using the V-belt drive system of Figure 9.13. Both shafts have pulleys that are split into two halves. The separation of the pulley halves is controlled by a linkage mechanism (not shown) that can smoothly accomplish the separation. Thus the radii (r_1 and r_2) of the two pulleys can be varied to provide an infinitely variable speed ratio (r_2/r_1). For example, if r_1 is increased, r_2 must decrease in order to maintain belt tension. The speed change can only be accomplished while the pulleys are turning, which allows the belt to shift its position. If the pulleys are stopped, trying to change speed will simply pinch the belt. Spindle speed changes on a lathe are frequently accomplished by this method.

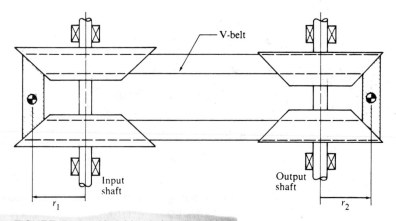

FIGURE 9.13   Variable-pitch pulley drive

1. flat
2. v-belt
3. cross section

9.7 MISCELLANEOUS BELTING

In addition to the standard flat and V-belts, many other types are available that have unique characteristics and application. Some of these designs are described below.

Round Belting. *Round belting* is used primarily for light load application, especially when the shafts are not parallel. One of the unique features of round belting is that it can be stretched by hand over the pulleys and snapped into place. Round belting can be made endless or come in any desired length. After cutting the belt to the required length, the ends are fastened with belt hooks. Standard size diameters for round belts range from ⅛ in. to ¾ in. Several very common applications are in sewing machines, vacuum cleaners, and tape reel mechanisms.

Positive-Drive Belting. A *positive-drive belt* (sometimes called a *timing belt*) is made of rubberized fabric and is reinforced internally with steel tensile wires. The belt has teeth that fit into grooves cut into the periphery of the pulley. Figure 9.14 shows fractional-horsepower and a 600 horsepower positive-drive belting system. Since the belt cannot slip, power is transmitted at a constant velocity ratio. Also, no initial tension is required. This means that the shafts can be mounted in fixed centers and that the bearing loads are reduced. Positive-drive belting can be run at much higher speeds

log drive

FIGURE 9.14   Fractional-hp and 600 hp positive-drive belting systems *(Courtesy of Maurey Manufacturing Corp., Chicago, Illinois)*

1 belt multigroove

FIGURE 9.15   Banded-together belts on multiple groove
sheave *(Courtesy of the Gates Rubber Co.
Denver, Colorado)*

that a standard V-belt and maintenance costs are much lower. Disadvantages are the
higher costs of the toothed belt and corresponding toothed pulley.

Banded-Together Belts.   Quite often, more than one belt is needed to transmit the
required horsepower. Either separate belts or a single unit called *banded-together belts*
can be used. Figure 9.15 shows a cutaway of a banded-together belt unit mounted on a
multiple groove sheave. The three V-belts are vulcanized with a reinforced tie band
that does not touch the land between the grooves of the sheave. The tie band gives the
belts lateral rigidity, which prevents the whipping of the belts from side to side. This
design is desirable for drives that are subjected to pulsating or heavy shock loads.
Under these loading conditions, individual V-belts are likely to whip laterally and even
come completely off the sheave.

Link-Type V-Belts.   A *link-type V-belt* is one that consists of a selected number of
easy-to-install links. Figure 9.16 shows a multiple groove sheave system using link-type
belts. This type of belt eliminates the need of splicing when endless belts cannot be
used. In Figure 9.17, we see two of the individual links which can be fastened together.
When any single link wears out, it can be readily replaced with a new link. Hence, the
stocking of various lengths of V-belts, as well as maintenance cost, is reduced. In
addition, since any desired length can be constructed, link belts can be used where the
shafts are mounted on fixed centers.

Variable-Pitch Multigroove Sheaves.   Figure 9.18 shows a *variable-pitch multigroove sheave*. The pitch diameter of the sheave can be changed, when the drive is

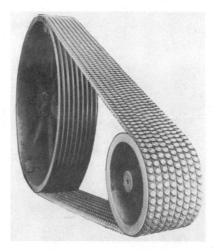

FIGURE 9.16   Multigroove sheave system using link-type belts *(Courtesy of the Gates Rubber Co. Denver, Colorado)*

FIGURE 9.17   Adjacent links of a link belt *(Courtesy of the Gates Rubber Co. Denver, Colorado)*

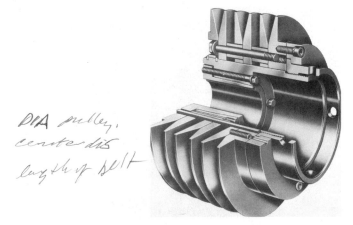

DIA pulley,
center dis
length of belt

cle center
distance

FIGURE 9.18   Variable-pitch multigroove sheave *(Courtesy of the Gates Rubber Co. Denver, Colorado)*

stationary, by means of movable discs that form the sheave grooves. Hence, this sheave provides a variable speed capability. To obtain the desired pitch diameter, a locking setscrew is loosened, and the adjusting nut is turned with a spanner wrench until the desired position is reached. Then the locking screw is retightened.

9.8   ROLLER CHAINS

A roller chain can be used to transmit power between parallel shafts and has features common to both gears and pulleys. Like gears, roller chains do not require an initial tension; hence, no slippage occurs. Because of their similarity to pulleys as wrapping connectors, roller chains are suitable for relatively long or short center distances. Compared to belts, chains are more compact, they can operate in a dirty and gritty environment, they are not affected by oil, and they can be used at the high temperatures encountered in ovens. The power transmission efficiency of a roller chain is high—98% to 99%.

Figure 9.19 shows the construction and nomenclature of roller chains. Each roller turns on a bushing that is swaged to an inner end plate. The pins that go through the center of the bushings are fixed to the outer end plates. Notice that a roller chain actually consists of two different types of links which are alternately connected:

Roller Links.   Each roller link contains two inner end plates, two rollers and two bushings.

Pin Links.   Each pin link contains two outer end plates and two pins.

One of the pins of a pin link goes through a bushing of one roller link, while the other pin of the same pin link goes through a bushing of an adjacent roller link. Thus any two adjacent roller links are fastened by an interconnecting pin link. The chain is

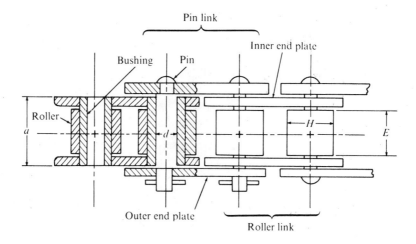

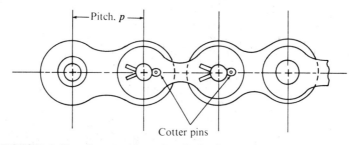

FIGURE 9.19   Construction of typical roller chain

made endless by the final attachment of a pin link whose two pins are held in place by either cotter pins as shown in Figure 9.19, or by a special end plate that serves as a cotter pin.

A whole number (preferably an even number) of links are required. If a chain contains an odd number of links, it is necessary that the last link, which is installed to make the chain endless, be an offset link. This requirement occurs because the last link must connect a roller link to a pin link, rather than a roller link to another roller link. An offset link has offset plates with one end containing only a pin and the other end containing a combination of a bushing and roller.

The pitch p of a roller chain is the distance between adjacent pins. The pitch is a very important roller chain parameter because each pin serves as a pivot between the adjacent interconnecting roller and pin links.

Figure 9.20 is a photograph showing a portable electric-powered roller chain hoist. The entire hoist weighs only 10 pounds and can lift up to 500 pounds. It has applications in automobile repair shops, refrigeration repair and installation, machine shops, light production lines, stockrooms and workshops in general.

Dimensions of standard roller chains are given in Table 9.3.

FIGURE 9.20   Portable electric-powered roller chain hoist
*(Courtesy of Eaton Corp. Hoisting Equip-
ment Div. Forrest City, Arkansas)*

TABLE 9.3   Dimensions of Standard Roller Chains

Chain Number	Pitch. p (in.)	Roller Diameter, H (in.)	Roller Width, E (in.)	Pin Diameter, d (in.)	Bushing Length, a (in.)
35	3/8	0.200	3/16	0.141	0.288
41	1/2	0.306	1/4	0.141	0.350
40	1/2	5/16	5/16	0.156	0.433
50	5/8	0.400	3/8	0.200	0.533
60	3/4	15/32	1/2	0.234	0.688
80	1	5/8	5/8	0.312	0.875
100	1 1/4	3/4	3/4	0.375	1.062
120	1 1/2	7/8	1	0.437	1.374
140	1 3/4	1	1	0.500	1.438
160	2	1 1/8	1 1/4	0.562	1.750
200	2 1/2	1 9/16	1 1/2	0.781	2.124

9.9 CHAIN GEOMETRICAL RELATIONSHIPS AND OPERATING CONDITIONS

Figure 9.21 shows a roller chain engaging two sprockets. The chain travels around the sprocket as a series of chordal links. Notice that the centerline of link AB is tangent to the pitch circle and that the centerline of link BC has moved radially inward by an amount e_1. The geometry produces two results:

— Impact between rollers and sprocket teeth, which causes wear and fatigue.
— A varying output (driven sprocket) torque and speed, even for a constant-speed driving sprocket.

To better visualize the varying torque and speed effect refer to Figure 9.22. Here we have an analogous situation in which we are pulling on a rope wrapped around a square pulley. As we pull with a smooth, constant force, the torque arm on the pulley changes from a maximum of r_a in Figure 9.22a to a minimum of r_b shown in Figure 9.22b. This changing radius varies the torque applied to the axle, which in turn varies the speed of the axle.

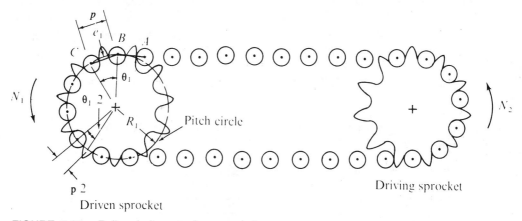

FIGURE 9.21   Roller chain engaging sprockets

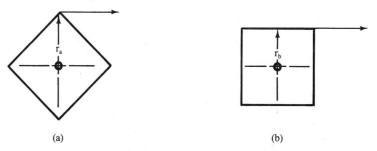

FIGURE 9.22   Demonstration of the changing torque arm of a sprocket

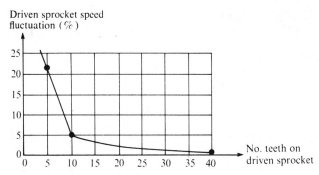

FIGURE 9.23   Driven sprocket speed variation versus number of teeth

To reduce impact, e_1 (in Figure 9.21) should be as small as possible. The ratio e_1/R_1 is a measure of the speed fluctuation of the driven sprocket. Figure 9.23 is a plot of the percent of speed variation versus the number of teeth on the driven sprocket. As the figure shows, a large number of teeth are desirable to reduce the speed variations and subsequent shock and wear. It is considered good practice to have a minimum number of 17 teeth unless space is severely limited or the speeds are very slow. The maximum recommended number of teeth is 120, because any elongation of the pitch from even slight wear causes the chain to ride high on the sprocket teeth.

Table 9.4 lists recommended maximum speeds of sprockets for standard roller chains. Exceeding these speeds will result in greatly reduced service life. Observe that

TABLE 9.4   Recommended Maximum Speeds of Sprockets of Standard Roller Chains

Chain No (Pitch) No. of Teeth.	35 (⅜)	41 (½)	40 (½)	50 (⅝)	60 (¾)	80 (1)	100 (1¼)	120 (1½)	140 (1¾)	160 (2)	200 (2½)
11	2260	1020	1690	1220	920	580	415	325	235	200	145
12	2590	1170	1940	1400	1050	670	475	375	270	230	165
13	2900	1310	2180	1570	1180	750	535	415	305	260	185
14	3170	1430	2380	1720	1290	820	585	455	335	280	205
16	3630	1630	2720	1960	1480	935	670	520	380	325	235
18	3970	1790	2980	2150	1610	1020	730	570	415	355	255
21	4210	1890	3160	2280	1720	1090	775	605	440	375	270
22	4380	1970	3290	2370	1780	1130	805	630	460	390	280
25	4510	2030	3380	2440	1830	1160	830	650	475	400	290
30	4490	2020	3370	2430	1830	1160	825	645	470	400	290
35	4290	1930	3220	2320	1740	1110	790	615	450	380	275
40	3970	1780	2970	2140	1610	1020	730	570	415	355	255
50	3110	1400	2330	1680	1270	805	575	450	325	275	200

the recommended maximum speed increases as the pitch of the chain decreases, and that the highest recommended maximum value occurs for 25 teeth. This is because a combination of high speed and a large number of teeth causes the chain to ride too far out on the sprocket teeth.

When you are designing a roller chain drive you will need an equation relating the pitch circle diameter D of the sprocket to the chain pitch p. Interestingly enough, we cannot find the circumference of the pitch circle by multiplying the chain pitch by the number of teeth or using the circumference to find D. Since the chain travels around the sprocket in a series of chordal links, we are actually obtaining the circumference of a polygon.

Since a higher accuracy is needed, the following explanation develops an equation relating chain pitch to the pitch diameter of the sprocket. Refer back to Figure 9.21. Each link on the sprocket subtends a central angle we will call theta θ. Thus

$$\theta = \frac{360°}{N} \tag{9.21}$$

where N = number of teeth on the sprocket

Also, from trigonometry, one-half the chain pitch subtends one-half the central angle. The link is perpendicular to the radial line that bisects it. A right triangle is formed that has 1/2 the pitch length as the opposite side to 1/2 the central angle, $\theta/2$, and the pitch circle radius is the hypotenuse. Thus

$$\text{Sin}\left(\frac{\theta}{2}\right) = \frac{p/2}{R} = \frac{p}{2R} = \frac{p}{D}$$

Rearranging and combining with Equation 9.21 above yields a useful relationship for the pitch diameter as a function of the chain pitch and the number of teeth on the sprocket:

$$D = \frac{p}{\text{Sin }(\theta/2)} = \frac{p}{\text{Sin }(360°/2N)}$$

$$D = \frac{p}{\text{Sin }(180°/N)} \tag{9.22}$$

where D = the pitch diameter of the sprocket

p = the chain pitch measured in inches

N = the number of teeth on the sprocket

The speed ratio of two sprockets is inversely proportional to their diameters and inversely proportional to the number of teeth on each, just as for mating gears. Speed ratios of 6:1 or less provide the most satisfactory chain life. If a ratio of greater than 6:1 is required, it should be made in two stages.

An approximate length L of chain in units of inches where C is the center distance (in inches) is given by

$$L = 2C + \frac{(N_1 + N_2)p}{2} + \frac{p^2(N_1 - N_2)^2}{39.4C}$$ (9.23)

where N_1 = the number of teeth on large sprocket 1

N_2 = the number of teeth on small sprocket 2

p = the pitch of the chain in inches

C = the center distance in inches

An ideal center distance is one that is about 40 times the pitch of the chain. The maximum center distance should be about 80 times the pitch and the minimum should be such that the angle of contact does not become less than 120°. The angle of contact will be greater than 120° for speed ratios less than or equal to 3:1. For ratios greater than 3:1 the center distance must be greater than the difference between the pitch diameters of the sprockets.

The type of loading and the operating conditions affect the life of a roller chain. Specifically, the following factors affect the horsepower capacity:

1.   Number of teeth on sprocket. The smaller the number of teeth, the greater the shock loading.
2.   Chain speed. The greater the chain speed, the greater the shock loading.
3.   Type of prime mover and driven loads. A combination of certain types of prime movers and loads can cause either steady, moderate-shock or heavy-shock loading.
4.   Lubrication. Poor lubrication can dramatically shorten the life of a roller due to premature wear.
5.   Environment. Dirty or dusty conditions tend to have an adverse affect on roller chain life.
6.   Duration of operation. A roller chain that operates 24 hours per day will have a shorter life (measured in years) than one which operates for shorter periods of time or infrequently.

9.10   HORSEPOWER CAPACITY OF ROLLER CHAIN

Quality roller chain is manufactured to specifications of the American National Standards Institute (ANSI). The steel roller chain and sprockets are designed and manufactured to provide long wear life and fatigue strength. When the manufacturer's recommendations are followed, approximately 15,000 hours of service life at full load may be expected.

Most quality roller chain manufacturers currently provide data tables to aid the engineer or technician in choosing the proper chain drive. Tables 9.5 and 9.6 are a sample of the tables available. Note that in Table 9.6 lubrication is specified:

— Type I is manual lubrication
— Type II is drip lubrication
— Type III is bath lubrication
— Type IV is oil stream lubrication

Ratings for intermediate numbers of sprocket teeth or rpm may be obtained by interpolation.

PROCEDURE

Roller Chain Selection

You must know the application horsepower and the rpm of the small sprocket.

Step 1 Multiply the application horsepower by the appropriate service factor listed here to obtain the desired design horsepower.

	10 hr/day	24 hr/day
Uniform load	1.0	1.2
Moderate shock	1.2	1.4
Heavy shock	1.4	1.7

Step 2 Go to Table 9.5 with the design horsepower and select an appropriate chain size (or number). This table is based on a sprocket with 17 teeth.

Step 3 Next, go to the horsepower rating table (Table 9.6) and determine the minimum size sprocket needed to provide, at the required speed, a rating equal to (or greater than) the design horsepower. Check Table 9.4 to be sure the maximum speed is not exceeded.

Step 4 Using the speed ratio calculate the number of teeth on the larger sprocket. Use Equation 9.22 and calculate the pitch diameters of the sprockets.

Step 5 Use Equation 9.23 and calculate the chain length. Adjust the length for an even number of pitches.

SAMPLE PROBLEM 9.4

Roller Chain Selection

PROBLEM: Design a roller chain drive that will transmit 3.5 hp. The chain drives a uniform load for 10 hr/day; the driving sprocket rotates at 1000 rpm; the center distance is 15 in.; and the speed reduction is 2:1. Determine **(a)** chain number and pitch, **(b)** pitch diameter and number of teeth for each sprocket, **(c)** percent speed fluctuation, **(d)** the length of chain adjusted for an even number of chain pitches.

Solution:

Step 1   The service factor is 1 so the design hp = 3.5.

Step 2   From Table 9.5 select a #40 chain with a ½ in. pitch.

Step 3   Table 9.6 indicates that a 17 tooth sprocket is capable of delivering over 4.98 hp. (This text does not supply data for chain numbers 25 and 35 otherwise we would check for a closer match to our design hp). From Table 9.4 we note that our sprocket's speed will be below the recommended maximum.

Step 4   A speed reduction of 2:1 indicates a driven sprocket with 17 × 2 = 34 teeth. Equation 9.22 is used to determine the pitch diameters of the sprockets.

$$D = \frac{p}{\sin\left(\dfrac{180°}{N}\right)}$$

Driving Sprocket $D_2 = \dfrac{0.5}{\sin\left(\dfrac{180°}{17}\right)} = 5.42$ in.

Driven Sprocket $D_1 = \dfrac{0.5}{\sin\left(\dfrac{180°}{34}\right)} = 2.72$ in.

Step 5   Equation 9.23:

$$L = 2C + \frac{(N_1 + N_2)p}{2} + \frac{p^2(N_1 - N_2)^2}{39.4C}$$

$$L = 2 \times 15 + \frac{(17 + 34)0.5}{2} + \frac{0.5^2(34 - 17)^2}{39.4 \times 15} = 42.87 \text{ in.}$$

Number of pitches = 42.87/0.5 = 85.74

Adjust to 86 pitches (an even number) and this gives a chain length of

$$L = 0.5 \times 86 = 43 \text{ in.}$$

Answers   **(a)** Chain number = 40. Chain pitch = 0.5 in.
                   (b) Drive sprocket: N_2 = 17 teeth. D_2 = 2.72 in.
                         Driven sprocket: N_1 = 34 teeth. D_1 = 5.42 in.
                   (c) From Fig. 9.23 speed variation is approximately 1%.
                   (d) Chain length = 43 in.

TABLE 9.5   Roller Chain Selection Table (Courtesy of the Boston Gear Co.)

RPM of Smaller Sprocket	DESIGN HORSEPOWER — CHAIN NUMBER												
	½	1	1-½	2	3	4	5	7-½	10	15	20	25	30
1800	25	25	35	35	35	40	40	40	50	80	60-2	80-2	-
1500	25	25	35	35	35	40	40	40	60	60	80	60-2	80-2
1200	25	35	35	35	40	40	40	50	60	60	60	80	100
1000	25	35	35	35	40	40	40	50	60	60	80	80	80
800	25	35	35	40	40	40	50	50	60	60	80	80	80
700	25	35*	35	40	40	50	50	50	60	80	80	80	80
600	35	35	35	40	50	50	50	60	60	80	80	80	100
500	35	35	40	40	50	50	50	60	80	80	80	100	100
400	35	35	40	40	50	50	60	60	80	80	100	100	100
350	35	40	40	40	50	50	60	80	80	80	100	100	100
300	35	40	40	50	60	60	60	80	80	80	100	120	100
250	35	40	50	50	60	60	80	80	80	100	120	120	120
200	40	40	50	50	60	80	80	80	100	100	120	120	120
175	40	40	50	60	60	80	80	80	100	100	120	120	140
150	40	50	50	60	80	80	80	100	100	120	120	140	140
125	40	50	60	80	80	80	100	100	120	120	140	140	140
100	40	50	60	80	80	80	100	120	120	140	140	160	160
80	50	60	60	80	80	80	100	120	120	140	160	160	160
70	50	60	60	80	80	80	100	120	140	160	160		
60	50	60	80	80	80	100	100	120	140	160	160		
50	50	60	80	80	80	100	120	120	160				
40	50	60	80	80	100	100	120	140	160				
30	60	80	80	100	100	120	140	140					
25	60	80	80	100	120	120	140	140					
20	60	80	100	100	120	120	140	160					
15	80	100	100	120	120	140	160						
10	80	100	120	120	140	140	160						

*Based on 17 Tooth Sprocket.

TABLE 9.6 Horsepower Ratings for ANSI Roller Chains (Courtesy of the Boston Gear Co.)

HP RATINGS—STANDARD SINGLE STRAND ROLLER CHAIN—NO. 40—½" PITCH

Teeth	P.D.	10	20	30	50	75	100	125	150	200	250	300	400	500	600	900	1200	1500	1800	2400	3000
11	1.77"	.054	.10	.15	.23	.33	.43	.53	.62	.80	.98	1.16	1.50	1.84	2.16	3.11	4.03	4.93	4.66	3.03	2.17
13	2.09	.065	.12	.17	.28	.40	.52	.63	.74	.96	1.18	1.39	1.80	2.20	2.59	3.73	4.83	5.91	5.99	3.89	2.79
15	2.40	.076	.14	.20	.32	.46	.60	.74	.87	1.12	1.37	1.62	2.10	2.56	3.02	4.35	5.64	6.89	7.43	4.82	3.45
17	2.72	.087	.16	.23	.37	.53	.69	.84	.99	1.29	1.57	1.85	2.40	2.94	3.45	4.98	6.45	7.89	8.96	5.82	4.17
19	3.04	.098	.18	.26	.42	.60	.78	.95	1.12	1.45	1.77	2.09	2.71	3.31	3.90	5.62	7.27	8.89	10.5	6.88	4.92
21	3.35	.109	.20	.29	.46	.67	.87	1.06	1.25	1.62	1.98	2.33	3.02	3.69	4.34	6.26	8.11	9.91	11.7	7.99	5.72
23	3.67	.120	.22	.32	.51	.74	.96	1.17	1.38	1.78	2.18	2.57	3.33	4.07	4.79	6.90	8.94	10.9	12.9	9.16	6.55
25	3.99	.132	.25	.35	.56	.81	1.05	1.28	1.51	1.95	2.38	2.81	3.64	4.45	5.24	7.55	9.78	12.0	14.1	10.4	7.43
Lubrication #		Type I						Type II						Type III					Type IV	

HP RATINGS—STANDARD SINGLE STRAND ROLLER CHAIN—NO. 50—⅝" PITCH

Teeth	P.D.	10	20	30	50	75	100	125	150	200	250	300	400	600	900	1200	1500	1800	2100	2400	2700
11	2.22"	.11	.20	.28	.45	.65	.84	1.03	1.21	1.56	1.91	2.25	2.92	4.21	6.07	7.86	7.44	5.58	4.42	3.62	3.04
13	2.61	.13	.24	.34	.54	.78	1.01	1.23	1.45	1.87	2.29	2.70	3.50	5.04	7.26	9.42	9.56	7.17	5.67	4.65	3.90
15	3.01	.15	.28	.40	.63	.90	1.17	1.43	1.69	2.19	2.67	3.15	4.08	5.88	8.48	11.0	11.9	8.89	7.03	5.76	4.83
17	3.40	.17	.32	.45	.72	1.04	1.34	1.64	1.93	2.50	3.06	3.60	4.67	6.73	9.70	12.6	14.3	10.7	8.48	6.95	5.83
19	3.80	.19	.36	.51	.81	1.17	1.51	1.85	2.18	2.82	3.45	4.06	5.27	7.59	10.9	14.2	16.9	12.7	10.0	8.22	6.89
21	4.19	.21	.40	.57	.90	1.30	1.69	2.06	2.43	3.15	3.85	4.53	5.87	8.46	12.2	15.8	19.3	14.7	11.6	9.55	8.01
23	4.59	.23	.44	.63	1.00	1.44	1.86	2.27	2.68	3.47	4.24	5.00	6.48	9.33	13.4	17.4	21.3	16.9	13.3	10.9	9.18
25	4.99	.26	.48	.69	1.09	1.57	2.04	2.49	2.93	3.80	4.64	5.47	7.09	10.2	14.7	19.1	23.3	19.1	15.1	12.4	10.4
Lubrication #		Type I						Type II						Type III					Type IV	

HP RATINGS—STANDARD SINGLE STRAND ROLLER CHAIN—NO. 60—¾" PITCH

Teeth	P.D.	10	20	30	50	75	100	125	150	200	250	300	400	600	900	1200	1400	1600	1800	2000	2200	2400
11	2.66"	.18	.34	.49	.78	1.11	1.44	1.76	2.07	2.69	3.29	3.87	5.02	7.23	10.5	11.9	9.45	7.70	6.49	5.51	4.78	4.20
13	3.13	.22	.41	.58	.93	1.33	1.72	2.11	2.48	3.22	3.94	4.64	6.01	8.65	12.5	15.3	12.1	9.89	8.34	7.08	6.14	5.39
15	3.61	.25	.47	.68	1.08	1.55	2.01	2.46	2.90	3.76	4.59	5.41	7.01	10.1	14.6	18.9	15.0	12.3	10.3	8.77	7.61	6.68
17	4.08	.29	.54	.78	1.24	1.78	2.30	2.82	3.32	4.31	5.26	6.20	8.03	11.6	16.7	21.7	18.2	14.8	12.5	10.6	9.18	8.06
19	4.56	.33	.61	.88	1.40	2.01	2.60	3.18	3.74	4.86	5.93	6.99	9.05	13.0	18.8	24.4	21.5	17.5	14.7	12.5	10.9	9.52
21	5.03	.37	.68	.98	1.56	2.24	2.89	3.54	4.17	5.41	6.61	7.78	10.1	14.5	21.0	27.2	24.9	20.3	17.1	14.5	12.6	11.1
23	5.51	.40	.75	1.08	1.72	2.47	3.19	3.91	4.60	5.97	7.29	8.59	11.1	16.0	23.2	30.2	28.6	23.3	19.6	16.7	14.4	12.7
25	5.98	.44	.82	1.18	1.88	2.70	3.49	4.28	5.04	6.53	7.98	9.40	12.2	17.5	25.4	32.9	32.4	26.4	22.3	18.9	16.4	14.4

Small Sprocket — RPM→

Lubrication #: Type I | Type II | Type III | Type IV

HP RATINGS—STANDARD SINGLE STRAND ROLLER CHAIN—NO. 80—1" PITCH

Teeth	P.D.	10	20	30	50	75	100	125	150	200	250	300	400	600	900	1000	1200	1400	1600	1800	2000	2200
11	3.55"	.42	.79	1.14	1.80	2.60	3.36	4.11	4.84	6.28	7.67	9.04	11.7	16.9	23.0	19.6	14.9	11.8	9.69	8.12	6.94	6.01
13	4.18	.51	.95	1.36	2.16	3.11	4.03	4.92	5.80	7.51	9.19	10.8	14.0	20.2	29.1	25.2	19.2	15.2	12.5	10.4	8.91	7.72
15	4.81	.59	1.10	1.59	2.52	3.63	4.70	5.75	6.77	8.77	10.7	12.6	16.4	23.6	34.0	31.2	23.8	18.9	15.4	12.9	11.0	9.57
17	5.44	.68	1.26	1.82	2.88	4.15	5.38	6.58	7.75	10.0	12.3	14.5	18.7	27.0	38.9	37.6	28.7	22.7	18.6	15.6	13.3	11.5
19	6.08	.76	1.43	2.05	3.25	4.68	6.07	7.42	8.74	11.3	13.8	16.3	21.1	30.4	43.8	44.5	33.9	26.9	22.0	18.4	15.7	13.6
21	6.71	.85	1.59	2.29	3.62	5.22	6.76	8.27	9.74	12.6	15.4	18.2	23.6	34.0	48.9	51.7	39.4	31.2	25.6	21.4	18.3	15.9
23	7.34	.94	1.75	2.52	4.00	5.76	7.46	9.12	10.7	13.9	17.0	20.0	26.0	37.4	53.9	59.2	45.1	35.8	29.3	24.6	21.0	18.2
25	7.98	1.03	1.92	2.76	4.38	6.30	8.17	9.98	11.8	15.2	18.6	21.9	28.4	40.9	59.0	64.9	51.1	40.6	33.2	27.8	23.8	20.6

Small Sprocket — RPM→

Lubrication #: Type I | Type II | Type III | Type IV

9.11   SI UNITS

The belt length and angle of contact equations can be used with SI units, but the formula for centrifugal force should not be used.

SAMPLE PROBLEM 9.5

Length and Angle of Contact in SI Units

PROBLEM: An open belt drive consists of two pulleys with diameters of 200 mm and 100 mm. The drive has a center distance of 600 mm. Find: **(a)** the angle of contact for both pulleys and **(b)** the belt length.

Solution

(a)
$$\text{Sin } \theta = \frac{\dfrac{D_1}{2} - \dfrac{D_2}{2}}{C} = \frac{100 - 50}{600} = 0.08333$$

$$\theta = 4.78°$$

$$\theta_{rad} = \frac{4.78}{57.3}$$

$$= 0.0834 \text{ rad}$$

$$\beta_1 = 180° + 2 \times 4.78° = 190°$$

$$\beta_2 = 180° - 2 \times 4.78° = 170°$$

(b) $L = 2X + Y + Z$

$$X = \sqrt{C^2 - \left(\frac{D_1}{2} - \frac{D_2}{2}\right)^2}$$

$$= \sqrt{600^2 - (100 - 50)^2} = 598 \text{ } mm$$

$$Y = \left(\frac{D_1}{2}\right)(\pi + 2\theta) = \left(\frac{200}{2}\right)(3.14 + 2 \times 0.0834)$$

$$= 331 \text{ mm}$$

$$Z = \left(\frac{100}{2}\right)(\pi - 2 \times 0.0834) = 149 \text{ mm}$$

$$L = 1680 \text{ mm} = 1.68 \text{ m}$$

9.12   SUMMARY

Power transmission between two shafts can be accomplished with belts and chains as well as with gears. Compared to gears, belts and chains are less expensive, and can accommodate large center distances more easily than gears. On the other hand, gears can operate at higher speeds and greater power and do not require the maintenance and periodic adjustments that belts and chains require.

The angle of contact is important when using belts and chains. In the case of belts, the total frictional force developed depends on the portion of the belt's length in contact with the pulley. The greater the arc of contact, the greater the frictional force available to develop torque. In the case of chains, the greater the angle of contact, the more teeth that engage the sprocket to absorb the driving force.

Flat belts are usually made from leather but may also be made of rubber or canvas. V-belts are designed to provide a wedging action on the pulley, thus greatly increasing the frictional force that provides the driving torque. Calculations of belt capacity must take into account the centrifugal force acting on the flat belts and V-belts if the belt speeds exceed 2100 fpm. Calculations must also consider the initial tension that must be placed on belts so that frictional forces can be developed between the belt and pulley.

There are various special purpose belts: round belts for very light loads (like sewing machines); timing belts, which have teeth that provide positive drive; multiple belts, which are banded together for increased horsepower transmission; and link-type V-belts, which can be easily assembled or installed.

This chapter discusses two types of variable-pitch pulleys. One type works in conjunction with the mating pulley. When one pulley's groove is opened (decreasing the pitch diameter) the other pulley must have its groove closed so that tension is maintained on the belt. The speed change is made while the pulleys are rotating. The other type of pulley can have its groove opened or closed by means of adjusting screws. This change must be made while the pulley is stationary and is suitable for semi-permanent speed changes.

Roller chains provide positive drive and do not require initial tension to transmit power. Shock loading on the chain increases if the number of sprocket teeth are reduced or if the speed is increased.

9.13   QUESTIONS AND PROBLEMS ▰▰▰▰▰▰▰▰

Questions

1.   Why are belts and chains called wrapping connectors?
2.   Name five factors that should be considered when selecting either a gear, belt, or chain drive system.
3.   Name one advantage of gears over belts and chains.
4.   Name one advantage of belts over gears and chains.
5.   Name one advantage of chains over belts and gears.
6.   What is the difference between slip and creep?
7.   How is a flat belt kept centered on a pulley?
8.   What is a serpentine drive?
9.   What is a quarter-turn drive?
10. Name three materials commonly used for belts.

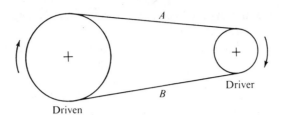

FIGURE 9.24   Sketch for question 12

11. What relative effect on the horsepower capacity does the centrifugal force have for a given allowable belt tensile force?

12. For the open-belt drive system of Figure 9.24, which side has the greater tension, A or B?

13. What advantages does a V-belt have over a flat belt?

14. True or false? A flat belt is a special case of a V-belt with a groove angle of 180°.

15. Why has a 40° included angle been established as the standard for V-belt cross sections?

16. What is round belting? Give one application.

17. What is a positive drive belt? Give one advantage.

18. What are banded-together belts? Give one advantage.

19. What are link-type belts? Give one advantage.

20. What is the purpose of a variable-pitch pulley?

21. Why do roller chains require lubrication?

22. Why is an even number of links preferred for a roller chain?

23. Give the reason for the rule of thumb: The recommended minimum number of teeth on a sprocket is 17.

Problems

1. Find the angle of contact on the small pulley of an open belt drive system. The pulley diameters are 6 in. and 9 in. and the center distance is 36 in.

2. Find the belt length for the system of Problem 1.

3. Solve problems 1 and 2 for a cross-belt drive.

4. In Problem 1, the 6 in. diameter drive pulley rotates at 1000 rpm. If combined slip and creep amount to 3 percent, find the speed of the 9 in. diameter pulley.

5. A ¼ in. thick flat leather belt is 4 in. wide. The driving pulley has a 9 in. diameter, and the driven pulley has an 18 in. diameter. The center distance is 36 in. Both pulleys are steel and the maximum belt stress is 300 lb/in.^{2} The driving pulley rotates at 1800 rpm. **(a)** Find the belt horsepower capacity and **(b)** find the belt length.

6. Compare the F_1/F_2 ratio of Problem 5 with a similar setup except the driving pulley rotates at 1000 rpm.

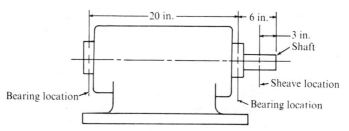

FIGURE 9.25   Sketch for problem 9

7. Find the width of a ¼ in. thick flat leather belt needed to transmit 10 hp at a pulley speed of 1800 rpm. The pulley is made of steel and has a 9 in. diameter, and the angle of contact is 160°. The maximum belt stress is 300 lb/in.^{2}

8. An 8 in. diameter V-belt sheave is used with a 10 in. diameter sheave, both of which have 36° groove angles. The center distance is 20 in. and the coefficient of friction is 0.35, using a belt whose weight density is 0.035 lb/in.^{3}. The belt speed is 3000 fpm and the allowable belt stress is 300 lb/in.^{2} If 20 hp is to be transmitted, how many belts are required?

9. Figure 9.25 shows the shaft extension of a 20 hp, 1800 rpm motor. The driving sheave has a 6 in. pitch diameter, while the driven sheave has a 12 in. pitch diameter. The center distance is 36 in. and the allowable belt stress is 300 lb/in.^{2} The belt is rubber and the weight density of the belt material is 0.04 lb/in.^{3}.
   (a) How many ¼ in. × 3 in. flat belts would be needed?
   (b) How many B-size V-belts would be needed? Groove angle = 40°.
   (c) What maximum radial load on the motor bearings occurs because of the use of flat belts?
   (d) What maximum radial load on the motor bearings occurs because of the use of B size V-belts?

10. Solve Sample Problem 9.4, except change the load to 12 hp and heavy shock for 24 hr/day.

11. An 1800 rpm motor drives a moderate shock load for 10 hr/day. The speed reduction ratio is 4:3. The power requirement is 8 hp and the center distance is 25 in. Select a suitable chain and find the length.

12. For the portable electric-powered roller chain hoist of Figure 9.20, the load capacity is 500 lb. The maximum chain velocity is 12 ft/min. Consider the sprocket pitch diameter to be about 6 in. Assuming heavy shock loads operating for no more than 10 hr/day, select a suitable roller chain.

SI Unit Problems

13. Solve Sample Problem 9.5, if D_1 = 400 mm.

14.   An open belt drive has pulley diameters of 80 mm and 450 mm. Find the angle of contact for the small pulley for center distances of 300 mm, 500 mm, and 750 mm.

15.   In Problem 13 the small pulley is the driver and rotates at 600 rpm. If the combined slip and creep is 3%, find the speed of the large pulley.

16.   (a) Solve Problem 14 for a cross-belt drive.
       (b) Find the belt length for the 500 mm distance.

10

Clutches and Brakes

Objectives

This chapter discusses a number of types of clutches and brakes, their operating characteristics, and the formulas used to determine the frictional torque that can be developed for transfer (in the case of clutches) or for braking. On completing this chapter, you will be able to

- Analyze the frictional forces and the torques developed for friction clutches and friction brakes.
- Explain the operating principles of a variety of clutches and brakes.
- Specify a block brake or band brake configuration so that the frictional force aids your efforts at braking rather than opposing you.

10.1   INTRODUCTION

A *clutch* is a device that connects and disconnects two collinear shafts. Hence, clutches are similar to couplings except that clutches must be capable of engaging and disengaging rapidly. A number of clutch designs can engage and disengage while the shafts are rotating. Couplings must be disassembled for disengagement. Friction plays a role in most types of clutches, and therefore heat dissipation is a factor that must be considered when designing a clutch.

Brakes are quite similar to clutches. The purpose of a brake is to stop the rotation of a shaft. For both brakes and clutches, initially there is one rotating shaft and one stationary shaft. After the clutch is engaged, both shafts rotate at equal speeds, and after the brake is applied, both shafts stop together. Therefore, a clutch brings a stationary shaft up to input speed and a brake brings a rotating shaft down to zero speed. Brakes and clutches should not only be fast-acting and easy to operate, but they should also not generate any significant shock loading during operation.

Braking action is produced by friction as a stationary part bears on a moving part. As a result, all the kinetic energy of the moving part is transformed into heat energy. This means that the dissipation of heat is a greater problem for brakes than for clutches. A brake material must have good wear properties, a high coefficient of friction, and the ability to withstand high temperatures. Most brake materials exhibit a rapid decrease in coefficient of friction if the temperature increases above a certain value. This phenomenon, called *brake fade,* frequently occurs in automobiles on downhill grades because the brakes cannot dissipate heat fast enough during sustained braking. Automotive asbestos brake linings begin to fade above 450° F.

10.2   JAW CLUTCHES

As shown in Figure 10.1, the *jaw clutch* is a positive-drive device with no slippage. One half of the clutch is pinned to one shaft, and the other half is free to slide along a spline of the second shaft. The shafts must be collinear. As a result of the positive drive and the resulting shock of engagement, this type of clutch is used for low-speed mechanisms. Figure 10.1a shows a *square-jaw clutch* that can transmit torque in either direction but does not engage very easily. A *spiral-jaw clutch* (Figure 10.1b) engages more readily but can transmit torque in only one direction. Normally, square-jaw clutches are not engaged while a shaft is turning, and the engagement of spiral-jaw clutches should be made only when the shaft speed is 50 rpm or less. Otherwise, heavy shock loads may be generated.

10.3   PLATE CLUTCHES

The *plate clutch* and the *cone clutch* (see Section 10.4 for a discussion of cone clutches) are not positive drive clutches because they rely on friction for engagement. The main advantage of a friction clutch is that engagement can be made smoothly by gradually increasing the normal force.

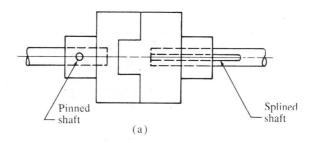

Pinned
shaft

Splined
shaft

(a)

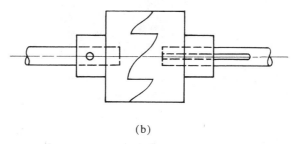

(b)

FIGURE 10.1   Jaw clutches

Figure 10.2 shows the construction of a plate friction clutch. When the input shaft rotates, its keyed cylinder and accompanying compression springs and pressure plate all rotate together. The compression springs push the pressure plate against the friction plate, which is splined to the output shaft. Because of the frictional forces on both sides of the friction plate, the output shaft is driven. A mechanism (not shown) is used to pull the pressure plate away from the friction plate, an action which compresses the spring a small amount. In this way, the clutch becomes disengaged.

Since the plate clutch (sometimes called disc clutch) depends on friction to transmit torque, it will slip if the torque is excessive. During slippage, heat is generated that must be dissipated. Normally, slippage takes place for only a short period of time

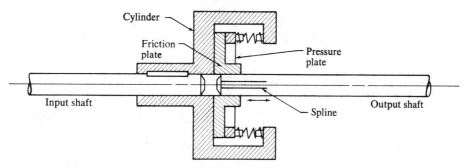

FIGURE 10.2   Plate friction clutch

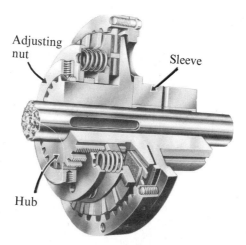

FIGURE 10.3   Slip clutch. *(Courtesy of the Hilliard Corp.,*
Elmira, New York)

during initial engagement. If the clutch is designed to limit the amount of torque that can be transmitted, it must be designed to dissipate heat on a continuous basis. Such a design is called a *slip clutch* and is illustrated in Figure 10.3. The slip clutch contains a steel plate that is lined with friction material on both sides. The steel plate (also called a *friction plate*) is mounted between two friction surfaces on the inner hub. The friction plate engages a gear tooth drive ring attached to the sleeve of the clutch. A series of compression springs squeezes the plate between the two friction surfaces. Notice the adjusting nut that is used to set the amount of compression on the springs. The compression determines normal force on the friction plate, and thus the amount of torque that can be transmitted.

The torque capacity of a plate friction clutch or slip clutch will now be derived by referring to Figure 10.2. The following nomenclature is applicable,

where D_o = outside diameter of the friction plate in inches

D_i = inside diameter of the friction plate in inches

R_m = mean radius of friction plate in inches

μ = coefficient of friction of mating surfaces

Q = axial force exerted on friction plate by springs in pounds

N = normal force acting on friction plate surface in pounds

F = frictional force acting on friction plate in pounds

T = torque capacity of clutch in pound-inches

p = maximum pressure acting on friction plate in pounds per inch^2

The frictional force equals the normal force multiplied by the coefficient of friction:

$$F = \mu N$$

For a plate clutch, the normal force equals the axial force. Thus we have

$$F = \mu Q$$

The friction torque equals the frictional force times the torque arm.

$$T = F \times R_m$$

The torque arm will be the **mean radius**, assuming that for worn, or soft surfaces the pressure is not uniform **but the wear is.** That is, the pressure is a maximum at the inner radius and zero at the outer radius. (This assumption is more conservative than assuming uniform pressure.)

For new clutches with rigid surfaces the pressure on the friction plate created by the force Q may be assumed to be uniformly distributed, and the torque arm will be 2/3 of the distance from the inner radius to the outer radius. However, the more conservative assumption is generally recommended and the one we will use.

Assuming uniform axial wear on the friction plate, the mean radius is defined

$$R_m = \frac{(D_o + D_i)}{4}$$

Combining the above equations, we have

$$T = \frac{\mu Q (D_o + D_i)}{4} \tag{10.1}$$

If maximum pressure is given instead of the normal force then the equation for torque with uniform wear becomes

$$T = \frac{\pi \mu p D_i (D_o{}^2 - D_i{}^2)}{8} \tag{10.2}$$

where p = the maximum pressure in pounds per inch^2

If more than one friction plate is used, the result is called a *multiple plate* or *multiple disc clutch*. The torque capacity is proportional to the number of pairs of mating surfaces. The number of plates that can be used is limited because heat dissipation and clutch engagement become more of a problem. Notice that the word *disc* is sometimes spelled *disk*.

SAMPLE PROBLEM 10.1 _____

Plate Clutch

PROBLEM: A single plate clutch has an outside diameter of 8 in. and an inside diameter of 6 in. If μ equals 0.35 and the allowable pressure is 30 lb/in.^{2}, how much torque can be transmitted?

Solution

Substitute directly into Equation 10.2.

$$T = \frac{3.14 \times .35 \times 30 \times 6(8^2 - 6^2)}{8} = 692 \text{ lb-in.}$$

10.4   CONE CLUTCHES

For a plate clutch, the axial force Q equals the normal force N acting on the friction plate. The normal force can equal many times the axial force by using the cone clutch which is illustrated in Figure 10.4. A cone clutch provides a "wedging action" similar to the V-belt pulley discussed in Chapter 9.

Let us analyze the torque capacity of a cone clutch by referring to a free-body diagram (FBD) of the cone in the completely engaged position (see Figure 10.5). The following nomenclature is applicable,

where Q = axial force applied while cone is fully engaged in pounds

N = normal force acting on cone surface in pounds

D_o = outside diameter of cone contacting surface in inches

D_i = inside diameters of cone contacting surface in inches

μ = coefficient of friction of mating conical surfaces

F = frictional force acting on conical surface in pounds

R_m = mean radius of conical surface in inches

α = the cone angle in degrees

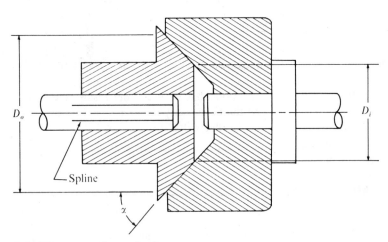

FIGURE 10.4   Cone clutch

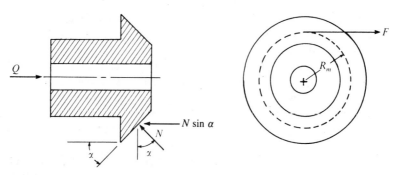

FIGURE 10.5   FBD of cone completely engaged

For the assumption of uniform wear, the frictional force F acts at the mean radius as shown in Figure 10.5. We now sum the horizontal forces in equilibrium on the cone in the engaged position (see Figure 10.5):

$$\sum F = 0 = Q - N \sin \alpha$$

Thus, we can find the frictional force:

$$F = \mu N = \frac{\mu Q}{\sin \alpha}$$

To find the torque capacity, we have

$$T = FR_m$$

where

$$R_m = \frac{(D_o + D_i)}{4}$$

Substituting for F and R_m, we have

$$T = \frac{\mu Q (D_o + D_i)}{4 \sin \alpha} \tag{10.3}$$

Thus a conical clutch can transmit more torque than a plate clutch with the same radial size, coefficient of friction, and axial force (compare Equations 10.3 and 10.1).

Specifically, the difference between the two equations is the factor $1/\sin \alpha$. This relationship means that the smaller the cone angle α, the greater the torque capacity. However, α should not be smaller than $8°$ because it makes the clutch difficult to disengage. Also, α should not be greater than $15°$ because a larger axial force Q (which is normally obtained by springs) would then be required, as is shown in Equation 10.3.

By solving the equation for Q, we observe that for a given torque capacity the axial force is directly proportional to sin α. Note that a plate clutch is a special case of a cone clutch where α equals 90° (sin 90° = 1).

During the process of clutch engagement, the conical surfaces slide relative to each other. This produces an axial frictional force F as shown in Figure 10.6. This axial frictional force, which occurs only during the engagement process, opposes the axial force required to engage the clutch. Let Q_E equal the axial force required to engage the clutch. Note that Q_E is not equal to Q because Q is the axial force required to transmit the torque T after the clutch has been engaged.

Let us evaluate Q_E. As shown in Figure 10.6, the horizontal component of frictional force F is added to our equilibrium equation. Summing the horizontal forces in equilibrium, we have

$$\sum F = 0 = Q_E - N \sin \alpha - F \cos \alpha$$

Solving for Q_E and substituting the expression μN for F, we obtain

$$Q_E = N(\sin \alpha + \mu \cos \alpha)$$

In deriving Equation 10.3, we noted that $N = Q/\sin \alpha$.
Substituting for N

$$Q_E = Q(1 + \mu \cot \alpha) \tag{10.4}$$

The results of Equation 10.4 are shown in Figure 10.7, where the ratio Q_E/Q is plotted against the cone angle α for values of μ equal to 0.3, 0.35, and 0.4. For a very typical cone clutch ($\mu = 0.35$ and $\alpha = 12°$), the force required to engage the clutch is 2.75 times the force required to transmit the torque.

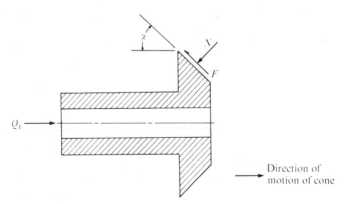

FIGURE 10.6   FBD of cone during engagement

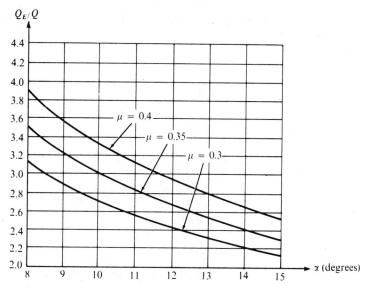

FIGURE 10.7 Q_E/Q versus α for cone clutch

SAMPLE PROBLEM 10.2

Cone Clutch Torque Transmission

PROBLEM: A cone clutch has an angle of 12° and a mean radius of 8 in. The clutch is to transmit 1000 lb-in. of torque and the coefficient of friction is 0.35.

(a) What minimum axial spring force is required to transmit the specified torque?

(b) What minimum axial force is required to engage the clutch?

(c) What horsepower can be transmitted at 1800 rpm?

Solution

(a) Use Equation 10.3.

$$Q = \frac{T \sin \alpha}{\mu R_m} = \frac{1000 \sin 12°}{0.35 \times 8} = 74.3 \text{ lb}$$

(b) Use Equation 10.4.

$$Q_E = Q(1 + \mu \cot \alpha) = 74.3(1 + .35 \times 4.70) = 197 \text{ lb}$$

(c) Use the horsepower-torque formula.

$$hp = \frac{T\omega}{63,000} = \frac{1000 \times 1800}{63,000} = 28.6 \text{ hp}$$

10.5 SPRING CLUTCHES

A *spring clutch* permits a unidirectional drive. It consists of two separate hollow cylindrical members, (cylinders 1 and 2 in Figure 10.8) located at the end of the input shaft. Cylinder 2 is pinned to the input shaft, while cylinder 1 is pinned to the output shaft and is free to rotate independently of the input shaft. A helical coil spring fits around the outside of the cylindrical members with a small radial clearance (approximately 0.002 in.).

If the input shaft is rotated in one direction, the inside diameter of the spring becomes smaller and the spring tightens around the outside surface of cylinders 1 and 2. This frictional grip causes the output shaft to rotate via the drive pin in cylinder 1. The spring has a wire of rectangular cross section to provide an increased surface area of contact. If the input shaft turns in the opposite direction, the inside diameter of the spring becomes larger and the frictional grip is lost. Thus the clutch is engaged in one direction and released in the other.

10.6 OVERRUNNING CLUTCHES

Sometimes it is desirable for an auxiliary power source to drive rotating machinery until the main power source takes over the job. For example, steam turbines must be brought up to speed by an auxiliary source before steam can be allowed into the turbine. Otherwise, the blades may be damaged because of uneven heating. One method of coupling is to use an *overrunning clutch* shown in Figure 10.9. The operation is as follows: Let's say member B of the clutch is driven by the auxiliary power source and member A is driven by the steam turbine. On start up, the auxiliary power source rotates member B at a designated speed. Member B drives member A counterclockwise as the two rollers are wedged to give positive drive. This of course turns the steam turbine. The spring keeps the roller in contact with both members at all times. As the turbine increases its own power it will cause member A to rotate at a faster rate than member B in the counterclockwise direction. Thus member A is said to overrun member B. This is equivalent, from a relative motion point of view, to holding A fixed and spinning B clockwise. In this latter case, the rollers do not become wedged and members A and B become disengaged.

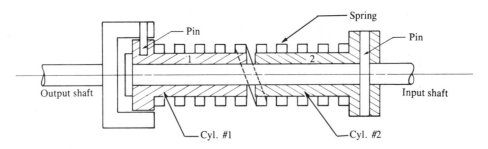

FIGURE 10.8   Spring clutch

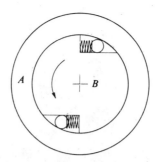

FIGURE 10.9   Overrunning clutch

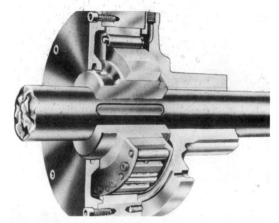

FIGURE 10.10   Industrial-type overrunning clutch. *(Courtesy of The Hilliard Corp., Elmira, New York)*

In Figure 10.10, we see a cutaway illustration of an overrunning clutch. It is used typically for two-speed drives in conveyors, paper processing machines and automatic machinery. In the larger sizes, it is capable of transmitting up to 300 hp.

10.7 ELECTRIC CLUTCHES

An electric clutch is one which uses an electromagnet for engagement. Figure 10.11 illustrates a multiple disc-type electric clutch. The inner driving member contains a coil that energizes an electromagnet. When the coil is energized, magnetic lines of force flow through the inner member and apply an attraction force that compresses the friction plates. Therefore, torque is transmitted from the inner member to the outer member. This particular clutch is capable of transmitting up to 900 lb-in. of torque.

Figure 10.12 shows an electromagnetic tooth clutch. In this design, the magnetic force of attraction causes positive engagement of the face teeth. When the coil is

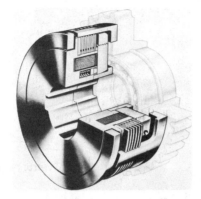

FIGURE 10.11   Electromagnetic   multiple-disc   clutch.
*(Courtesy of Motor Components Division,
The Bendix Corp.)*

FIGURE 10.12   Electromagnetic tooth clutch. *(Courtesy
of Motor Components Division, The Ben-
dix Corp.)*

deenergized, spring-loaded plungers separate the teeth. This type of clutch is positive
drive and is capable of transmitting up to 15,000 lb-in. of torque.

10.8   THE FLUID CLUTCH

In Figure 10.13, we see a *fluid clutch*, which consists of the following components:

1. A leakproof enclosure that is free to rotate on bearings. Vanes are attached to
the inside of the enclosure, and the output shaft is attached to the outside.

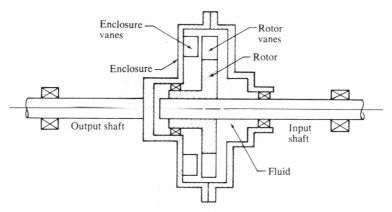

FIGURE 10.13 Fluid clutch

2. A rotor which is mounted on the input shaft and is free to turn inside the enclosure. Vanes are mounted around the periphery of the rotor.

3. A specified amount of fluid, which is contained in the enclosure.

As the input shaft rotates, the rotor vanes push fluid against the vanes attached to the enclosure. The momentum of the fluid and viscous effects causes the enclosure, and thereby the output shaft, to rotate.

A fluid coupling provides good protection against overloads, and if the output shaft locks, the rotor vanes merely slosh through the fluid. Power transmission is smooth and can be controlled by changing either the level or the viscosity of the fluid.

10.9 DRY FLUID CLUTCHES

Unlike the fluid clutch, the dry fluid clutch does not use a liquid but uses a specified amount of heat-treated steel shot instead. The steel shot is placed inside the housing, which is keyed to the driving shaft. (See Figure 10.14 for this configuration [the steel shot is not shown.] When the driving shaft rotates, it throws the steel shot to the outer surface of the housing by centrifugal force. The steel shot becomes wedged between the housing and rotor, and torque is transmitted to the driven shaft without slippage once a certain speed is reached. Notice the valve at the outer surface of the housing. This valve is used to introduce the steel shot.

10.10 BLOCK BRAKES

A *block brake* in its simplest form is a properly shaped block pressed against a rotating drum as shown in Figure 10.15. The applied force is usually delivered by means of a lever to obtain a mechanical advantage. The block is usually covered with a friction-

FIGURE 10.14   Dry fluid clutch. *(Courtesy of Dodge Manufacturing Division, Reliance Electric Co. Mishawaka, Indiana)*

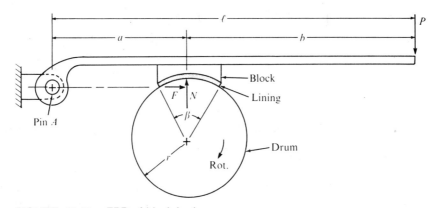

FIGURE 10.15   FBD of block brake

material lining that is replaced when worn out. The following nomenclature will be used for an analysis of the block brake,

where P = applied force in pounds

N = normal force exerted on block by drum in pounds

μ = coefficient of friction

F = frictional force exerted on block by drum in pounds

r = radius of drum in inches

T = retarding torque on drum pound-inches

β = angle of contact of block on drum (degrees)

K = correction factor for angle of contact (see Figure 10.16)

e = eccentricity in inches—the distance that the pivot is placed above or below the line of action of the frictional force (see Figure 10.17)

l = the distance from the lever arm pivot to the applied force in inches

We will now analyze the forces acting on the free-body diagram (FBD) of the block and lever in Figure 10.15. The forces shown in the figure are those acting on the block and lever. The block and lever are assumed to be in an equilibrium condition. Analysis is accomplished by a summation of moments. By taking moments about the pivot point of the lever (pin A), the normal force is found:

$$\sum M = 0 = +Pl - Na$$

and

$$N = \frac{Pl}{a}$$

Notice that for the block brake of Figure 10.15, the line of action of the frictional force F goes through the center of pin A and therefore has a zero torque arm. If the pin location were raised or lowered, it would change the operation of the brake, and the frictional force would have to be included in the equation. Changing the pin location will be discussed later in this section.

The frictional force F is found next:

$$F = \mu N = \frac{\mu Pl}{a}$$

The retarding torque is

$$T = Fr = \frac{\mu r Pl}{a}$$

This equation assumes that the line of action of the frictional force goes through the center of pin A. It also assumes that the pressure is uniformly distributed between the block and drum. If the angle of contact β is less than 60°, such assumptions are correct, but for β greater than 60°, the nonuniform pressure distribution must be taken into account. The factor K (from Figure 10.16) accounts for the nonuniform pressure and is placed in the torque equation, as shown below.

$$T = KFr = \frac{K\mu r Pl}{a} \tag{10.5}$$

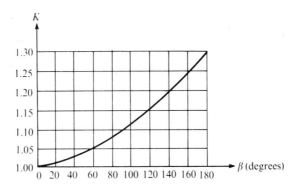

FIGURE 10.16   Torque correction factor for block brake

Equation 10.5 applies only to the configuration in Figure 10.15. This equation is presented primarily to show you how to develop the torque equation and it is recommended that you start with the summation of moments equation when analyzing block brakes and band brakes, which are discussed in the next section.

Please note one other point. For the configuration in Figure 10.15, it makes no difference which way the drum rotates. However, in the configurations in Figure 10.17, the direction of drum rotation does make a difference because it determines the direction of the frictional force and the amount of retarding torque.

As mentioned before, it is possible for pin A to be located above or below the line of action of the frictional force by an amount e. The four additional configurations for

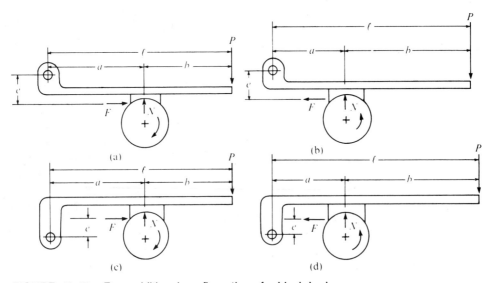

FIGURE 10.17   Four additional configurations for block brake

block brakes are illustrated in Figure 10.17. Let us analyze how the eccentricity e and the direction of rotation affect the amount of retarding torque. Observe that the configurations of Figure 10.17b and c require less applied force P than those in Figures 10.17a and d in order to develop a specified retarding torque T. This is because the frictional force F attempts to rotate the lever in the same direction as the applied force P. However, this configuration can cause the brake to grab, which can be dangerous. Just as the systems of Figures 10.17b and c are mathematically identical, so are those of Figures 10.17a and d. Note that in these latter two figures the frictional force attempts to rotate the lever in the opposite direction to the applied force.

PROCEDURE

Finding the Retarding Torque or the Applied Force of a Block Brake

Step 1 Consider that the brake block and lever arm are in equilibrium. Draw a free-body diagram (FBD) of the block and lever and indicate the position and direction of the forces acting on the block and lever.

Step 2 Write a summation of moments equation. Sum the moments about the pivot point.

Step 3 Rearrange the equation to find the required information. Remember to include the torque correction factor K. Use these equations to substitute into the summation equation:

$$T = KFr \quad \text{and} \quad F = \mu N.$$

If you are trying to find P, you generally know the torque from which you can find the numerical values of F and N to substitute into the summation equation. If you are trying to find the torque, then you probably know P, so substitute F/μ for N in the summation equation and solve for F. Then solve for the torque.

The following sample problems will demonstrate the analysis of block brakes.

SAMPLE PROBLEM 10.3

Block Brake With Zero Eccentricity

PROBLEM: A block brake has the configuration of Figure 10.15. The following data are given: $r = 6$ in., $a = 6$ in., $b = 10$ in., $\beta = 80°$, $\mu = 0.35$, and the drum rotates at 500 rpm and transmits 10 hp. Find the force P required to stop the drum.

Solution

Steps 1, 2, and 3 have already been done for us in this section. Therefore, from Figure 10.16, $K = 1.08$. Use the horsepower-torque equation to find torque.

Step 4 $hp = \dfrac{T\omega}{63,000}$

$T = 10hp \times \dfrac{63,000}{500rpm}$

$T = 1260$ lb-in.

$P = \dfrac{Ta}{\mu r/K} = \dfrac{1260 \times 6}{0.35 \times 6 \times 16 \times 1.08} = 208$ lb

SAMPLE PROBLEM 10.4

Block Brake With Eccentricity

PROBLEM: Solve Sample Problem 10.3 for the configuration shown in Figure 10.17*b*. The pin eccentricity equals 3 in.

Solution

Step 1   This step has been done for us in Figure 10.17*b*.

Step 2   Moment summation about the pivot point

$$\sum M_A = 0 = +Fe + Pl - Na$$

Step 3 $F = \dfrac{T}{Kr} = \dfrac{1260}{(1.08 \times 6)} = 194.4$ lb

$N = \dfrac{F}{\mu} = \dfrac{194.4}{0.35} = 555.5$ lb

Substituting in the summation equation

$0 = (194.4 \times 3) + (P \times 16) - (555.5 \times 6)$

$P = \dfrac{(555.5 \times 6) - (194.4 \times 3)}{16}$

$= 172$ lb

As expected, the value of P is less than that obtained in Sample Problem 10.3.

10.11   BAND BRAKES ▰▰▰▰▰▰▰▰▰▰▰

A band brake consists of a steel band wrapped partially around a rotating drum (see Figures 10.18–10.20). The steel band has a friction-material lining that is replaced when worn out. When a force P is applied to the lever as shown, it tightens the band around the drum and thus offers resistance to motion. We will analyze two classes of band brakes: the simple band brake and the differential band brake.

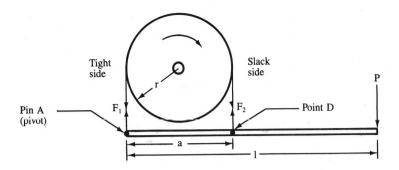

(a) Simple Band Brake Schematic

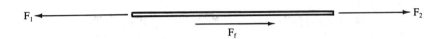

(b) Band Stretched Straight

FIGURE 10.18   Simple band brake (a) Simple band brake schematic (b) Band stretched straight

Simple Band Brake.   The *simple band brake* is shown in Figure 10.18a. Band brakes operate in a fashion that is similar to a pulley-belt system. Thus F_1 and F_2 are called the tight and slack band tensions, respectively. Note that for this configuration, F_1 is tied to the pivot point of the lever arm. Figure 10.18b shows the steel band straightened out so that it is easy to see that F_1 is equal to the sum of the frictional force, F_f, and F_2. The resisting torque is supplied by the frictional force.

That is,

$$T = F_f r = (F_1 - F_2)r$$

where r = the drum radius in inches

The relationship of F_1 to F_2 is the same as that used for a flat belt pulley system without centrifugal force factored in (see Equation 9.13 in Chapter 9).

$$\frac{F_1}{F_2} = e^{\mu\beta}$$

where β = the angle of contact of the band on the drum

μ = the coefficient of friction

The problems in this text will be restricted to finding either the applied force, P or the torque. Our approach will be to solve the above equations for F_1 and F_2 and substitute these (as needed) into a moment summation equation for the forces acting on the lever arm.

A summation of moments will be made about pin A in Figure 10.18a. Note that F_1 has a zero moment arm.

$$\sum M_A = 0 = Pl - F_2 a$$

$$P = \frac{F_2 a}{\ell}$$

If the drum rotation were reversed to a counterclockwise direction, the frictional force would be in the opposite direction to that shown in Figure 10.18b and the forces F_1 and F_2 would be reversed. The larger force, F_1, would be applied at point D and this means that P would have to be larger. This condition will be left as a student problem.

In actual practice, the dimension a is seldom equal to the diameter because the angle of wrap is usually greater than 180° and can be as great as 270°. For lever arm proportions, the length l is usually made about ten times the length a.

PROCEDURE

Finding the Force Applied to a Simple Band Brake

Step 1 Construct a free body diagram of the brake lever.

Step 2 Solve the torque equation and the force ratio equation to find F_1 and F_2.

Step 3 Sum the moments of the forces about the pivot point. Substitute F_1 and F_2 as needed and solve for the force P.

SAMPLE PROBLEM 10.5

Simple Band Brake

PROBLEM: A band brake has the configuration of Figure 10.18a. The following data are given: $r = 10$ in., $a = 20$ in., $\ell = 30$ in., $\mu = 0.35$ and $\beta = 180°$ (this could be determined by inspection). If the retarding torque is to be 2000 lb-in., find the required applied force for clockwise drum rotation.

Solution

Step 1 has been done for us in Figure 10.18a.

Step 2 $T = F_f r$.

$$F_f = \frac{2000}{10} = 200 \text{ lb}$$

$$F_1/F_2 = e^{\mu\beta} = e^{0.35\pi} = 3.00$$

$$F_1 - F_2 = F_f = 200 \text{ lb}$$

$$3.00F_2 - F_2 = 200$$

$$F_2 = 100 \text{ lb} \quad \text{and} \quad F_1 = 300 \text{ lb}$$

Step 3 The moment summation about pin A has already been done for us in the text above, therefore

$$P = \frac{F_2 a}{\ell} = \frac{100 \times 20}{30} = 66.7 \text{ lb}$$

Differential Band Brake. The *differential band brake* is shown in Figure 10.19 and 10.20. Our discussion will be directed to Figure 10.19. At first glance, it might appear that this configuration is not as effective as the simple band brake because as the lever is moved in a clockwise direction by force P, section b of the lever moves up. This tends to slacken the tight side, F_1. On viewing the free body diagram (FBD) of the lever in Figure 10.19, we know that F_1 aids P and F_2 opposes it, so one might think a larger force P is required for a smaller F_1. However, the opposite effect occurs; that is, a smaller force P is actually required. In order to understand this seeming paradox we must realize that the ratio F_1/F_2 remains the same. Therefore, the effect is that the frictional force remains the same but both F_1 and F_2 become smaller.

The ratio of a/b cannot be arbitrary, so let us carefully define these two dimensions of the lever arm before proceeding. In the Figure 10.19, the following lever arm symbols are used,

where　　a = the distance from the pivot (pin A) to the slack side force, F_2 (the same as for the simple band brake in Figure 10.18)

　　　　b = the distance from pin A to the tight side force, F_1

　　　　l = the distance from pin A to the applied force, P (the same as for the simple band brake)

The ratio of dimension a/b should be greater than the ratio of the forces F_1/F_2 or $e^{\mu\beta}$. Otherwise, the brake may lock up. Usually, this is an undesirable condition,

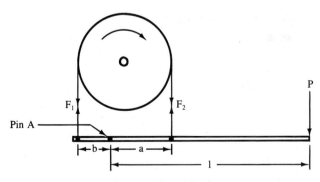

FIGURE 10.19 Differential band brake

although it could be used to prevent reverse rotation of a shaft. To understand how the brake can lock, we will write a moment summation equation for the brake in Figure 10.19 and then apply the F_1/F_2 ratio from Sample Problem 10.5.

The moment summation (about pin A) for the forces on the lever arm in Figure 10.19 is as follows:

$$\sum M_A = 0 = P\ell + F_1 b - F_2 a$$

Therefore

$$P = \frac{F_2 a - F_1 b}{\ell}$$

The force ratio from Sample Problem 10.5 tells us that F_1 is three times as large as F_2: $F_1 = 3F_2$. Now, if a were three times as large as b (let $a = 3$ in., $b = 1$ in., $F_1 = 300$ lb, and $F_2 = 100$ lb), we can see that the numerator of the above equation would becomes zero. Of course, this means $P = 0$. Thus, a frictional force is obtained with a zero force applied to the lever.

PROCEDURE

Finding the Force Applied to a Differential Band Brake

The procedure is the same as the one used to solve simple band brakes. Sample Problem 10.6 follows this procedure for a differential band brake.

SAMPLE PROBLEM 10.6

Differential Band Brake

PROBLEM: A band brake has the configuration of Figure 10.19. The following date are given: $r = 10$ in., $a = 18$ in., $b = 2$ in., $\ell = 30$ in., $\mu = 0.35$, $\beta = 180°$, and the retarding torque = 2000 lb-in. Find the required applied force, P, if the drum rotates clockwise, and determine if the brake will lock.

Solution

Step 1   is already presented to us in Figure 10.19.

Step 2 $e^{\mu\beta}$ and F_1 and F_2 have already been calculated in Sample Problem 10.5. $F_1 = 300$ lb and $F_2 = 100$ lb

Step 3   The moment equation has already been written in the text material above.

$$P = \frac{F_2 a - F_1 b}{\ell} = \frac{100 \times 18 - 300 \times 2}{30} = 40 \text{ lb}$$

The ratio $a/b = 18/2 = 9$. This is greater than the $e^{\mu\beta}$ ratio of 3 and therefore the brake will not lock.

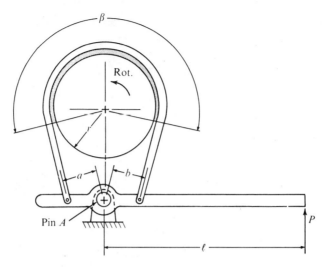

FIGURE 10.20   Band brake

Figure 10.20 represents a typical differential band brake design. Note that force *P* is now applied in an upward direction and the drum rotates counterclockwise, resulting in the reversal of the tight and slack sides of the brake band. This reversal is indicated by the positions of *a* and *b,* as compared to Figure 10.19 (Remember: *a* is the distance to the slack side, and *b* is the distance to the tight side of the band).

10.12 AUTOMOTIVE DRUM BRAKES

The most common use of brakes is in the wheel brakes of the automobile. This section will cover the expanding-shoe drum brake; the caliper disc brake will be presented in Section 10.13.

Figure 10.21 is a sketch of a Duo-Servo drum brake. This design actually uses the automobile's momentum to stop itself. The construction and operation are described as follows:

1.   Only one anchor pin is used and is located near the actuating cylinder.
2.   A hinged floating link connects the bottom ends of the two brake shoes to which the lining material is attached.
3.   When the cylinder moves the leading shoe against the rotating drum, friction tends to pull the shoe around with the drum.
4.   The rotational frictional force goes through the floating link to the other shoe, forcing it against the drum.
5.   The motion through the floating link is called *servo action* because the friction from the one shoe increases the effectiveness of the other shoe. Servo action occurs when braking in either forward or reverse directions.

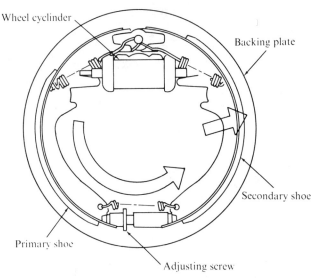

FIGURE 10.21   Schematic of Duo-Servo drum brakes. *(Courtesy of The Bendix Corp.)*

The hydraulic system of an automobile normally consists of one master cylinder and the four wheel cylinders. When the brake pedal is depressed, the master cylinder sends pressurized oil to each wheel cylinder. The wheel cylinders extend and actuate the pivoted brake shoes, forcing the shoes against the drum in each wheel and producing the retarding torque. When the brake pedal is released, the tension springs retract the cylinders to the original position. Notice the adjusting screw, which accommodates for brake lining wear by shifting the brake shoes closer to the drum.

Figure 10.22 is a photograph of a Duo-Servo drum brake. Observe the thickness of the brake lining on the shoes and the gap between the lining and drum. The four inner holes are for the wheel bolts.

One modern safety feature on automobiles is the use of two independent master cylinders (in one housing). One master cylinder operates the two rear wheel cylinders and one operates the two front wheel cylinders. This is called a dual hydraulic braking system. If one system fails (for example, if there is an oil leak with resulting loss of hydraulic fluid), braking action is still available at two of the four wheels.

The Duo-Servo brake is available with automatic adjusting capability as shown in Figure 10.23. Such a design adjusts itself in small increments in proportion to the wear of the linings, and therefore the brake pedal travel is constant for the life of the lining.

The construction and operation of this automatic adjuster system are as follows:

1. A steel cable is attached at one end to the anchor pin and at the other to a hinged steel adjusting lever.
2. The cable goes over a cable guide that is connected to the secondary brake shoe.
3. This system works only when braking in the reverse direction of automobile travel.

FIGURE 10.22   Photograph of Duo-Servo drum brakes. *(Courtesy of The Bendix Corp.)*

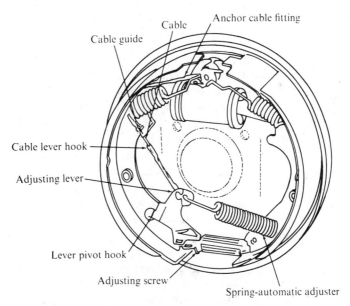

FIGURE 10.23   Automatic adjustment Duo-Servo brake. *(Courtesy of The Bendix Corp.)*

4. Wear on the lining allows the secondary shoe to move far enough from the anchor during reverse braking to force the cable to move the adjusting lever into engagement with the next tooth on the wheel of the adjusting screw.

FIGURE 10.24   Photograph of automatic adjustment Duo-
Servo brake. *(Courtesy of The Bendix
Corp.)*

5.   Upon release of the brake pedal, a spring returns the lever to its previous position. This moves the adjusting screw several degrees, resulting in about 0.0005
in. adjustment on each brake shoe. Figure 10.24 is a photograph of an automatic
adjustment Duo-Servo brake.

One of the main problems encountered with brakes is excessive temperature
buildup when the heat is not dissipated rapidly enough. Most brake materials exhibit a
rapid decrease in the value of the coefficient of friction if the temperature increases
above a certain value (see Figure 10.25 for a typical relationship). As a result, a given
brake pedal force does not produce as much retarding torque. Let us say the temperature increases from 400° F to 1000° F during prolonged braking. According to Figure
10.25, it takes five times as much brake-pedal force to produce the same braking action. This phenomenon is called *brake fade* and usually occurs in automobiles on downhill grades where frequent and prolonged braking is necessary. Automobile
manufacturers recommend downshifting on long downhill grades to reduce brake temperature buildup and brake fade.

Asbestos has been a useful component of brake shoes for many years. However,
because of health hazards the EPA is requiring that asbestos be phased out completely
by 1997. The new asbestos-free materials are referred to as organics and semi-
metallics. Brake shoe materials are now engineered for specific cars and for front or
rear wheels. Compared to organics, semi-metallics are more stable, perform better at
hot temperatures and have a good life. Therefore, semi-metallics are generally used on
the front wheels because the front brakes typically absorb 60% of the braking action on
rear wheel drive vehicles and may absorb as much as 75% to 80% of the braking action
on front wheel drive vehicles.

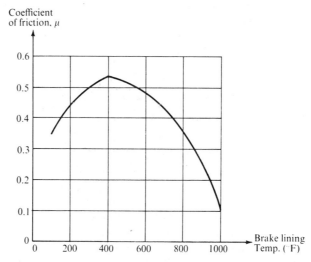

FIGURE 10.25   Brake fade effect resulting from excessive
temperatures

10.13 AUTOMOTIVE DISC BRAKES

Figure 10.26 is a photograph of an automotive caliper disc brake. The disc diameter
and thickness can be any size within specified space requirements. Typical dimensions
for automobiles are a disc diameter of 11.5 in. and a thickness of 3/8 in. Both sides of
the disc are used thus providing a large frictional surface as well as a large cooling
surface. The thickness allows the disc to not only distribute the heat more evenly but
also to act as an effective heat sink. Such a combination means that a disc brake can be
designed to have high torque capacity and good heat dissipation characteristics. The

FIGURE 10.26   Automotive caliper disc brake. *(Courtesy
of The Bendix Corp.)*

disc brake is therefore capable of absorbing large values of kinetic energy, for which the stopping of a high-speed automobile is a good application.

An assembly of the disc brake on the front wheel spindle of an automobile is shown in Figure 10.27. Notice that there is a stationary friction pad located at each side of the brake disc, which rotates with the wheel. During braking, each friction pad is pressed against the side of the disc by a piston actuated inside a hydraulic cylinder. The cylinder is typically about 2 1/8 in. in diameter. The hydraulic pressure will vary up to about 600 psi for power brakes and about 300 psi without power assistance. Springs are used to retract the pistons, and thus reestablish the clearance between the disc and friction pads when the brake pedal is released. The caliper can be readily removed for inspection and the lining replaced without breaking any hydraulic connections.

An analysis will now be made for the braking torque from the frictional force applied by the brake pads. We will also make the analysis for the stopping distance of automobiles. (These equations may also be applied to drum brakes.)

The frictional torque that each brake pad can develop equals frictional force times the radius arm:

$$T = F \times r$$

where r = the radius arm and is measured from the center of the wheel spindle to the center of the brake pad.

The frictional heat energy of the brakes equals the kinetic energy of the automobile prior to braking. This relationship assumes a level road, no other significant sources of friction (such as bearings), and ignores wind resistance.

Frictional heat energy = kinetic energy of automobile

$$F \times \text{distance} = \left(\frac{1}{2}\right)mV^2$$

$$F2\pi rN = \left(\frac{1}{2}\right)\left(\frac{W}{g}\right)V^2$$

Since $T = Fr$ we can also write

$$2\pi TN = \left(\frac{1}{2}\right)\left(\frac{W}{g}\right)V^2 \tag{10.6}$$

where F = frictional force in pounds

m = the mass of the automobile in slugs

r = torque arm of friction pad in feet

N = number of revolutions of wheel during braking to a stop

W = weight of automobile in pounds

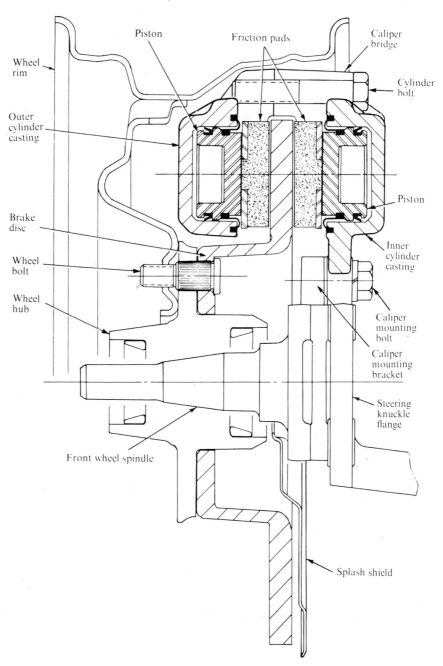

FIGURE 10.27 Assembly of automobile disc brake on front wheel spindle. *(Courtesy of The Bendix Corp.)*

g = acceleration of gravity (32.2 ft/s²)

V = maximum velocity of automobile in feet per second

T = torque in pound-feet

Also, the distance traveled by the automobile is

$$S = \pi DN \tag{10.7}$$

where   S = distance traveled by automobile in feet

D = diameter of tires in feet

Combining the above equations yields this usable equation for braking distance

$$S = \frac{WDV^2}{129T} \tag{10.8}$$

SAMPLE PROBLEM 10.7

Disc Brakes

PROBLEM: A 4000 lb automobile uses tires with a diameter of 28 in. and travels on a level road at 60 mph. Each wheel has disc brakes. The radius arm of the brake pads is 5.25 in., the force applied to each pad is 1500 lb, and the coefficient of friction = 0.36. Find the braking torque for each wheel and the stopping distance.

Solution

$F = \mu N = 0.36 \times 1500 = 540$ lb (per brake pad)

Torque per wheel is

$$T = 2 \text{ pads} \times F \times r = 2 \times 540 \text{ lb} \times \frac{5.25 \text{ in.}}{12 \text{ in/ft}}$$

$T = 472.5$ lb-ft

$$S = \frac{WDV^2}{129T} = \frac{4000 \times (28/12) \times (88 \text{ ft/s})^2}{129 \times 472.5 \times (4 \text{ wheels})} = 296 \text{ ft}$$

10.14 SI UNITS

The equations in this chapter can be used with SI units. To be consistent and avoid unit errors, dimensions must be in meters, forces in newtons, pressure in pascals (1 Pa = 1 N-m), and power in watts. The equation relating power to torque and rpm is

$$P_W = \frac{T\omega}{9.55} \tag{10.9}$$

where P_w = the power in watts

T = the torque in newton-meters

ω = the rotational speed in rpm

SAMPLE PROBLEM 10.8

Plate Clutch in SI Units

PROBLEM: A single plate clutch has an outside diameter of 250 mm and an inside diameter of 120 mm. The coefficient of friction = 0.3 and the allowable pressure on the clutch = 200 kPa. How much torque can be transmitted?

Solution

$$T = \frac{\pi \mu p D_i (D_o^2 - D_i^2)}{8}$$

$$T = \frac{3.14 \times 0.3 \times 200\,000 \times 0.120(0.250^2 - 0.120^2)}{8}$$

$$T = 136 \text{ N-m}$$

10.15 SUMMARY

Clutches are similar to couplings except that clutches must be capable of engaging and disengaging rapidly. Frequently, this action must be done while the equipment is running. For most of the clutches discussed in this chapter, friction is the driving force that produces torque. Jaw clutches do not rely on friction but on compressive and shear forces to provide the positive driving torque. Fluid clutches containing oil rely on the momentum and the viscosity of the fluid to transmit the driving torque. Dry fluid clutches rely on the wedging action of steel balls under centrifugal force to transfer the driving torque.

Plate clutches rely on friction to transfer torque. Cone clutches are similar to plate clutches except that a wedging action increases the torque that can be transferred with the same axial force of engagement as the plate clutch. The smaller the cone angle, the more difficult it becomes to disengage the clutch. The larger the cone angle, the larger the axial force needed to engage the clutch. A good compromise is to make the cone angle, α, between 8° and 15°.

A spring clutch drives in only one direction. Its method of operation is such that the inside diameter of the helical spring tends to contract when the spring is rotated in one direction. The spring then pinches down on a cylinder attached to a mating shaft. .The pinching action produces the friction that provides the driving torque. If the helical spring is rotated in the opposite direction, the diameter of the spring expands and no friction, therefore no driving torque, is produced.

Overrunning clutches work on the principle that steel balls become wedged between two adjacent clutch parts to transmit torque. One part is the driving cylinder with slots for the steel balls and the other part is a sleeve that fits over the cylinder. If the sleeve starts to rotate faster than the cylinder, the steel balls become unwedged and no torque is transmitted.

The term block brake refers to a friction block that is pressed against a rotating drum by means of a lever. The frictional force provides a stopping torque. The term band brake refers to a band wrapped around a rotating drum and actuated by means of a lever. The advantages of the band brake over the block brake are that the angle of contact can be made greater than 180° and the leverage can be increased by the way the tight side and slack side of the band are attached to the brake arm.

There are two standard types of automotive brakes, the drum type and the disc type. The drum brake is designed so that the frictional force on one shoe not only provides a braking torque but also forces a second shoe against the drum to provide additional braking. Above 400° the coefficient of friction tends to drop off, causing brake fade. Disc brakes are designed to radiate heat more effectively than drum brakes, thus they can be used for longer periods without brake fade.

10.16 QUESTIONS AND PROBLEMS ▬▬▬▬▬▬▬▬

Questions

1. What is the basic difference between (a) a clutch and a coupling and (b) a clutch and a brake?
2. Why are jaw clutches used only for slow-speed mechanisms?
3. What advantage does a friction clutch have over a positive-drive clutch?
4. What is the purpose of a slip clutch?
5. What advantage does a cone clutch have over a plate clutch?
6. Why shouldn't the cone angle of a cone clutch be less than 8°?
7. Make a sketch of an overrunning clutch and explain how it works.
8. Briefly explain how electric clutches operate.
9. What is the difference between a fluid clutch and a dry fluid clutch?
10. Explain why the dissipation of heat is a greater problem for brakes than for clutches.
11. For the band brake of Figure 10.20, what determines whether the force P is applied in the upward or downward direction?
12. Why do some automotive hydraulic brake systems contain two master cylinders?
13. What is the difference between a drum brake and a disc brake?
14. Name an advantage of a disc brake over a drum brake.
15. Explain the phenomenon of a brake fade. Do you think that the use of magnesium wheels can reduce brake fade? Explain your answer.

Problems

1. (a) If, in Sample Problem 10.1, the outside diameter is reduced by 10%, what is the percent change in torque?

(b) Repeat **(a)** but instead of changing the outside diameter, increase the inside diameter by 10%.

2. A single plate clutch has an inside diameter of 5 in. The allowable pressure is 25 lb/in.^{2} and μ equals 0.35. If 1000 lb-in. of torque are to be transmitted, what should be the outside diameter?

3. Solve Sample Problem 10.2 but change the cone clutch angle to 8°.

4. Solve Sample Problem 10.2 if the cone angle is 90° (making it a plate clutch).

5. A cone clutch has an outside diameter of 30 in., an inside diameter of 10 in., and a 14° cone angle. The clutch is to transmit 1500 lb-in. of torque and the coefficient of friction is 0.38. **(a)** What minimum axial spring force is required to transmit the torque? **(b)** What minimum axial force is required to engage the clutch?

6. For the block brake of Figure 10.15, the following data are given: $r = 8$ in., $a = 8$ in., $b = 12$ in., $\beta = 100°$ and $\mu = 0.40$. If the applied force P equals 250 lb, what retarding torque can be expected?

7. Solve Problem 6 above using $e = 4$ in. The other data are the same. Solve for all four configurations of Figure 10.17. What conclusion can be made by designing an eccentricity into the configuration of a block brake?

8. Derive the following equation for the block brake of Figure 10.17a:

$$T = K\mu r \, \frac{Pl}{a + \mu e}$$

Is this equation also valid for the block brake of Figure 10.17d? Explain your answer.

9. Solve Sample Problem 10.5 for the simple band brake except reverse the rotation of the drum.

10. Figure 10.28 shows a simple band brake.

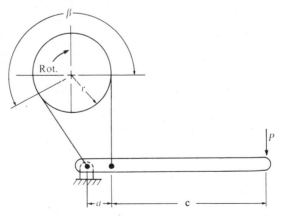

FIGURE 10.28   Sketch for problem 10

Find the torque capacity assuming $\mu = 0.30$, $\beta = 240°$, $a = 4$ in., $c = 36$ in., $r = 5$ in., and $P = 25$ lb.

11. Solve Sample Problem 10.6 except reverse drum rotation and direction of force P. (Remember, distance "a" is the measure to the slack side.)

12. Solve Sample Problem 10.6 except
    (a) Let $a = 15$ in. and $b = 5$ in.
    (b) Let $a = 17$ in. and $b = 3$ in.
    Do you think the brake will lock in either case? What happens to P when a is increased and b is decreased?

13. A differential band brake has the configuration of Figure 10.20. The following data are given: $r = 10$ in., $a = 5$ in., $b = 1$ in., $l = 10$ in., $\mu = 0.35$ and $\beta = 200°$. If the retarding torque is to be 2500 lb-in., find the force P.

14. The friction plate of an automobile clutch has a 6 in. inside diameter and a 9 in. outside diameter. There are ten springs, each of which is compressed 1/2 in. to give a total spring force of 2600 lb. This clutch is used with an engine having the characteristics shown in Figure 10.29. The coefficient of friction is 0.35.
    (a) What factor of safety does the clutch have?
    (b) How much wear of the friction lining can take place before the clutch slips? Remember, the spring force is directly proportional to the amount of compression

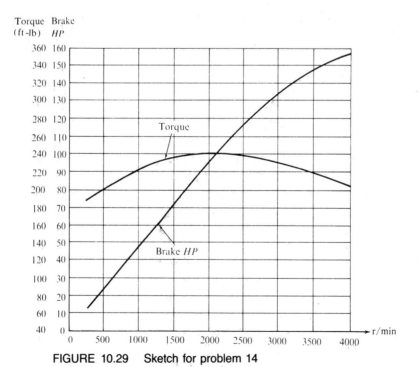

FIGURE 10.29   Sketch for problem 14

15. The Uniform Vehicle Code gives a maximum permissible braking distance (not including reaction time) of 300 ft for an automobile traveling at 60 mph (88 ft/s). A 3000 lb car has 23 in. dia. wheels, and all four wheels have disc brakes. The center of the brake pad is 4.25 in. from the center of the spindle. The pistons that actuate the brake pads are 2 1/4 in. in diameter. The coefficient of friction = 0.38. Will a hydraulic pressure of 600 psi stop the car within the required distance?

SI Unit Problems

16. Solve Sample Problem 10.8 except change the outside diameter to 320 mm.
17. How much power is transmitted by the clutch in Problem 16 if the clutch is turning at 500 rpm?
18. A block brake has the configuration of Figure 10.17a. The following data are given: $r = 150$ mm, $a = 150$ mm, $b = 250$ mm, $\beta = 90°$, $\mu = 0.35$ and $e = 30$ mi. The drum rotates at 400 rpm and transmits 2 kW of power. Find P required to stop the drum.
19. A simple band brake has the configuration of Figure 10.18. The following data are given: $r = 125$ mm, $a = 250$ mm, $l = 650$ mm, $\mu = 0.40$, $\beta = 180°$, the retarding torque = 250 N-m. The drum rotates clockwise. Find the required applied force P (in newtons).

Mechanical Fasteners and Power Screws

Objectives

This chapter discusses the characteristics and advantages of a variety of mechanical fasteners and the design of power screws. On completing this chapter, you will be able to recommend suitable fasteners for a variety of applications, . . . you also will be able to determine the amount of applied torque required to raise or lower a given load with a power screw.

11.1   INTRODUCTION

Fasteners are devices that permit one machine part to be joined to a second part. Hence, fasteners are involved in almost all designs. The acceptability of any product depends not only on the selected components but also on the means by which they are fastened together. The principal purposes of fasteners are to provide the following design features:

— Disassembly for inspection and repair
— Modular design, where a product consists of a number of subassemblies. (Modular design aids manufacturing as well as transportation.)

There are three main classifications of fasteners, which are described below.

Removable.   This type permits the parts to be readily disconnected without damaging the fastener. An example is the ordinary nut-and-bolt fastener.

Semipermanent.   For this type, the parts can be disconnected, but some damage usually occurs to the fastener. One such example is a cotter pin.

Permanent.   When this type of fastener is used, it is intended that the parts will never be disassembled. Examples are riveted and welded joints.

The following factors should be taken into account when selecting fasteners for a given application:

— Primary function
— Appearance
— A large number of small size fasteners versus a small number of large size fasteners (an example is bolts)
— Operating conditions such as vibration, loads and temperature
— Frequency of disassembly
— Adjustability in the location of parts
— Types of materials to be joined
— Consequences of failure or loosening of the fastener

The importance of fasteners is realized when referring to any complex product. In the case of the automobile, thousands of parts are fastened together to produce the total product. The failure or loosening of a single fastener could result in a simple nuisance, such as a door rattle, or a serious occurrence, such as a wheel coming off. Such possibilities must be taken into account in the selection of the type of fastener for the specific application.

11.2   SCREW THREAD TERMINOLOGY

The most widely used type of fastener is the screw thread configuration. Figure 11.1 shows the design and terminology of a screw thread on the end of a bolt. Such a thread

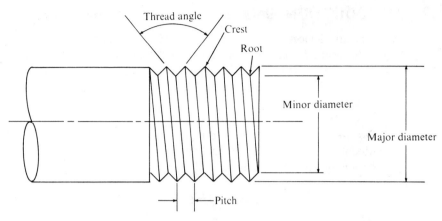

FIGURE 11.1   Screw thread terminology

is called an external thread, whereas the thread of a mating nut is an internal thread. Screw threads may be either right-hand (Figure 11.1) or left-hand. A right-hand bolt will advance into a nut when turned clockwise. The converse happens for a left-hand thread.

The following terminology applies to screw threads:

Major Diameter.   The *major diameter* is the largest diameter of the thread; it determines the nominal thread size.

Minor Diameter.   The *minor diameter* is the smallest diameter of the thread. For a external thread, the minor diameter is frequently called the root diameter.

Pitch p.   The *pitch* is the axial distance between any point of one thread and the corresponding point of an adjacent thread. If the number of threads per inch equals n, the pitch can be expressed mathematically as follows:

$$p = \frac{1}{n}$$

Lead L.   The distance a bolt advances into a nut in one revolution is called the *lead*. On a single-thread screw, the pitch and lead are equal. Sometimes a bolt has two or three unique threads running along separate helical paths. For multiple-thread fasteners, the relationship between lead and pitch is defined by

$$L = kp$$

where $k = 1$ for single thread

$k = 2$ for double threads

$k = 3$ for triple threads

Double and triple threads are used to provide a large amount of threaded engagement for a given number of turns.

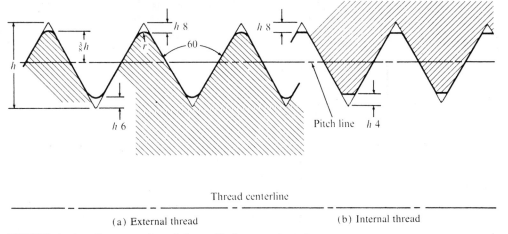

FIGURE 11.2   Cross section of the unified screw thread

The most popular thread form for threaded fasteners is specified in an accord signed in 1948 by the United States, Britain, and Canada. It is the *Unified Thread Standard,* and the thread form is shown in Figure 11.2. The Unified Thread is available in a number of thread series. The first three listed below are probably the most familiar to you. The next four series are a sampling of what is referred to as ''constant pitch'' threads. In other words the pitch stays the same for all diameters of screws and bolts from roughly 1 in. in diameter up to about 6 in. in diameter.

Coarse (UNC).   This series, Unified National Coarse, is used primarily for ordinary assembly applications because it is easy to assemble. Also, for some conditions, the threads have a greater stripping strength compared to the fine thread series.

Fine (UNF).   This series finds applications where vibration is of concern. For a given diameter, fine thread screws and bolts are stronger in tension than coarse and the threads are easier to tap in hard materials (less material is removed).

Extra Fine (UNEF).   This series finds applications where fine adjustments must be made, such as in the use of precision instruments.

4-Pitch (4N).   This is a continuation of the coarse (UNC) series, but in larger sizes.

8-Pitch (8N).   This is also a coarse series option in the larger sizes and applications include the bolting of high-pressure pipe flanges.

12-Pitch (12N).   This is used in boiler practice and is also considered a continuation of the fine thread series for the larger sizes.

16-Pitch (16N).   This is a fine thread series suitable for adjusting collars and retaining nuts where fine adjustments are required.

Table 11.1 provides the number of threads per inch for common sizes of UNC, UNF, and UNEF series threads.

TABLE 11.1   Unified Screw Threads

Nominal Diameter	Threads per Inch		
	UNC	UNF	UNEF
1/4	20	28	32
5/16	18	24	32
3/8	16	24	32
7/16	14	20	28
1/2	13	20	28
9/16	12	18	24
5/8	11	18	24
3/4	10	16	20
7/8	9	14	20
1	8	12	20
1 1/8	7	12	18
1 1/4	7	12	18
1 3/8	6	12	18
1 1/2	6	12	18
1 5/8	—	—	18
1 3/4	5	—	16
2	4 1/2	—	16

There are three fits available in the Unified Thread System. These are given as follows, where letter A denotes an external thread and letter B denotes an internal thread:

Fit 1A or 1B.   This fit provides the loosest fit and thus, facilitates assembly.

Fit 2A or 2B.   This fit provides an intermediate amount of looseness and hence is the most commonly used fit.

Fit 3A or 3B.   This type provides the tightest fit and, as a result, is used primarily in precision equipment.

There is a standard method to specify Unified Screw threads. For example, the designation 1/2 in.-20 UNF-2A indicates an external thread with a 1/2 in. major diameter, a pitch of 1/20 in. (20 threads per inch), and a class 2 fit.

We will not discuss metric threads because standards for metric threads have not yet become commonplace.

11.3   COMMON SCREW FASTENERS ▬▬▬▬▬

There are numerous types of common screw fasteners which can be utilized to connect one member to another. The following is a brief description of the more popular ones:

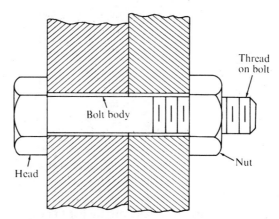

FIGURE 11.3   Through bolt assembly

Through Bolt.   The *through bolt* goes through both pieces that are held together and is made with either square or hexagonal heads and nuts (see Figure 11.3). The holes into which the bolts fit must have adequate clearance. Notice that the bolt thread is drawn using a conventional schematic representation.

Stud.   The *stud* holds parts together where the hole can be drilled only through one of them. The thread extends along almost the entire length of the stud shank and there is no head on the stud. As shown in Figure 11.4, one end of the stud is screwed into one of the parts while the other extends through the second part. Then the nut is screwed on and tightened. Automobile engine cylinder heads are usually attached to the engine blocks by means of studs.

Cap Screw.   A *cap screw* fastens a plate to another component as shown in Figure 11.5. When the parts are assembled, the cap screw passes through a clearance hole in one part and screws in a threaded hole in the other.

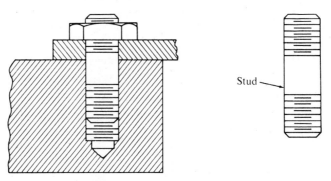

FIGURE 11.4   Stud and stud assembly

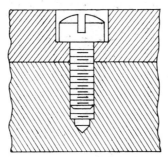

FIGURE 11.5   Cap screw assembly

Machine Screw.   *Machine screws* are simliar to cap screws, but are used mainly for plates having thin sections as shown in Figure 11.6.

Carriage Bolt.   The *carriage bolt* has a square section directly under the head as illustrated in Figure 11.7. the shape of the head is such that it cannot be held by a wrench or screwdriver. The square section is held by a square hole in one of the parts. Thus, as the nut turns, the bolt is prevented from turning by the square section. This arrangement produces a flush surface, which is a desirable characteristic.

Eyebolt.   An *eyebolt* (see Figure 11.8) is frequently used for providing a ring on a part for lifting purposes.

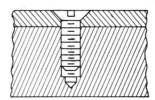

FIGURE 11.6   Machine screw assembly

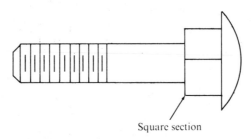

Square section

FIGURE 11.7   Carriage bolt

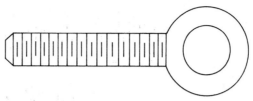

FIGURE 11.8   Eyebolt

U-Bolt.   A *U-bolt* is shown in Figure 11.9. It uses two nuts for the fastening of parts. A popular application is the fastening together of the leaves of an automotive leaf spring.

Turnbuckle.   A *turnbuckle* contains a central member which is threaded to two separate rods as illustrated in Figure 11.10. One end of the central member contains a right-hand thread, whereas the other end contains a left-hand thread. As the central member is rotated, the rods are pulled closer together or pushed further apart depending on the direction of rotation of the central member. A typical application is for the tightening of cables. Another is in the bow compass used for the drawing of circular arcs. In this application, the compass itself is an adjustable central member and turning the threads adjusts the relative positions of the two legs of the compass.

Setscrews.   The *setscrew* is used to locate parts on a shaft and to prevent axial or rotary motion between the shaft and mounted component. A common application is the mounting of light-duty pulleys on shafts. For example, the rotating drum in a clothes

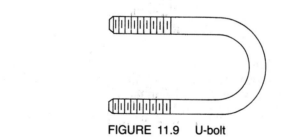

FIGURE 11.9   U-bolt

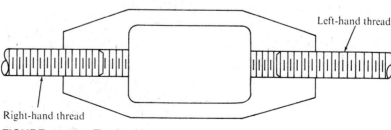

FIGURE 11.10   Turnbuckle

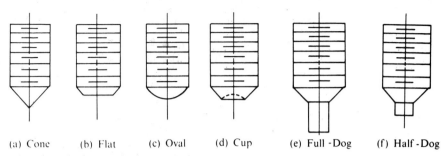

(a) Cone   (b) Flat   (c) Oval   (d) Cup   (e) Full-Dog   (f) Half-Dog

FIGURE 11.11   Types of setscrews

dryer has a pulley mounted on its axle by the use of setscrews. The pulley is connected to the motor by a V-belt. Figure 11.11 shows the various types of setscrews and Figure 11.12 shows an application.

Since the torque-carrying capacity of a setscrew is small, two or three setscrews are sometimes used for a single component. The primary advantage of a setscrew is that it can be removed very easily.

The *cone point* setscrew design is used when the mating parts are to have a precise position relative to each other. The *flat point* design does virtually no damage to the shaft and is used with a hardened shaft that has a recess to accept the flat end as shown in Figure 11.12. The *oval-type* setscrew is used with a mating hole in the shaft. It is used when frequent adjustments are anticipated. The *cup point* design is the most popular type, since it requires no special preparation of the shaft. Thus components may be assembled with the minimum amount of preparation. The cup point will, however, result in some damage to the shaft. *Full-dog* and *half-dog* setscrews are used when maximum holding power is required. The shaft contains a drilled hole that accepts the cylindrical ends of the setscrew. Since the half-dog type can be used in a flat-type setscrew application, it is more popular than the full-dog design.

Socket-Head Screws.   Figure 11.13 is a photograph showing socket-head type screws. The head of each screw contains a hexagonal opening, or socket. Each screw is assembled using the hex wrench shown in the foreground of the photograph.

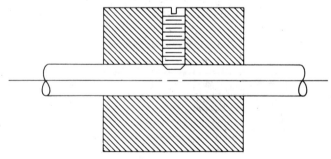

FIGURE 11.12   Use of flat-type setscrew

FIGURE 11.13   Hexagonal socket-head screws (*Courtesy of Standard Pressed Steel Co., Jenkin-town, Pennsylvania*

11.4 WASHERS

Washers are frequently used with bolted fasteners. The most common type is the plain washer, which is shown in Figure 11.14*a*. Such a washer increases the bearing area under the nut or head of a bolt; hence, plain washers are used to protect mating

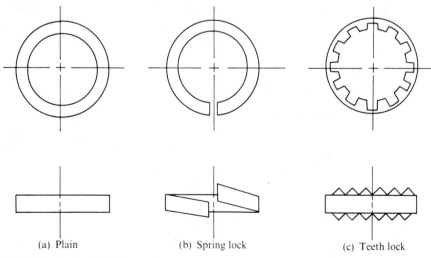

(a) Plain   (b) Spring lock   (c) Teeth lock

FIGURE 11.14   Washers

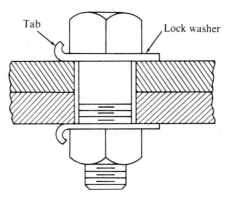

FIGURE 11.15   Positive-type lock washer

surfaces of bolted parts. They are especially needed when the parts are made of soft metals that can be easily damaged by a turning nut or bolt head during tightening.

Lock washers are used to keep bolted fasteners from loosening. The type shown in Figure 11.14*b* provides a spring force that helps maintain the desired bolt tension. In Figure 11.14*c*, we see lock washer that has protruding teeth that dig into the fastener and mating parts and prevent loosening.

Sometimes it is absolutely necessary that a bolted fastener not come loose, even as a result of vibration. In such cases, a *positive-type lock washer* like that shown in Figure 11.15 is used. This type of washer contains a tab that is bent up against the flat portion of the nut or bolt head. The tab prevents rotational motion of the nut or bolt head.

11.5 RETAINING RINGS

Retaining rings are spring-type fasteners which are used to secure components onto shafts or inside the bores of housings. There are two basic types, as shown in Figure 11.16. The *external-type retaining ring* is expanded, using a special tool, over a shaft;

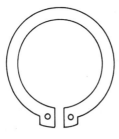

(a) External type   (b) Internal type

FIGURE 11.16   Retaining rings

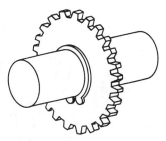

FIGURE 11.17   External-type retaining ring positioning sprocket on shaft (*Courtesy of Industrial Retaining Ring Co., Irvington, New Jersey*)

FIGURE 11.18   Internal-type retaining ring holding bushing inside bore of housing (*Courtesy of Industrial Retaining Ring Co., Irvington, New Jersey*)

whereas the *internal-type retaining ring* is compressed and then inserted into the bore of a housing. In either case, the retaining ring snaps into a groove that locks it into place. Figure 11.17 shows an external-type retaining ring locking a sprocket on a shaft. Figure 11.18 illustrates the use of the internal-type retaining ring, which holds a bushing inside the bore of a housing. A typical pair of industrial pliers used to apply and remove retaining rings is shown in Figure 11.19.

Typical materials for retaining rings are carbon-spring steel, beryllium copper, and stainless steel, all of which permit repeated reuse without loss of resiliency. Advantages of retaining rings include faster assembly and simpler design, since more complicated fasteners using screw-type collars and threaded nuts are eliminated.

11.6   PIN FASTENERS

Pins are frequently used to secure parts to each other and to hold against shear forces. The simplest type of pin is the *dowel pin*, which is solid and cylindrical. Dowel pins are hardened and ground to a tolerance of ±0.0001 in. This close tolerance permits

FIGURE 11.19   Industrial pliers for installing retaining
rings (*Courtesy of Industrial Retaining
Ring Co., Irvington, New Jersey*)

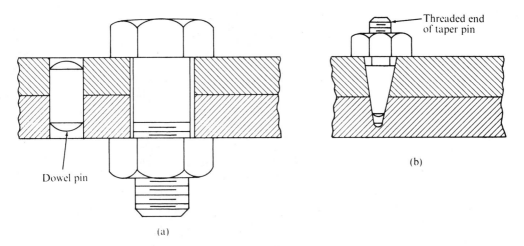

FIGURE 11.20   Application of dowel and taper pins

dowel pins to locate one part very precisely relative to another during machining
or assembly. To achieve this precision, dowel pins are pressed into holes that have
been reamed or bored to very close tolerances. See Figure 11.20a for a dowel pin
application.

Taper pins serve the same function as dowel pins, but rather than cylindrical, they
are conical. The conical taper equals 0.250 in./ft. of length of pin. The two parts must
have their holes reamed simultaneously for precise taper alignment. The taper pin is
placed into position and then driven against the hole (female taper) to produce a fric-

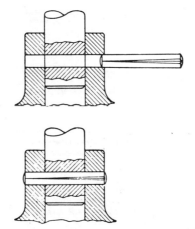

FIGURE 11.21 Application of groov-pin *(Courtesy of Groov-Pin Corp., Ridgefield, New Jersey)*

tional taper lock. Sometimes taper pins are threaded at one end. A nut is then applied as shown in Figure 11.20*b*.

A Groove-Pin is a cylindrical pin having three longitudinal grooves. The pin diameter next to the grooves is slightly larger than the nominal diameter of the pin before insertion into a hole. When the pin is forced into a drilled hole of the same diameter, the constraining action (of the hole wall) causes the material near the grooves to be displaced, setting up a locking fit. Advantages of Groove-Pins are:

1. They can withstand severe shock and vibration.
2. They eliminate failures due to loose or disengaged pins.
3. They require only a drilled hole without the need for close tolerances.
4. They can be reused.

Figure 11.21 shows an application where a Groove-Pin is used to fasten a shaft inside a bore of a housing. Figure 11.22 illustrates various types of Groove-Pins having different body and groove shapes to suit particular applications.

11.7 MISCELLANEOUS FASTENERS

One of the fascinating aspects of the field of fasteners is the seemingly infinite variety of different types available. The following is a brief presentation of some of the more popular ones.

Palnut® Lock Nuts. These fasteners are single-thread lock nuts made of spring-tempered steel which provides resilient locking action. Figure 11.23*a* shows the regular type of Palnut®, which requires the least amount of space for installation. An application is illustrated in Figure 11.23*b*, where the lock nuts are applied on top of ordinary

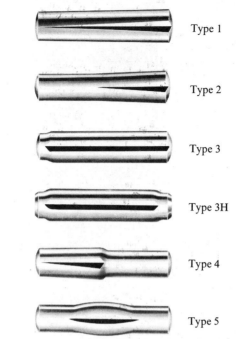

Type 1

Type 2

Type 3

Type 3H

Type 4

Type 5

FIGURE 11.22   Various types of groov-pins (*Courtesy of Groov-Pin Corp., Ridgefield, New Jersey*)

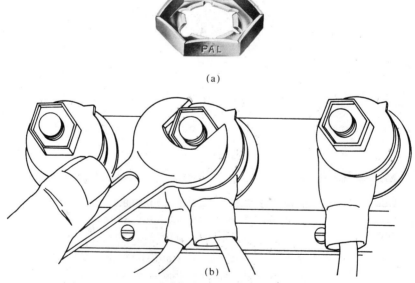

(a)

(b)

FIGURE 11.23   Regular-type Palnut® with application in electrical systems (*Courtesy of The Palnut Co., Mountainside, New Jersey*)

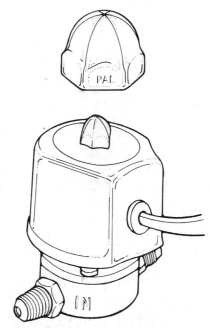

FIGURE 11.24 Acorn-type Palnut® with application in re-frigeration systems (*Courtesy of The Pal-nut Co., Mountainside, New Jersey*)

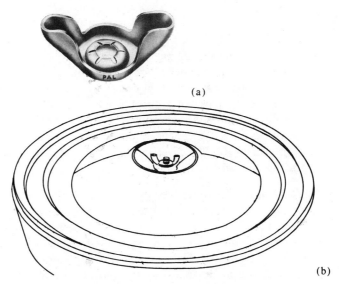

(a)

(b)

FIGURE 11.25 Wing-type Palnut® with application in auto-motive systems (*Courtesy of the Palnut Co., Mountainside, New Jersey*)

FIGURE 11.26   Assortment of speed nuts (*Courtesy of Eaton Corp., Cleveland, Ohio*)

nuts to keep electrical connections tight. Lock washers interfere with conductivity and thus increase resistance. In addition, lock washers often cause overheating and corrosion. The upper portion of Figure 11.24 shows the *acorn-type lock nut*. A typical application is to attach the cover of a direct-acting solenoid valve used in refrigeration systems of vending machines (see lower portion of Figure 11.24). In Figure 11.25*a*, we see the *wing-type lock nut*. The application in Figure 11.25*b* is quite familiar: a wing-type lock nut is used to hold the cover and filter assembly on an automobile air cleaner firmly in place despite vibration.

Eaton Speed Nuts. There are literally thousands of types of speed nuts in existence (see Figure 11.26). Since the nut is actually pushed over a stud (speed nuts are not rotated onto a stud as is the case for regular nuts), they are quickly installed. Figure 11.27*a* shows one type of speed nut. Its spring-like characteristics resist loosening as illustrated in Figure 11.27. Two distinct forces are exerted on the screw as the fastener is tightened. An inward thread lock is provided by two arched prongs because, as the prongs are compressed, they move inward to engage against the root of the screw thread. Also, the arched spring lock is made by the compression of the arch in

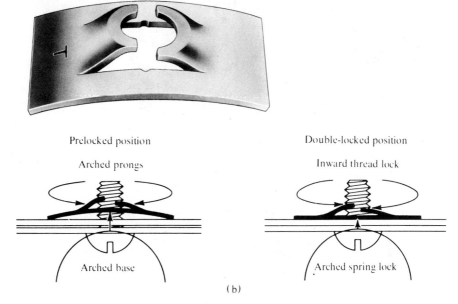

Prelocked position   Double-locked position

Arched prongs   Inward thread lock

Arched base   Arched spring lock

(b)

FIGURE 11.27   Speed nut with spring-like locking action (*Courtesy of Eaton Corp., Cleveland, Ohio*)

both prongs as the fastener is tightened. Thus a strong upward force is developed against the screw threads.

Elastic Stop Nuts.   Figure 11.28 shows a nut with a nylon insert that will not rotate whether it is tight or loose except when wrenched. The nylon insert is a very effective energy absorber and hence prevents loosening during severe vibrations and shock.

Heli-Coil Inserts.   Figure 11.29 shows a Heli-Coil insert, which is used to repair threaded holes that have been damaged. First a larger hole is drilled and tapped, and then the insert is installed into the tapped hole. Special repair kits are available for

FIGURE 11.28   Elastic stop nut containing nylon insert (*Courtesy of Amerace-Esna Corp., Union, New Jersey*)

FIGURE 11.29   Heli-Coil insert (*Courtesy of Heli-Coil Products, Danbury Connecticut*)

tapping and installation of the Heli-Coil insert. Figure 11.30 shows installation of the Heli-Coil insert into a newly tapped spark plug hole.

11.8   POWER SCREWS

A power screw consists of a screw and nut and is used to transmit power or motion in machines. As illustrated in Figure 11.31, the screw turns and the nut (which is prevented from turning by a guide not shown) moves along the axis of the screw. The axial movement of the nut is used to drive a load.

FIGURE 11.30   Installation of Heli-Coil insert (*Courtesy of Heli-Coil Products, Danbury Connecticut*)

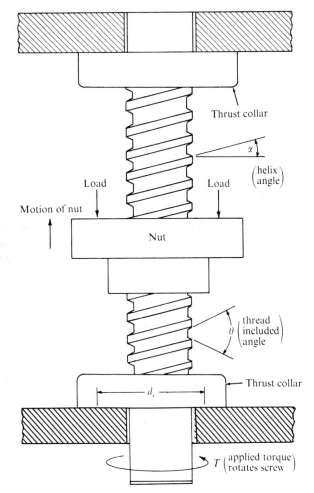

FIGURE 11.31 Power screw

Before proceeding further, we will identify the symbols used in power screw equations,

where T = applied torque in pound-inches

Q = resistive load in pounds

F = horizontal force acting on nut at distance r from the centerline of the screw in pounds

N = the force normal to the helix surface in pounds

F_f = the frictional force acting on the nut in pounds

R = the resultant of the normal force N and the frictional force F_f

r = pitch radius in inches

d = pitch diameter in inches

α = helix angle of thread in degrees

β = angle of friction in degrees

μ_n = coefficient of friction between nut and thread

μ_c = coefficient of friction between thrust collar and support

d_c = mean diameter of thrust collar in inches

It should be noted that considerable friction exists in a conventional power screw between the nut and thread and also between the thrust collar and its support. This friction is a consequence of sliding contact.

The two types of threads used for power screws are the *square* and *Acme threads* (see Figure 11.32 for tooth profiles). The square thread is the most efficient for transmitting power, but is normally not used because it is very expensive to manufacture. The Acme thread is usually made with a 29° included angle as shown in Figure 11.32b. Table 11.2 shows standard Acme screw dimensional data.

The thread on a square or Acme screw travels in the path of a helix. If the helix were unwrapped from the screw, it would form an inclined plane as depicted in Figure 11.33. The length of the base of the inclined plane equals the circumference of a circle whose diameter equals the pitch diameter d of the thread. The height of the inclined plane equals the lead ℓ of the thread. the helix angle α can therefore be mathematically defined, referring to Figure 11.34, as

$$\text{Tan } \alpha = \frac{\ell}{\pi d}$$

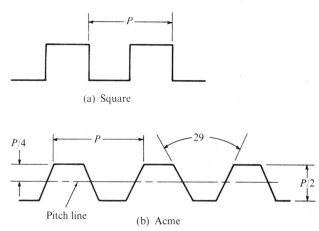

(a) Square

(b) Acme

Pitch line

FIGURE 11.32   Power screw tooth profiles (p = pitch)

TABLE 11.2   Standard Dimensions of Acme Threads

Outside Diameter	Threads/in.	Outside Diameter	Threads/in.
1/4	16	1 1/2	4
3/8	12	1 3/4	4
1/2	10	2	4
5/8	8	2 1/4	3
3/4	6	2 1/2	3
7/8	6	2 3/4	3
1	5	3	2
1 1/4	5	4	2

This will be used in the equation relating torque to load and friction. But, before demonstrating how the equation for torque is obtained, some ground rules and concepts must be established.

1. We will be dealing only with the square thread, as this is simpler.

2. The equation will be for the condition of raising the load. You will be expected to develop the equation for lowering the load, if needed.

3. We will use the angle of friction (identified as β) instead of the coefficient of friction. Remember that the coefficient of fricton μ equals the frictional force divided by the normal force. Since this is a tangent function, we can say that the coefficient of friction equals the tangent of angle β:

$$\mu = \tan \beta$$

As the screw is rotated (in Figure 11.31) to raise the load, our understanding of the process's dynamics must be based on Figure 11.34. That is, our effort is to push a block up a circular incline (a helix). The torque that is developed equals the horizontal force F applied at a distance of the pitch radius r from the centerline of the screw:

$$T = F \times r$$

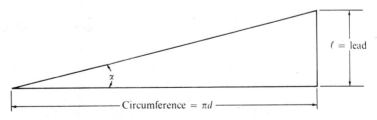

FIGURE 11.33   Definition of thread helix angle

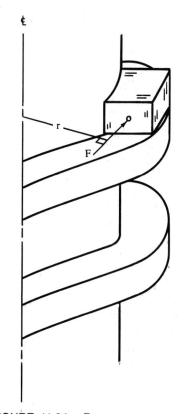

FIGURE 11.34   Power screw concept

We have an equilibrium condition, and we want to find the force F, sufficient to overcome the friction and start the sliding action. There are three concurrent forces acting on the block (see Figure 11.35a): the horizontal force F, the load Q, and a force R. R is the vectorial sum of the normal force N and the frictional force opposing the impending motion. The easiest thing to do now is draw the force triangle and solve for F. The force triangle is drawn in Figure 11.35b; we can see that

$$F = Q \times \tan (\alpha + \beta)$$

Thus, the torque required for a square thread screw to raise a load is found by

$$T = rQ \tan (\alpha + \beta) = \frac{d}{2} Q \tan (\alpha + \beta) \tag{11.1}$$

Equation 11.1 can also be obtained by summing the components of the forces parallel to the incline and perpendicular to the incline, setting the two equations equal to zero, and solving simultaneously. Refer to Figure 11.35a.

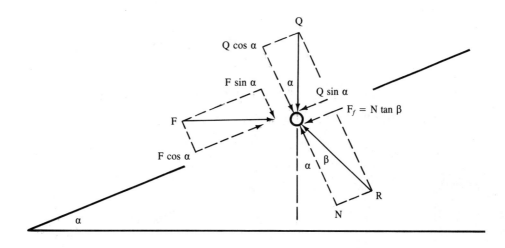

(a)

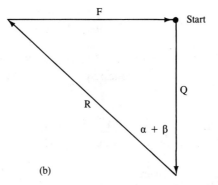

(b)

FIGURE 11.35   Forces acting on a power screw when lifting a load Q

$$\sum F_{\text{normal}} = N - F \sin \alpha - Q\cos \alpha = 0$$

$$\sum F_{\text{parallel}} = F \cos \alpha - Q \sin \alpha - F_{\text{f}} = 0$$

Substitute $N \tan \beta$ for F_{f} in the second equation. We need an equation with F and Q but not N, so multiply the first equation by $\tan \beta$ and solve simultaneously. The $N \tan \beta$ terms in each equation cancel each other, leaving

$$F \cos \alpha - (F \sin \alpha)(\tan \beta) - Q \sin \alpha - (Q \cos \alpha)(\tan \beta) = 0$$

Rearranging

$$F = \frac{F \sin \alpha \tan \beta}{\cos \alpha} + \frac{Q \sin \alpha}{\cos \alpha} + \frac{Q \cos \alpha \tan \beta}{\cos \alpha}$$

$$F = (F \tan \alpha)(\tan \beta) + Q \tan \alpha + Q \tan \beta$$

Collecting terms and rearranging

$$F - (F \tan \alpha)(\tan \beta) = Q(\tan \alpha + \tan \beta)$$

$$F = \frac{Q(\tan \alpha + \tan \beta)}{1 - (\tan \alpha)(\tan \beta)} \tag{11.2}$$

If we now look up trigonometry identities in a handbook we will see that

$$\frac{\tan \alpha + \tan \beta}{1 - (\tan \alpha)(\tan \beta)} = \tan (\alpha + \beta)$$

which is one of the factors in Equation 11.1. However, Equation 11.2 is needed to help write an equation for the Acme thread.

The equation for the Acme thread is more complicated. The reason is that, unlike the square thread, where all of the forces are acting in the same vertical plane, the forces on the Acme thread are not acting in the same vertical plane. The normal force on an Acme thread must be perpendicular to the helix surface, which is inclined 14 1/2° (1/2 of 29°) to the vertical. This means that the normal force in Figure 11.35a is now a component of the normal force for an Acme thread. Of course, the frictional force is based on the normal force. The cosine of 14 1/2° (0.968) must be factored into Equation 11.2, as shown below. The torque required for an Acme thread to raise a load is found by

$$T = \left(\frac{Qd}{2}\right) \left(\frac{0.968 \tan \alpha + \tan \beta}{0.968 - (\tan \alpha)(\tan \beta)}\right) \tag{11.3}$$

The torque required to overcome friction of the thrust collar (see Figure 11.31) uses Equation 10.1 for plate clutches (Chapter 10). Some symbols are changed to conform with our present discussion:

$$T = \frac{\mu_c Q d_c}{2} \tag{11.4}$$

or

$$T = \frac{\mu_c Q(D_o + D_i)}{4}$$

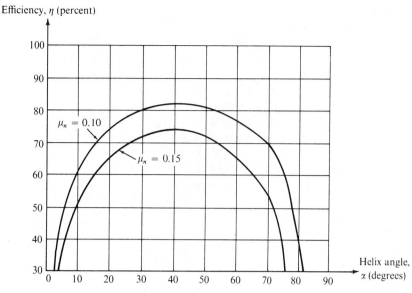

FIGURE 11.36   Efficiency of Acme power screw ($\mu_c = 0$)

If a ball thrust bearing is used instead of a sliding contact thrust collar, μ_c becomes negligible and can be ignored. The coefficient of friction (for sliding contact between nut and screw as well as between thrust collar and support) usually ranges between 0.10 and 0.15. The amount of friction present can be evaluated by the efficiency of a power screw which is defined

$$\eta = \frac{T'}{T} \times 100 \tag{11.5}$$

where η = power screw efficiency (percent).

T' = torque required if there is no friction in pound-inches

T = torque required with friction in pound-inches

If μ_c and μ_n both equal zero, there is no friction, and both Equation 11.1 and 11.3 become

$$T' = \left(\frac{d}{2}\right) Q \tan \alpha \tag{11.6}$$

Inspection of equations 11.1, 11.3, 11.5 and 11.6 reveals that the efficiency of a power screw depends on the coefficient of friction and the helix angle. Figure 11.36 is a graph of efficiency versus helix angle for two different coefficients of friction. The graph disregards the thrust collar friction. Observe that maximum efficiency equals

82% and 73% for $\mu_n = 0.10$ and 0.15, respectively. However, these maximum values of efficiency occur at a helix angle of approximately 40°, which is an extremely large value. A typical helix angle for a single-threaded screw (5 threads per inch) is about 4°, which produces efficiency values of about 40% and 30% for $\mu_n = 0.10$ and 0.15. Using a double-threaded screw increases the helix angle to about 8°. Although this increases the efficiency, the mechanical advantage decreases as the helix angle is increased because a greater value of torque is required to overcome a given load.

High efficiency, in addition to causing a loss of mechanical advantage, may also prevent the power screw from being self-locking when lifting a load. In a large number of applications we want the self-locking feature, for example, in automotive screw jacks. The importance of the helix angle (for a given coefficient of friction) in providing the self-locking feature can be explained with the aid of Figure 11.35a. For the condition when the load is stationary in a raised position, no force F is applied, and impending motion is down the plane. F_f reverses its direction and opposes the parallel component of Q, that is $Q \sin \alpha$. As the helix angle increases, F_f decreases (because N decreases) and $Q \sin \alpha$ increases. At some angle α, F_f will be overcome and the load will force rotation of the screw as it moves to the lowest position. This condition is sometimes called *overhauling*.

PROCEDURE

Solving Power Screw Problems

The procedure is rather straight-forward. Find the helix angle and the angle of friction, or their tangents, and then substitute these into the appropriate torque equation. Add the torque required to overcome collar friction, if required.

SAMPLE PROBLEM 11.1

Acme Power Screw

PROBLEM: A single-threaded screw jack has Acme threads. The screw pitch diameter is 1 in. and there are 5 threads/in. The thrust collar has a 2 in. outside diameter and a 1 in. inside diameter. $\mu_n = 0.15$ and $\mu_c = 0.10$. How much force P must be applied to a 12 in. radius jack wrench to raise a 1000 lb load?

Solution

$$\text{Tan } \alpha = \frac{\ell}{\pi d} = \frac{1/(\text{no. threads per in.})}{\pi d}$$

$$= \frac{1/5}{\pi 1} = 0.0637$$

$$\text{Tan } \beta = \mu_n = 0.15$$

Using Equation 11.3 and 11.4 together.

$$T = \left(\frac{1000 \times 1}{2}\right)\left(\frac{0.968 \times 0.0637 + 0.15}{0.968 - 0.0637 \times 0.15}\right) + \frac{0.10 \times 1000(2 + 1)}{4}$$

$$T = 110 + 75 = 185 \text{ lb-in.}$$

$$P = \frac{185 \text{ lb·in.}}{12 \text{ in. torque arm}} = 15.4 \text{ lb}$$

SAMPLE PROBLEM 11.2

Efficiency of a Power Screw

PROBLEM: For the screw jack in Sample Problem 11.1, find (a) the efficiency η, ignoring thrust collar friction (b) the overall efficiency, including the thrust collar.

Solution

(a) $T' = \left(\dfrac{d}{2}\right) Q \tan \alpha = \left(\dfrac{1}{2}\right) 1000 \times 0.0637$

$T' = 31.9$ lb-in.

T (without collar) $= 110$ lb-in.

$\eta = \left(\dfrac{T'}{T}\right)100 = \left(\dfrac{31.9}{110}\right)100 = 29.0\%$

(b) $\eta = \left(\dfrac{31.9}{185}\right)100 = 17.2\%$

11.9 FORCE ANALYSIS OF FASTENERS ▬▬▬▬▬▬▬

Bolted or riveted fasteners are frequently subjected to loads that can produce shear or tensile forces in the body of the bolts or rivets. Possible failure of the bolts or rivets must be considered in the design of such a fastener. If a bolted fastener absorbs a tensile force, it is also possible for the threads to be stripped by excessive shear stresses.

When more than one bolt is used, sometimes each bolt does not absorb the same amount of load. However, it is common practice to make each bolt the same size. This reduces inventory and also the possibility of installing the wrong size bolt at a given location of the fastener.

Let us initially examine bolted (or riveted) connectors in which shear is the type of bolt load. Then we will analyze bolted brackets in which the bolt experiences tensile stresses and the threads absorb shear stresses. Finally, bolt torsional stresses induced during tightening will be investigated.

Fastener With Single Bolt Loaded in Shear. Figure 11.37a shows two plates fastened together using a single bolt. An external load P is applied to each plate as shown. A free-body diagram of the upper portion of the fastener is illustrated in Figure 11.37*b*. From mechanics, we conclude that a shear force of magnitude P acts on the

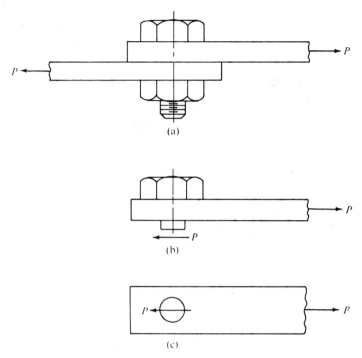

FIGURE 11.37   Fastener with single bolt loaded in shear

cross-sectional area of the bolt. Figure 11.37*c* shows a second view of this shear force P. The bolt shear stress τ can be obtained from

$$\tau = \frac{P}{A} \tag{11.7}$$

where A is the cross-sectional area of the bolt.

Fastener With Multiple Bolts Loaded in Shear (Zero Eccentricity Design).

Figure 11.38 shows two plates bolted together using three bolts. Notice that the centroid of the bolt cross-sectional areas lies on the line of action of the applied load P. Such a configuration has zero eccentricity, and each bolt carries an equal share of the load. Equation 11.8 yields the shear stress τ in each bolt:

$$\tau = \frac{P}{NA} \tag{11.8}$$

where   N = number of bolts.

            A = cross-sectional area of each bolt.

Eccentrically Loaded Bolted Connector Loaded in Shear.

When the line of action is applied load P does not go through the centroid of the bolts, a couple or

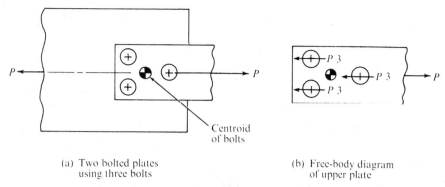

(a) Two bolted plates
    using three bolts

(b) Free-body diagram
    of upper plate

FIGURE 11.38   Fastener with multiple bolts loaded in shear (zero eccentricity design)

torque about the centroid is generated. This produces a secondary shear force in each bolt. Figure 11.39 shows such a bolted connector. The eccentricity e is the distance from the centroid of the bolts measured perpendicularly to the line of action of the applied load P. This eccentricity results in secondary shear forces in addition to the direct shear forces experienced in Figure 11.38. This can be demonstrated as follows:

Place equal and opposite forces of magnitude P (dashed line in Figure 11.39) at the centroid of the bolts. The direction of one force is parallel to the applied load P, while the other force is opposite in direction. Since the two forces are equal and opposite in direction, the actual total system loading is not affected. Notice that a couple is produced by the applied load P and the equal but opposite-in-direction force P at the centroid. The magnitude of the couple or torque about the centroid can be found using

$$T = Pe \tag{11.9}$$

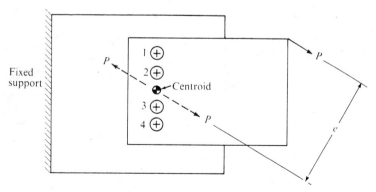

FIGURE 11.39   Eccentrically loaded bolted connector loaded in shear

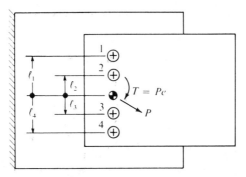

FIGURE 11.40   Dual loads on bolts as a result of eccentricity

In addition to the couple Pe, we still have the other force P at the centroid. Thus we really have two loads, as shown in Figure 11.40. The two loads are a couple and a force passing through the centroid. Notice that the distances ℓ_1, ℓ_2, ℓ_3 and ℓ_4 are the distances from the centroid to the center of the corresponding bolt.

A dual load system can be readily handled by using the concept of superposition: *The total effect of two or more loads equals the vector sum of the effects of the individual loads assumed to be acting alone.* This technique is illustrated in Figure 11.41 where the resulting shear force on each bolt is the vector sum of the direct shear due to P and the secondary shear due to the couple Pe.

Figure 11.41a shows the direct shear of $P/4$ acting on each bolt. the direction of the direct shear forces is opposite to the load P. Notice that the secondary shear forces have the following characteristics (see Figure 11.41b):

1.   Direction is perpendicular to the line running from the center of the bolt to the centroid of all the bolts.

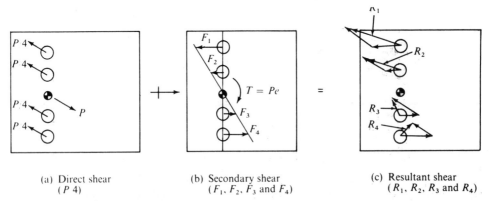

(a) Direct shear
(P 4)

(b) Secondary shear
(F_1, F_2, F_3 and F_4)

(c) Resultant shear
(R_1, R_2, R_3 and R_4)

FIGURE 11.41   Vector addition of direct and secondary shear forces

2. Direction opposes applied torque T.
3. Magnitude is proportional to the distance from the center of the bolt to the centroid of all the bolts as is graphically illustrated in Figure 11.41b, where F_1, F_2, F_3, and F_4 are the secondary shear forces. This characteristic results in the following:

$$\frac{F_1}{F_2} = \frac{\ell_1}{\ell_2} \tag{11.10}$$

$$\frac{F_1}{F_3} = \frac{\ell_1}{\ell_3} \tag{11.11}$$

$$\frac{F_1}{F_4} = \frac{\ell_1}{\ell_4} \tag{11.12}$$

Also from mechanics, the summation of moments about the centroid must equal zero:

$$F_1\ell_1 + F_2\ell_2 + F_3\ell_3 + F_4\ell_4 - Pe = 0 \tag{11.13}$$

Since we have four equations and four unknowns (F_1, F_2, F_3, and F_4), a mathematical solution is available by substituting directly into Equation 11.13:

$$F_1\ell_1 + \left(F_1 \frac{\ell_2}{\ell_1}\right)\ell_2 + \left(F_1 \frac{\ell_3}{\ell_1}\right)\ell_3 + \left(F_1 \frac{\ell_4}{\ell_1}\right)\ell_4 - Pe = 0$$

The final result is

$$F_1 = \frac{Pe\ell_1}{\ell_1^2 + \ell_2^2 + \ell_3^2 + \ell_4^2} \tag{11.14}$$

After the value of F_1 is obtained from Equation 11.14, F_2, F_3, and F_4 can be evaluated using Equations 11.10, 11.11 and 11.12. As a result, the direct and secondary shear forces are solvable. Figure 11.42c shows the graphical method for adding the direct and secondary shear forces to obtain the resultant shear forces, R_1, R_2, R_3, and R_4.

SAMPLE PROBLEM 11.3 _____

Forces on Bolts

PROBLEM: For the bolted connector of Figure 11.42, find (**a**) the direct and secondary shear force on each bolt; (**b**) the maximum total resulting shear force.

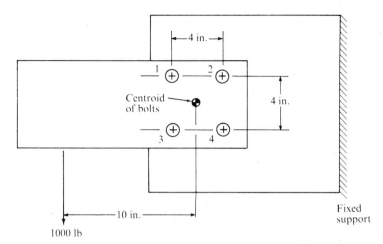

FIGURE 11.42   Bolted connector

Solution

(a) Figure 11.43a shows the direct shear forces, whereas Figure 11.43b illustrates the secondary shear forces. The direct shear force on each bolt equals 1000/4 = 250 lb.

The applied torque from the 10 in. eccentricity equals 1000 × 10 = 10,000 lb-in. The distance ℓ from the centroid to the center of each bolt is the same for each bolt. Therefore, each bolt receives the same magnitude of secondary shear force P as shown in Figure 11.43b. Summing moments about the centroid yields

$$P\ell + P\ell + P\ell + P\ell - 10,000 = 0$$

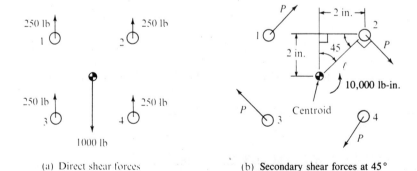

(a) Direct shear forces   (b) Secondary shear forces at 45°

FIGURE 11.43   Direct and secondary shear forces (a) Direct shear forces (b) Secondary shear forces at 45°

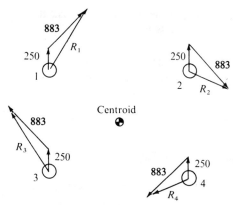

FIGURE 11.44 Graphical addition of direct and secondary shear forces

where $\ell = \sqrt{2^2 + 2^2} = 2.83$ in.

Therefore $P = 883$ lb

(b) As shown in Figure 11.44, bolts 1 and 3 each receive the maximum
total resulting shear force. The result is

$R_1 = R_3 = 1074$ lb

Bolted Bracket (Bolt Tension and Thread Shear).   Figure 11.45 shows a
bracket bolted to a fixed frame using four bolts. An external load P is applied to the top
of the bracket as shown. The purpose of the rib is to give the bracket adequate bending
stiffness. The flange is designed to take all the shear loading away form the bolts. The

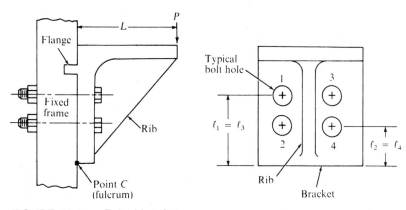

FIGURE 11.45 Bolted bracket

load P attempts to rotate the bracket clockwise about fulcrum point C. This means that a very tiny gap opens up between the fixed frame and bracket as the bolts stretch. The gap is zero at the fulcrum and increases linearly, reaching a maximum at the top of the bracket. Therefore, the tensile force that a given bolt experiences is proportional to its distance from the fulcrum. This relationship is represented by

$$\frac{F_1}{F_2} = \frac{\ell_1}{\ell_2} \tag{11.15}$$

Summing moments about the fulcrum,

$$PL - F_1\ell_1 - F_3\ell_3 - F_2\ell_2 - F_4\ell_4 = 0 \tag{11.16}$$

For the system of Figure 11.45, $\ell_1 = \ell_3$ and $\ell_2 = \ell_4$. Therefore, $F_1 = F_3$ and $F_2 = F_4$. Combining Equations 15.11 and 11.16,

$$PL - F_1\ell_1 - F_1\ell_1 - \left(F_1\frac{\ell_2}{\ell_1}\right)\ell_2 - \left(F_1\frac{\ell_2}{\ell_1}\right)\ell_2 = 0$$

Solving for F_1 produces

$$F_1 = \frac{PL\ell_1}{2(\ell_1^2 + \ell_2^2)} \tag{11.17}$$

The tensile stress S_1 in bolt 1 can be found using

$$S_1 = \frac{F_1}{A_t} \tag{11.18}$$

A_t is the area that the tensile stress acts on for an external thread. Experiments have shown that the tensile stress acts on an area slightly larger than the root area (or minor diameter area). Information on A_t is supplied in Table 11.3, which provides basic dimensions for coarse thread screws and bolts.

The shear stress in the threads can be determined by referring to Figure 11.46, which shows the tensile force F_1 acting on the nut. The thread shearing area (stripping area), A_s, is a cylindrical area defined by

$$A_s = \pi dt \tag{11.19}$$

where t equals the thickness of the nut and d is the root diameter (or minor diameter) of a screw or bolt. Thus the thread shear stress, S_s, can be found:

$$S_s = \frac{F_1}{\pi dt} \tag{11.20}$$

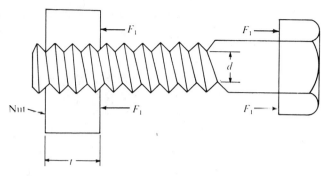

FIGURE 11.46   Stripping of threads

If the allowable shear stress for the thread is known, the required nut thickness based on stripping of the threads becomes

$$t = \frac{F_1}{\pi S_s d} \qquad\qquad (11.21)$$

SAMPLE PROBLEM 11.4

Tensile and Shear Forces on a Bolt

PROBLEM: For the bolted bracket of Figure 11.45, P = 1000 lb, L = 60 in., ℓ_1 = ℓ_3 = 20 in., and ℓ_2 = ℓ_4 = 10 in. Find (a) the maximum bolt tensile force; (b) the required nut thickness if the root diameter of the bolt is 0.25 in. and the allowable shear stress is 6000 lb/in.^{2}

Solution

(a) Substituting into Equation 11.17,

$$F_1 = \frac{PL\ell_1}{2(\ell_1^2 + \ell_2^2)} = \frac{(1000)\,(60)\,(20)}{2[(20)^2 + (10)^2]} = 1200 \text{ lb}$$

(b) From Equation 11–21, the required nut thickness is

$$t = \frac{F_1}{\pi S_s d} = \frac{1200}{\pi(6000)\,(0.25)} = 0.255 \text{ in.}$$

Bolt Torsional Stress Induced During Tightening.   When a bolt is tightened with a torque wrench, a tensile force is induced, producing friction between the nut or bolt head and the mating surface being clamped. As the tensile force increases, so does the friction. The applied torque must overcome the frictional force. After the bolt has been tightened, the external torque is removed and the torsional stresses reduce to zero,

TABLE 11.3 Basic Dimensions for Coarse Thread Series (UNC/UNRC) (*From American Society of Mechanical Engineers, document ANSI B1.1–1982*)

Nominal size	Basic major diameter D, in	Threads/ in, n	Basic pitch diameter* E, in	Root Dia. UNR design (minor diameter) external† (Ref.), k_s, in	Basic minor diameter internal, K, in	Section at minor diameter at $D-2h_b$, in²	A_t Tensile stress area,‡ in²
1(0.073)§	0.0730	64	0.0629	0.0544	0.0561	0.00218	0.00263
2(0.086)	0.0860	56	0.0744	0.0648	0.0667	0.00310	0.00370
3(0.099)§	0.0990	48	0.0855	0.0741	0.0764	0.00406	0.00487
4(0.112)	0.1120	40	0.0958	0.0822	0.0849	0.00496	0.00604
5(0.125)	0.1250	40	0.1088	0.0952	0.0979	0.00672	0.00796
6(0.138)	0.1380	32	0.1177	0.1008	0.1042	0.00745	0.00909
8(0.164)	0.1640	32	0.1437	0.1268	0.1302	0.01196	0.0140
10(0.190)	0.1900	24	0.1629	0.1404	0.1449	0.01450	0.0175
12(0.216)§	0.2160	24	0.1889	0.1664	0.1709	0.0206	0.0242
1/4	0.2500	20	0.2175	0.1905	0.1959	0.0269	0.0318
5/16	0.3125	18	0.2764	0.2464	0.2524	0.0454	0.0524
3/8	0.3750	16	0.3344	0.3005	0.3073	0.0678	0.0775
7/16	0.4375	14	0.3911	0.3525	0.3602	0.0933	0.1063
1/2	0.5000	13	0.4500	0.3334	0.4167	0.1257	0.1419

Size		Threads					
9/16	0.5625	12	0.5084	0.4633	0.4723	0.162	0.182
5/8	0.6250	11	0.5660	0.5168	0.5266	0.202	0.226
3/4	0.7500	10	0.6850	0.6309	0.6417	0.302	0.334
7/8	0.8750	9	0.8028	0.7427	0.7547	0.419	0.462
1	1.000	8	0.9188	0.8512	0.8647	0.551	0.606
1 1/8	1.1250	7	1.0322	0.9549	0.9704	0.693	0.763
1 1/4	1.2500	7	1.1572	1.0799	1.0954	0.890	0.969
1 3/8	1.3750	6	1.2667	1.1766	1.1946	1.054	1.155
1 1/2	1.500	6	1.3917	1.3016	1.3196	1.294	1.405
1 3/4	1.7500	5	1.6201	1.5119	1.5335	1.74	1.90
2	2.000	4½	1.8557	1.7353	1.7594	2.30	2.50
2 1/4	2.2500	4½	2.1057	1.9853	2.0094	3.02	3.25
2 1/2	2.5000	4	2.3376	2.2023	2.2294	3.72	4.00
2 3/4	2.7500	4	2.5876	2.4523	2.4794	4.62	4.93
3	3.000	4	2.8376	2.7023	2.7294	5.62	5.97
3 1/4	3.2500	4	3.0876	2.9523	2.9794	6.72	7.10
3 1/2	3.500	4	3.3376	3.2023	3.2294	7.92	8.33
3 3/4	3.7500	4	3.5876	3.4523	3.4794	9.21	9.66
4	4.000	4	3.8376	3.7023	3.7294	10.61	11.08

*British; effective diameter
†See formula under definition of tensile stress area in Appendix B of ANSI B1.1-1987.
‡Design form. See Fig. 2B in ANSI B1.1-1982.
§Secondary sizes.

SOURCE: ANSI B1.1-1982, reproduced by permission.

leaving only the tensile stresses. Thus, if a bolt is not overtorqued, it should not fail during the tightening procedure.

The torque necessary to produce the required bolt tensile force can be approximated by

$$T = KdF \tag{11.22}$$

where T = the torque in pound-inches

d = minimum bolt diameter in inches

F = required bolt tensile force in pounds

K = lubrication factor. For lubricated bolts and nuts, $K = 0.15$; for unlubricated bolts and nuts, $K = 0.2$.

The torsional shear stress τ induced during tightening can be found using

$$\tau = \frac{16}{\pi} \frac{T}{d^3} \tag{11.23}$$

SAMPLE PROBLEM 11.5

Torsion on Bolt

PROBLEM: A lubricated bolted joint is to be tightened so that a 100 lb. tensile force is induced. The bolt diameter is 1/2 in. Find the minimum allowable shear stress for the bolt.

Solution.

$$T = KdF = (0.15)\,(0.5)(100) = 7.5 \text{ lb-in.}$$

$$\tau = \frac{16}{\pi}\left(\frac{T}{d^3}\right) = \frac{16(7.5)}{\pi(0.5)^3} = 306.0 \text{ lb/in.}^2$$

11.10 RIVETED JOINTS

Riveting is a popular method for permanently joining two parts together. Rivets can be used to readily join parts having dissimilar materials and unequal thicknesses. The two basic types of riveted joints are lap joints and butt joints as shown in Figure 11.47.

A rather unique type of rivet called a Rivnut is illustrated in Figure 11.48. It is a tubular rivet with internal threads. As shown in Figure 11.49, a hand or power tool pull-up stud engages the threads of the Rivnut. This exerts a pull that causes the shank to expand tightly against the material being fastened. After the Rivnut is installed, various parts such as hinges, brackets or knobs can be readily attached, since the internal threads of the Rivnut are exposed. Rivnuts are widely used in stoves, kitchen cabinets, appliances, storm doors and windows, vending machines and automobiles. The main advantages are simplified production and reduced asszmbly time.

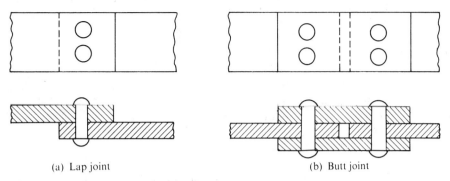

(a) Lap joint   (b) Butt joint

FIGURE 11.47 Types of riveted joints

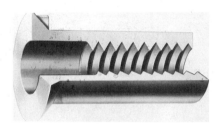

FIGURE 11.48 Rivnut (*Courtesy of the B.F. Goodrich Co.*)

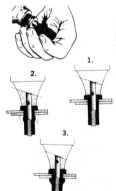

INSTALLATION

Rivnut is threaded onto pull-up stud of a hand or power-heading tool.

Rivnut on header tool mandrel, inserted in drilled hole ready for installation. Arrow indicates direction of mandrel movement as tool is operated.

Mandrel retracts pulling threaded portion of Rivnut shank toward blind side of work, forming bulge around unthreaded shank area. Rivnut is clinched securely in place. Unthreading the tool mandrel leaves internal Rivnut threads intact, unharmed.

BLIND NUT PLATE

Installed Rivnuts also serve as blind nut plates for simple screw attachments. When imperative that attached part be flush fit, countersunk Rivnut heads are used instead of flat heads.

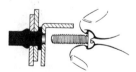

FIGURE 11.49 Installation and application of rivnuts (*Courtesy of the B.F. Goodrich Co.*)

Figure 11.50 shows the critical areas stressed in a riveted joint. The cross section of each rivet is stressed in shear, as in Figure 11.50b. Each half of a rivet has a bearing (compressive) stress applied, as shown in Figure 50c. The critical bearing area is considered to be the plane area represented by $t \times d$. An equal bearing stress appears

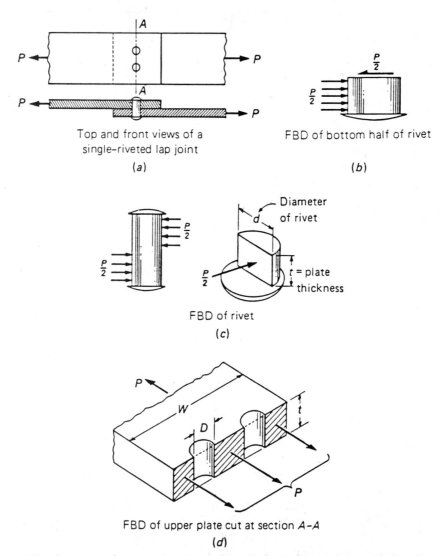

Top and front views of a
single-riveted lap joint

(a)

FBD of bottom half of rivet

(b)

FBD of rivet

(c)

FBD of upper plate cut at section *A-A*

(d)

FIGURE 11.50   Critical areas in riveted joints (a) Top and front views of a
single-riveted lap joint (b) FBD of bottom half of rivet (c) FBD
of rivet (d) FBD of upper plate cut at section A-A

opposite the rivet in the plate. The net cross-sectional areas of plates are stressed in
tension or compression, as portrayed in Figure 11.50d. The shaded area in the sketch is
the critical area, or net area, in tension.

The formula for the net area, with two holes in the plate, is as follows:

$$A = (W - 2D)t.$$

Therefore, in order to analyze a riveted joint, we must compare the stresses and forces appearing on

— The circular cross section of each rivet
— The projected plane area t × d of the rivet and
— The net cross section of the main plate

$$\text{eff(in\%)} = \frac{\text{safe joint load}}{\text{safe gross plate load}} \times 100$$

The efficiency of a riveted joint is equal to 100 times its safe load divided by the tensile (compressive) load the gross plate area will take. A riveted joint can never be 100% efficient because of the holes in the plates. The following assumptions and facts apply to riveted joint problems.

1. A basic assumption is that each rivet takes its fair share of the load as determined by the number of shear areas. Experiments have proven the soundness of this assumption.

2. To prevent the possibility of tear out and to simplify calculations, rivet holes should not be placed closer to the end of the plate than about two times the rivet diameter. *Tear out* occurs when the plate material ruptures in a direction toward the end of the plate instead of across the plate.

3. Boiler and pressure vessel fabrication is done in a shop where the rivet holes are drilled and matched and the rivets are set by machine. Because of these methods and close control, the rivet hole is taken as the same size as the rivet when making calculations.

4. Structural members may have their rivet holes punched out; this step weakens the material slightly around the hole. Also, conditions cannot be controlled as closely in the field as in the shop. For these reasons, it is standard practice to add 1/16 in. to the hole diameter to obtain the ''effective'' hole diameter when solving structural rivet problems. However, we will **not** do this for problems in this text. In order that you may concentrate on the basics, consider all hole or rivet diameters to be effective hole diameters.

5. There are concentrated stresses in the plate area around the rivet holes. However, specifications and codes frequently allow the technician or engineer to avoid this computation by lowering the allowable stress in this plate area. For the purpose of simplification, this text will use the same stress on plate cross sections with or without rivet holes.

PROCEDURE

Solving Riveted Lap Joint Problems

Ideally, the three critical areas (net plate area, rivet shear area, and rivet bearing area) should all be stressed to their maximum safe values. But the use of standard sizes and materials means that these values probably will not be reached. Therefore, you must determine the safe load a joint can take when each critical area is stressed to its allowable value. Compare the loads you have calculated, and choose the lowest value as the maximum safe load for the joint.

Step 1   Refer to the free-body diagrams in Figure 11.50 for each of the steps below where a critical area must be determined.

Step 2   Determine the number of shear areas on the rivets, and analyze for stress or force as specified.

Step 3   Determine the number of bearing areas on the rivets, and analyze.

Step 4   Determine the net cross-sectional area of the smallest plate, and then solve for the stress or force as called for in the problem. If there is more than one row of rivets, the net cross-sectional area for each row may have to be analyzed.

Step 5   If the problem is to determine the load the joint can take, then the smallest of the loads calculated above is the answer.

Sample Problem 11.6 illustrates a riveted lap joint.

SAMPLE PROBLEM 11.6

Riveted Lap Joint

PROBLEM: Given the following sketch and facts, determine (**a**) the allowable load P and (**b**) the efficiency of the joint,

where   plates = 1/2 in. thick

rivets = 3/4 in. in diameter

allowable S_t = 20,000 pounds per inch^2

allowable S_c = 32,000 pounds per inch^2

allowable τ = 15,000 pounds per inch^2

Solution (a)

Step 1   Refer to the free-body diagram.

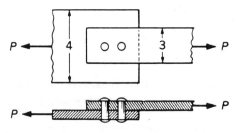

FIGURE 11.51   Sketch for Sample Problem 11.6

Step 2 The load as determined by shear areas is

$$P = \tau A = 15\,000 \times \frac{\pi(3/4)^2 \times 2}{4} = 13{,}250 \text{ lb}$$

Step 3 The load as determined by bearing areas is

$$P = S_c A = 32{,}000 \times 2 \times \frac{3}{4} \times \frac{1}{2} = 24{,}000 \text{ lb}$$

Step 4 The load as determined by tension area is

$$P = S_t A = 20{,}000 \times \left(3 - \frac{3}{4}\right) \times \frac{1}{2} = 22{,}500 \text{ lb}$$

Step 5 Thus, the allowable load is

$$P = 13{,}250 \text{ lb}$$

Solution (b)

The 3 in. plate with no holes can take a tensile load of

$$S_t A = 20{,}000 \times 3 \times \frac{1}{2} = 30{,}000 \text{ lb}$$

Thus, the efficiency of the joint is

$$\text{efficiency} = \frac{13{,}250 \times 100}{30{,}000} = 44\,\%$$

PROCEDURE

Solving Riveted Butt Joint Problems

The steps for solving riveted butt joint problems are essentially the same as for lap joints, with the added condition that the cover plates must be considered. If a free-body diagram is drawn for each half of a butt joint, you can see that the loads on each half are identical and that half of the joint is resisting the other half. Therefore, only half of the joint need be considered. Also, if 2 cover plates are used, the rivets will each have 2 shear areas.

11.11 WELDED JOINTS

Welding is less expensive than riveting as a means of joining parts. In addition, from an appearance point of view, welding is superior to riveting because welding can result in an invisible joint. Since arc welding is the most popular type, it is the only one which will be discussed here.

Arc welding is the joining of two metallic parts by using a sustained electric arc formed between the work and a metal rod electrode. The high temperature developed melts the metal of the work and the tip of the electrode, causing metal from the electrode to cross the arc to the weld joint.

Today, many machine frames are fabricated using weldments (an assemblage of welded pieces) rather than by casting. This eliminates the need for patterns, molds, and many finishing processes. Instead, components such as plates, channels, angles, and rods are cut to the correct length and welded together to produce the desired frame. Advantages of weldments over castings are lighter weight and lower costs. In addition, once the welds have been lightly ground, it is usually easier to apply protective coatings such as paint for corrosion resistance and appearance purposes. Figure 11.52 shows the most common types of welded joints.

The strength of a welded joint depends on the properties of the parent (base) metal as well as on the weld metal and the geometry of the weld. The most popular type of arc weld is the *fillet weld,* because it is easy to apply, as well as to design. A strength analysis of a fillet weld will therefore be made by referring to Figure 11.53 which shows a fillet weld absorbing a load P. The minimum weld area (A_{min}) resisting the load depends on the throat of the weld, which is shown in Figure 11.53b. Since a fillet weld cross section is essentially a 45° triangle, we have the following relationship for A_{min}:

$$A_{min} = WL \cos 45° = 0.707WL \qquad (11.24)$$

where W = weld leg size in inches

   L = total weld length in inches

Standard weld sizes are 1/8, 3/16, 1/4, 5/16, 3/8, 1/2, 5/8, 3/4, 7/8 inches and 1 inch. The weld size used depends on the plate thickness. Normally, the weld leg size equals approximately 75% of the plate thickness.

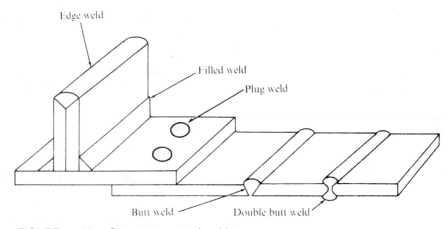

FIGURE 11.52   Common types of welds

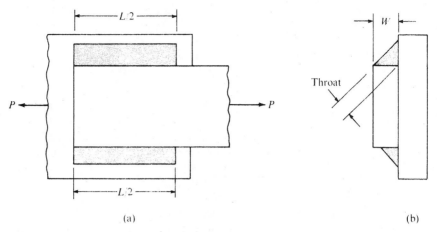

FIGURE 11.53   Loaded fillet weld joint

Virtually all fillet welds are subjected to shear stresses. The shear stress τ equals the shear force divided by the minimum shear area. Thus we have

$$\tau = \frac{P}{0.707WL} \tag{11.25}$$

The allowable shear stress for steel weldments is 14,000 lb/in^2 for static loads and 5000 lb/in.^{2} for dynamic loads. Sample Problems 11.7 and 11.8 show the principles involved in the strength analysis of welded joints.

SAMPLE PROBLEM 11.7

Welded Joint

PROBLEM: Two plates are joined by three segments of 1/2 in. fillet welds as illustrated in Figure 11.54. What allowable static load P can be sustained?

Solution

From Equation 11.27, we have

$$P = 0.707WL\tau$$

$$= (0.707)\left(\frac{1}{2}\right)(12)(14,000) = 59,400 \text{ lb}$$

SAMPLE PROBLEM 11.8

Welded Joint With Eccentric Load

PROBLEM: A 4 × 3 × 5/8 in. standard steel angle is fillet-welded to a plate using a 1/2 in. size weld as shown in Figure 11.55. Find the required weld lengths L_1 and L_2, assuming a static load.

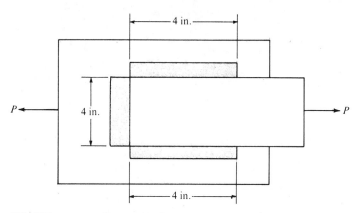

FIGURE 11.54   Sketch for Sample Problem 11.7

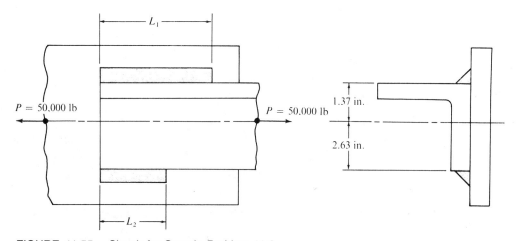

FIGURE 11.55   Sketch for Sample Problem 11.8

Solution

In order to prevent twisting of the two parts, the load should be applied at the centroid of the angle iron as shown. Also, the weld lengths L_1 and L_2 should be such that the centroid of the original angle iron is not changed. Thus we have the following mathematical relationship:

$$\frac{L_1}{L_2} = \frac{2.63}{1.37} = 1.92$$

Also, from Equation 11.27, we have

$$L_1 + L_2 = \frac{P}{0.707W_\tau} = \frac{50,000}{(0.707)(0.5)(14,000)} = 10.1 \text{ in.}$$

Solving the preceding two equations simultaneously yields the following:

$$L_1 = 6.64 \text{ in.}$$

$$L_2 = 3.46 \text{ in.}$$

TABLE 11.4   International Standard (Metric) Threads (*From Kent's Mechanical Engineers Handbook, 12th Edition, Published by J. Wiley & Sons Inc.*)
(Dimensions in millimeters)

Outside Diameter	Pitch	Nut Minor Diameter	Outside Diameter	Pitch	Nut Minor Diameter	Outside Diameter	Pitch	Nut Minor Diameter
6	1.0	4.70	22	2.5	18.75	48	5.0	41.50
7	1.0	5.70	24	3.0	20.10	52	5.0	45.50
8	1.25	6.38	27	3.0	23.10	56	5.5	48.86
9	1.25	7.38	30	3.5	25.45	60	5.5	52.86
10	1.5	8.05	33	3.5	28.45	64	6.0	56.21
12	1.75	9.73	36	4.0	30.80	68	6.0	60.21
14	2.0	11.40	39	4.0	33.80	72	6.0	64.21
16	2.0	13.40	42	4.5	36.15	76	6.0	68.21
18	2.5	14.75	45	4.5	39.15	80	6.0	72.21
20	2.5	16.75						

11.12 SI UNIT FASTENERS

Metric threads are manufactured with a 60° included angle as is the Unified Screw Thread (see Figure 11.2). Basic sizes of metric threads are listed in Table 11.4. The standard designation for metric fasteners is as the follows M6 × 1 - 6H. The M is the thread system symbol for metric; the 6 indicates the nominal size of the fastener—6mm in diameter; the 1 represents a pitch of 1 mm; and 6H refers to the tolerance class.

We will need to find the root diameter (minor diameter) of a fastener. The thread depth = 0.7 × p where p is the pitch. Therefore the minor diameter = the nominal diameter − (2 × 0.7 × pitch), or

$$d = D - (2 \times .7p) \tag{11.26}$$

SAMPLE PROBLEM 11.9

Metric Fastener

PROBLEM: Refer to Figure 11.37. An M20 × 2.5 fully threaded bolt is used to hold the plates together. The allowable shear stress for the fastener is 35 MPa. What allowable load can be placed on the plates if the bolt is most likely to fail in shear stress?

Solution.

We need to find the root area of the bolt.

$$d = D - 2 \times 0.7p = 20 - 2 \times .7 \times 2.5 = 16.5 \text{ mm}$$

$$A = \frac{\pi d^2}{4} = \frac{3.14 \times 0.0165^2}{4} = 213.7(10)^{-6} \text{ m}^2$$

$$F = SA = 35(10)^6 \times 213.7(10)^{-6} = 7480 \text{ N}$$

11.13   SUMMARY

This chapter discusses a large variety of removable and semipermanent fasteners and power screws. Fasteners are an important part of almost all manufactured products.

The most widely used type of fastener is the screw thread configuration. Screw thread terminology includes the terms major diameter, minor (root) diameter, pitch, pitch diameter, lead, thread angle, and crest. The most popular standardized thread sizes are specified by the Unified Thread Standard. Standards for metric threaded fasteners have not yet become common.

Some common screw fasteners discussed are the nut-and-bolt, stud, cap screw, machine screw, carriage bolt, eyebolt, U-bolt, turnbuckle, and setscrew.

Torsion formulas for power screws were developed. Inspection of the formulas indicated that the efficiency of a power screw depends on the helix angle and the coefficient of friction. The efficiency of a power screw is the ratio between the torque required without friction and the torque required with friction.

Other fastener devices mentioned in this chapter are washers, retaining rings, pins (dowels, tapers, and groove), acorn nuts, wing nuts, speed nuts, threaded inserts called Heli-Coil inserts, and permanent fasteners such as rivets and welds.

11.14   QUESTIONS AND PROBLEMS

Questions

1. Briefly discuss the three main classifications of fasteners.
2. Name six factors which should be taken into account when selecting fasteners.
3. Relative to screw threads, what is the difference between pitch and lead?
4. Relative to screw threads, what does the following designation denote: 1/4-20UNC-2B?
5. What is the difference between a bolt and a stud?
6. What is a turnbuckle? Name one application.
7. What is a setscrew? Name one application.
8. What is the purpose of a power screw?
9. The efficiency of a power screw depends on what two parameters?

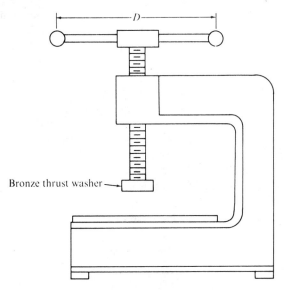

FIGURE 11.56   Sketch for Problem 8

10. What advantage does a square thread screw have over an Acme lead screw in power applications?
11. Give two purposes for using washers with bolted fasteners.
12. Show by sketches an application for an external and an internal retaining ring. What advantage does a retaining ring offer as compared to collars with setscrews?
13. What is a dowel pin? When would it be used?
14. What is a speed nut? What advantage does it offer?

Problems

1. Solve Sample Problem 11.1 except change the Acme threads to square threads.
2. What is the overall efficiency of Problem 1?
3. A triple-threaded Acme power screw with a 2 1/2 in. pitch diameter and 2 threads/in. is used with a collar having a 4 1/2 in. outer diameter and a 3 in. inner diameter. The coefficient of friction is 0.15 for both the threads and the collar. Find the force required to raise a 2500 lb load if the force is applied at a radius of 24 in.
4. What would be the answer to Problem 3 if a single-threaded screw were used?
5. Find the efficiency for Problem 3 if (**a**) collar friction is ignored; (**b**) collar friction is taken into account.
6. Find the efficiency for Problem 4 if (**a**) collar friction is ignored; (**b**) collar friction is taken into account.
7. A small screw press, like that shown in Figure 11.56, is to be designed for a 4500 lb capacity.

(a) Select a steel screw with Acme threads. The screw has an outside diameter of 1 inch (obtain other data from Table 11.2 and Figure 11.32). The bronze thrust washer has an outside bearing diameter of 0.75 in. and an inside diameter of 0.25 in.

(b) Find the torque required to raise a 4500 lb load. The coefficient of friction is 0.10 for both the threads and thrust washer.

(c) What handwheel diameter D is required if the operator is to exert a 45 lb force with each hand to develop full output capacity?

8. Develop an equation for the torque required to lower a load on a square thread screw. Your equation will be similar to Equation 11.1.

9. (a) Solve Sample Problem 11.3 if the load is 4000 lb.

   (b) Choose a suitable bolt from Table 11.3. (Bolts are fully threaded, shear stress = 14,400 psi

10. Find the load P that can be carried by the bolted fastener of Figure 11.57 if the allowable shear stress is 10,000 psi and the bolts are 7/8-9 UNC (see Table 11.3). Bolts are in shear at the shank area.

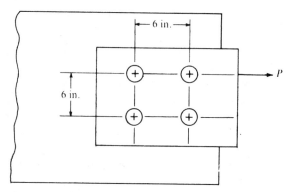

FIGURE 11.57   Sketch for Problem 10

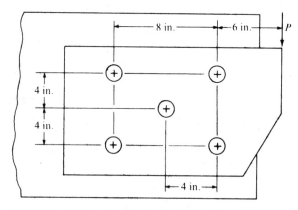

FIGURE 11.58   Sketch for Problem 11

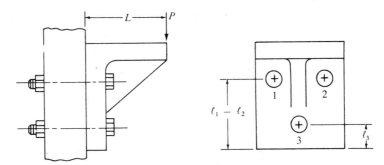

FIGURE 11.59   Sketch for Problem 12

11.   For the bolted fastener of Figure 11.58, find the design load P assuming a maximum load of 5000 lb for any one bolt.
12.   For the bolted bracket of Figure 11.59, $P = 1000$ lb, $L = 60$ in., $l_1 = l_2 = 20$ in., $l_3 = 10$ in. The allowable tensile stress $= 18,000$ psi. and the allowable shear stress $= 6,000$ psi.
    (a) Choose a bolt from Table 11.3.
    (b) Find the required nut thickness.
13.   A 3/4-10 UNC nut and bolt may be torqued to a shear stress of 6000 psi. For an unlubricated condition,
    (a) What tensile force can be developed?
    (b) What is the tensile stress developed?
14.   In an effort to increase the efficiency of Sample Problem 11.6, the rivets are changed to $\frac{7}{8}$ in. in diameter. Find (a) the allowable load P and (b) the efficiency of the joint.
15.   The butt joint in Figure 11.47b has all of the plates 4 in. wide and 3/8 in. thick. The rivets and holes are 3/4 in. in diameter. Find (a) the allowable load for the joint and (b) the efficiency. Use the allowable stresses in Sample Problem 11.6.
16.   For the welded joint of Figure 11.60, what maximum dynamic load P can be sustained?

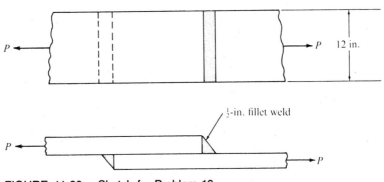

FIGURE 11.60   Sketch for Problem 16

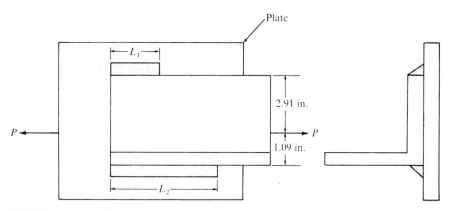

FIGURE 11.61 Sketch for Problem 17

17. Figure 11.61 shows a 4 in. $\times$ 4 in. $\times$ 1/2 in. angle welded to a 1/2 in. plate, using 1/2 in. fillet welds. A load P of 30,000 lb is applied through the centroid of the angle as shown. Find the required lengths of weld L_1 and L_2 for a static load.

SI Unit Problems

18. Solve Sample Problem 11.9, except use an M33 by 3.5 bolt.
19. Refer to Figure 11.57. Change the 6 in. dimensions to 150 mm each. M16 by 2.0 bolts are used. The bolts are fully threaded. The allowable shear stress = 70 MPa. Find load P.
20. Two steel plates, each 15 mm thick and 90 mm wide, are lap-joined with one row of two rivets (see Fig. 11.47). Rivets and rivet holes are 20 mm in diameter; allowable tensile stress = 150 MPa; allowable compressive stress = 600 MPa; and the allowable shear stress = 99 MPa. Determine the load the joint can take.
21. Refer to Figure 11.53. A steel bar is lap-welded to a plate. The bar is 40 mm wide by 6 mm thick. The two fillet welds are 45 mm long and the weld leg is 6 mm. If the allowable shear stress on the weld = 99 MPa, what load P can the welds take?

12

Spring Design

Objectives

This chapter describes various types of springs and their applications, develops the design equations for helical coil and leaf springs and lists a number of factors to consider in spring design. On completing this chapter, you will be able to

- Decide on the most appropriate type of spring for a given application.
- Choose a satisfactory spring material.
- Analyze and design coil and leaf springs.

12.1   INTRODUCTION

Springs are mechanical members which are designed to give a relatively large amount of elastic deflection under the action of an externally applied load. **Hooke's Law,** which states that deflection is proportional to load, is the basis for the behavior of springs. However, some springs are designed to produce a nonlinear relationship between load and deflection. The following is a list of the important purposes and applications of springs.

Control of Motion in Machines.   This category represents the majority of applications, such as providing operating forces in clutches and brakes. Also, springs are used to maintain contact between two members such as a cam and its follower. Keys in typewriters are returned to their normal positions by the use of springs.

Reduction of Transmitted Forces as a Result of Impact or Shock Loading. Applications here include automotive suspension system springs and bumper springs. Most elevator chutes contain springs to protect passengers if the cable breaks and the elevator crashes into the bottom of the chute.

Storage of Energy.   Applications in this category are found in clocks, movie cameras, and lawn mowers having recoil starters. Many parking meters use spring mechanisms that perform the timing function. Other common applications are the take-up reels for lubrication hoses in gasoline service stations and for the electrical cords in home vacuum cleaners.

Measurement of Force.   Scales used to weigh people is a very common application for this category.

Most springs are made of steel, although phosphor bronze, silicon bronze, brass, monel, inconel, and beryllium copper are also used. Springs are universally made by companies that specialize in their manufacture. The helical coil spring is the most popular type of spring; torsion bars and leaf springs are also widely used. If the wire diameter (assuming a helical coil spring) is less than about 5/16 in., the spring will normally be cold-wound from hard-drawn or oil-tempered wire. For larger diameters, springs are formed using hot-rolled bar. Figure 12.1 illustrates a finished hot-wound spring (still red-hot, right off the mandrel) ready for submerging into a quenching bath.

It is good practice to consult with a spring manufacturing company when selecting a spring design, especially if high loads, high temperatures, or stress reversals are anticipated or if corrosion resistance is required. To properly select a spring, a complete study of the spring requirements, including space limitations, must be undertaken. Many different types of special springs are available to satisfy unusual requirements or applications. These include Belleville springs, volute springs, constant force springs, power springs, and garter springs. Spring wires and flat strips come in standard sizes in both the U.S. customary units (inches) and metric units (millimeters). Because of space limitations, we will not discuss standard sizes in either category. As a note of interest, though, standard practice in industry at the present time is to make all calculations in U.S. customary units and then convert the dimensions to the nearest standard metric dimensions.

FIGURE 12.1 Finish hot wound spring. *(Courtesy of Associated Spring Corp., Bristol, Connecticut)*

DifFerenc – distance between wraps

12.2 COMPRESSION AND EXTENSION COIL SPRINGS

The helical coil spring is the most popular type of spring. Figure 12.2 shows the configuration of a compression-type coil spring having four different types of ends. In each case, when there is no load, the coils are separated. As the load is applied, the coils move closer together, but do not touch. It is desirable that compression coil springs have as much contact as possible with the mating parts at the ends of the springs. The four types of ends used on compression coil springs are described as follows:

Plain End (Figure 12.2a). The *plain-end* coil spring is the least expensive to produce, but the spring tends to bow sideways under load, resulting in increased stresses on one side of the spring because the load is not transmitted in an axial direction.

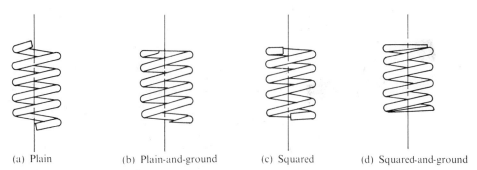

(a) Plain   (b) Plain-and-ground   (c) Squared   (d) Squared-and-ground

FIGURE 12.2 Types of ends for helical compression springs

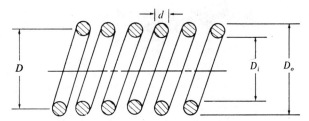

FIGURE 12.3   Sectional view of compression coil spring.

Plain-and-Ground End (Figure 12.2b).   The *plain-and-ground-end* coil spring will seat against its mating machinery better than the plain end coil spring. However, during storage, handling, and assembly, the ends tend to become tangled. This tangling leads to time-consuming sorting and sometimes even distortion of critical shape dimensions.

Squared End (Figure 12.2c).   The *squared-end* coil spring does not tend to become tangled during the manufacturing process as readily as the plain-and-ground end.

Squared-and-Ground End (Figure 12.2d).   The squared-and-ground-end coil spring enjoys the advantage of not becoming readily tangled during manufacturing; in use the load is transmitted in a perfectly axial direction (no sideways bowing).

Regardless of the type of end used, a partially dead or inactive coil exists at each end of a compression spring. This inactive coil must be subtracted from the actual total number of coils to find the number of active coils. The following is an approximate rule for finding the active number of coils:

—   Plain end - subtract a total of ½ turn
—   Plain-and-ground end - subtract a total of 1 turn
—   Squared end - subtract a total of 1½ turns
—   Squared-and-ground end - subtract a total of 2 turns

Figure 12.3 is a sectional view of a compression coil spring, showing its nomenclature. All measurements are in inches.

The following is a list of spring terms which apply to helical coil springs in general.

Outside Diameter D_o.   The *outside diameter* D_o is the maximum diameter of the coil spring. This parameter is important when radial space is restricted.

Inside Diameter D_i.   The *inside diameter* D_i is the minimum diameter of the coil spring. This parameter is of concern when a part such as a pin must fit inside the spring.

Mean Diameter D.   The *mean diameter* D is the average diameter of the coil spring. Stress and deflection analysis depend on the mean diameter.

$$D_o - D_i = 2d \qquad d = \frac{D_o - D_i}{2}$$

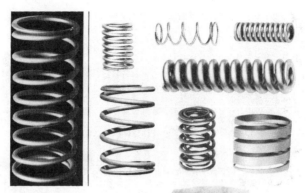

FIGURE 12.4   Miscellaneous compression coil springs. *(Courtesy of Associated Spring Corp., Bristol, Connecticut)*

Wire Diameter *d*.   This parameter *d* refers to the diameter of the wire that is wound to create the coils.

Free Height L.   The term *free height L* refers to the overall length of the spring in the free or unloaded configuration.

Solid Height L_0.   The term *solid height L_0* refers to the length of the spring when all the coils are touching.

Deflection δ.   The deflection δ refers to the change in length of a spring resulting from an applied load.

Direction of Coiling.   This term refers to the direction in which the coils are wound: either right-hand or left-hand. The springs of Figure 12.2 are coiled right-hand.

Spring Index *C*.   The *spring index C* refers to the ratio of the coil mean diameter to the coil wire diameter: *D/d*.

Active Number of Turns *n*.   This term refers to the effective number of coil turns supporting the external load. It equals the total number of actual turns minus the inactive turns at the ends.

Figure 12.4 is a photograph that shows various types of compression coil springs. Notice that one of the springs has a rectangular-wire cross section. When space is limited radially or when the maximum solid height is restricted, the use of square or rectangular cross sections permits greater energy storage in less space. However, the use of square or rectangular wire should be avoided when possible because of cost.

Figure 12.5 is a photograph of various types of extension coil springs, which are characterized by the loop at each end. Observe the different variations in the design of the end loops. Notice that the coils touch each other until an axial load is applied; thus extension coil springs exert a tensile force on mating parts.

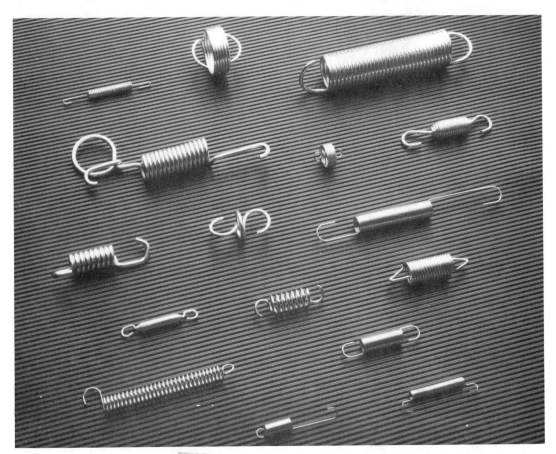

FIGURE 12.5   Miscellaneous extension coil springs *(Courtesy of Hardware Products Co., Inc., Chelsea, Massachusetts)*

12.3   STRESSES IN HELICAL COIL SPRINGS WITH CIRCULAR-WIRE CROSS SECTIONS ▬▬▬

A helical coil spring experiences a shear stress when acted upon by an external load. The magnitude of stress can be determined by referring to Figure 12.6, which depicts a free-body diagram of a compression coil spring undergoing an axial load of magnitude P. Notice that, in order to have equilibrium along the axis of the spring, the wire cross section must experience a transverse shear force P. Rotational equilibrium requires that the wire cross section absorb a torque T of magnitude $P(D/2)$. This torque induces a torsional shear stress.

Let us now evaluate the maximum resultant shear stress, τ_{max}, which is the vector sum of the transverse and torsional shear stresses:

$$\tau_{max} = \text{transverse shear stress} + \text{torsional shear stress}.$$

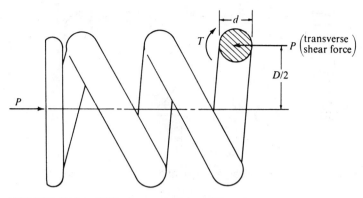

FIGURE 12.6   FBD of compression coil spring

$$\tau_{max} = \frac{P}{A} + \frac{Tc}{J}$$

where A = the wire cross-sectional area in inches^2
             ($\pi \ (d^2/4)$ for circular area)

c = the radius of wire cross section in inches (d/2)

J = the polar moment of inertia of wire cross section in inches^4 = ($\pi \times$ $d^4/32$)

Thus,

$$\tau_{max} = \frac{P}{\pi d^2/4} + \frac{P(D/2) \ (d/2)}{\pi d^4/32} = \frac{4P}{\pi d^2} + \frac{8PD}{\pi d^3}$$

This equation is represented graphically in Figure 12.7, which shows the stresses superimposed on an enlarged view of the wire cross section. Note that the maximum resultant shear stress occurs at the inner surface of the wire cross section. This is where the transverse and torsional shear stresses are additive. However, in most spring designs, the transverse shear stress is very small compared to the torsional shear stress. So the transverse shear stress part of the equation is dropped during initial design calculations. This gives us

$$\tau_{max} = \frac{8PD}{\pi d^3} \qquad\qquad\qquad\qquad (12.1)$$

Let's just analyze this equation for a moment. We note that the shear stress is directly proportional to the load P and to the coil diameter D. The stress is inversely proportional to the cube of the wire diameter. This means that for small percentage changes in wire diameter there is a tripling effect on the stress. For example, a 5% decrease in the wire diameter results in the stress increasing by almost 15%.

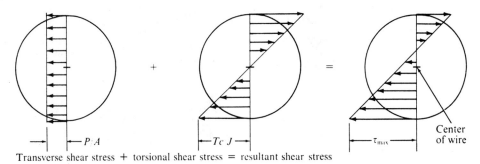

Transverse shear stress + torsional shear stress = resultant shear stress

FIGURE 12.7   Graphical representation of coil shear stress

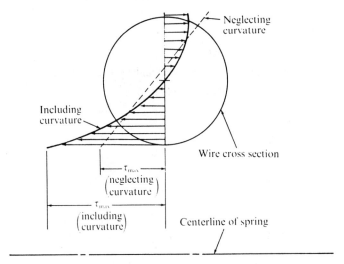

FIGURE 12.8   Effect of coil curvature on shear stress

There is another factor to consider before we complete stress equation for springs. The torsional stress equation is based on the strain developed in a straight member rather than a coiled one. This strain is discussed again in Section 12.5 (and illustrated in Figure 12.12) in relation to deflection. Strain is affected by length: the inside length of the wire in a coil is shorter than the outside length. This fact means that although the deformations may be the same on the inside and outside of the cross section of the wire the strains and stresses will not be the same. This inequality is shown graphically in Figure 12.8.

A stress factor, called the Wahl factor (symbol—K), was developed by Mr. A. M. Wahl to account for this curvature. Arthur M. Wahl was a research engineer for Westinghouse Electric Co. and authored a text entitled *Mechanical Springs* in 1944. This stress factor also accounts for the transverse shear stress, so that Equation 12.1 can be used with just this factor added. The formula for the Wahl factor is given below. Figure 12.9 shows the Wahl Factor K plotted against the Spring Index C.

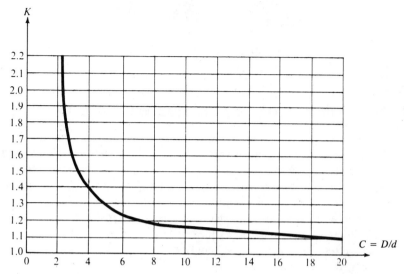

FIGURE 12.9   Wahl factor versus spring index

$$K = \frac{(4C - 1)}{(4C - 4)} + \frac{0.615}{C}$$

As can be seen from Figure 12.9, the value of K becomes significant when the spring index C is less than 18. This is because the curvature effect becomes appreciable.

Incorporating the Wahl factor, Equation 12.1 is modified to read

$$\tau_{max} = \frac{8KPD}{\pi d^3} \qquad\qquad (12.2)$$

Equation 12.2 is the one in general use because it has been demonstrated to give values accurate within 2% of more exact equations for spring indexes of 3 or greater.

SAMPLE PROBLEM 12.1

Helical Coil Spring Stress

PROBLEM: A helical coil spring has a 1 in. diameter wire and a mean coil diameter of 4 in. The spring supports a 2000 lb load. Find the maximum shear stress using Equation 12.2.

Solution

$$C = \frac{D}{d} = \frac{4}{1} = 4$$

From Figure 12.9,

$$K = 1.4$$

$$\tau_{max} = \frac{8KPD}{\pi d^3} = \frac{8 \times 1.4 \times 2000 \times 4}{3.14 \times 1^3}$$

$$\tau_{max} = 28{,}500 \text{ lb/in.}^2$$

12.4 SPRING MATERIALS AND ALLOWABLE STRESSES ▬▬▬

The selection of a suitable spring material is just as important a consideration as the design of the spring. A wide variety of spring materials are available to meet strength and cost requirements. Table 12.1 shows typical properties of common spring materials.

The physical characteristics of any given spring material may be enhanced by cold drawing (also called cold working), or heat treating. Most metals can be hardened and strengthened by cold drawing. However, excessive residual stresses may be created be the cold-drawing process and by the bending of the material when forming the springs. These residual stresses can be removed by a process called stress relieving. In this process, the material is heated to 400–800° F and held there for a prescribed period of time. This not only removes the excessive residual stresses but stabilizes the springs geometrically. High-carbon and alloy steels may be hardened by a heat treating process that includes heating to a temperature in the range of 1500° F, quenching, and then tempering to remove residual stresses. Some materials such as beryllium copper and 17-7 PH stainless steel are *age hardenable* (or *precipitation hardenable*). This term means that the materials can be hardened simply by allowing them to stay at normal ambient temperatures for a specified period of time. Music wire (ASTM A228) is one of the hardest and strongest materials used for springs.

Figure 12.10 provides minimum ultimate tensile strength values of various spring materials. Note that the tensile strength levels vary with wire diameter. This is usually because cold working and heat treating are less effective on the larger diameters. Table 12.1 has information for converting the minimum tensile stress to an allowable shear stress for the various spring materials listed.

The surface treatment of a spring can greatly increase the fatigue strength. One such surface treatment is *shot peening,* in which small steel shot is propelled at the spring at high velocity. This produces very tiny dents in the spring surface, in effect smoothing out larger surface defects such as tiny cracks, pits, and notches. The result is a very thin layer of surface with increased strength. Although shot peening tends to decrease the corrosion resistance of the wire, subsequent surface treatment such as plating provides good corrosion resistance. Zinc or cadmium plate is commonly used. In addition to plating, springs can be protected with many types of coatings, such as paint, phosphate, oil, grease, wax, or plastic materials.

12.5 DEFLECTION AND DESIGN OF COIL SPRINGS ▬▬▬

The deflection δ of a compression coil spring is shown in Figure 12.11, where the springs are represented schematically. With no load, the length of the spring equals its free height L. Under a load P the spring length foreshortens by an amount δ.

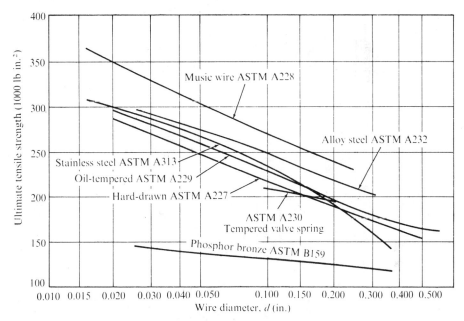

FIGURE 12.10   Minimum tensile strength of spring wires *(Courtesy of Associated Spring Corp., Bristol, Connecticut)*

An equation defining δ as a function of spring parameters can be derived by noting that a coil spring (under axial load) deforms because of twisting of the wire cross section. Thus we can visually unwrap a coil spring having n turns. This produces a shaft (torsion bar) absorbing a torque as illustrated in Figure 12.12. The length L of the shaft can be found from

$$L = \pi D n$$

where n = the number of coils

From Figure 12.6, it was previously determined that the torque absorbed by the wire of a coil spring equals the product of the load P and the mean radius of the coil $D/2$.

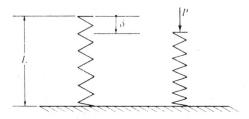

FIGURE 12.11   Deflection of compression coil spring

TABLE 12.1 Typical Properties of Common Spring Materials (Courtesy of Associated Spring Corp., Bristol, Connecticut)

Common Name Specification	E (10^6 lb/in.^{2})	G (10^6 lb/in^2)	Density, ρ (lb/in^3)	Sizes Normally Available, (in.) Min	Max	Fatigue Applications*	Allowable Shear Stress as a % of Minimum Tensile Strength†	Max Service Temp (°F)	°C
High-carbon steel wires									
Music ASTM A228	30	11.5	0.284	0.004	0.025	E	45	250	121
Hard-drawn ASTM A227	30	11.5	0.284	0.028	0.625	P	40	250	121
Oil-tempered ASTM A229	30	11.5	0.284	0.020	0.625	P	45	250	121
Valve-spring ASTM A230	30	11.5	0.284	0.050	0.250	E	45	250	121
Alloy-steel wires									
Chrome-vanadium AISI 6150, ASTM A231	30	11.5	0.284	0.032	0.438	E	45	425	218.5
Chrome-silicon AISI 9254, ASTM A401	30	11.5	0.284	0.035	0.375	F	45	475	246

Material								
Stainless-steel wires								
Austenitic								
AISI 302, ASTM A313	28	10	0.286	0.005	0.375	G	30–40	550	288
Precipitation-hardening								
17–7 PH	29.5	11	0.286	0.030	0.500	G	45	650	343
Nickel-chrome								
A286	29	10.4	0.290	0.016	0.200	—	35	950	510
Copper-base alloy wires								
Phosphor-bronze								
ASTM B159	15	6.3	0.320	0.004	0.500	G	40	200	93.3
Beryllium-copper								
ASTM B197	18.5	7.0	0.297	0.003	0.500	E	45	400	204

Note: E and G can vary with cold work, heat treatment and operating stress.

*E = Excellent; G = Good; F = fair; P = poor.

†The allowable bending stress (for torsion & flat springs) is taken as 75% of minimum tensile strength

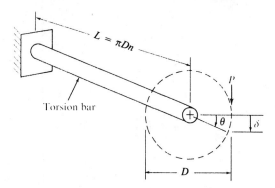

FIGURE 12.12   Conversion of coil spring into a torsion
bar

Therefore P acts with a moment arm value of $D/2$, as shown in Figure 12.12. As the shaft twists through an angle θ, P moves through a distance δ:

$$\delta = \frac{D}{2}\theta$$

θ equals the angle of twist in units of radians. The above equation is simply the geometric relationship that states: The arc length equals the radius times the central angle in radians. Since $\theta = TL/GJ$ (from shaft theory), we have upon substitution

$$\delta = \frac{D}{2}\left(\frac{TL}{GJ}\right)$$

where $T =$ the torque in pound-inches

             $L =$ the straight length of the spring wire in inches

             $G =$ the modulus of rigidity in pounds per inch^2

             $J =$ the polar moment of inertia of the wire in inches^4

In order to put this equation in its final form, the following relationships will be used: $T = P(D/2)$, $L = \pi Dn$ and $J = \pi d^4/32$. Substituting and rearranging:

$$\delta = \frac{8PD^3n}{Gd^4} \qquad\qquad\qquad (12.3)$$

Equation 12.3 is again rearranged to show that there is a linear relationship between load and deflection:

$$P = \left(\frac{Gd^4}{8D^3n}\right)\delta = k\delta$$

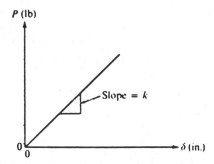

FIGURE 12.13   Linear relationship between P and δ

(Note that for a given spring, everything in parenthesis is a constant.)

Thus, we have

$$k = \frac{P}{\delta} = \frac{Gd^4}{8D^3n} = \frac{Gd}{8C^3n} \tag{12.4}$$

Specifically, k is a measure of the stiffness of a spring and is measured in pounds per inch. The terms *spring gradient* or *spring rate* are used interchangeably to identify the parameter k. You must guard against confusing the above terms and their symbol with the term spring index (symbol—C) and the symbol for the Wahl factor—K. Equation 12.4 indicates that for a given load, as the stiffness of the coil decreases, the number of coils increases.

Figure 12.13 shows the straight-line relationship of Equation 12.4. Note that the slope of the line equals the spring rate k. In some instances, the spring must operate between two loads, a maximum and some smaller load other than zero. In this case the loads must still be on the curve in Figure 12.13. The spring rate can be calculated by finding the difference between the two loads and dividing by the difference in deflection for the two loads, as indicated by the triangle in Figure 12.13.

SAMPLE PROBLEM 12.2

Spring Deflection

PROBLEM: A compression coil spring must foreshorten 5 in. under a 50 lb load. The following parameters are known:
spring index $C = 8$ and $G = 12 \times 10^6$ lb/in.^{2}

$$\tau_{max} = 80,000 \text{ lb/in.}^2$$

Find the number of active coils.

Solution

When looking at Equations 12.1 through 12.4, we see that we must know D and d. But the spring index $(C = 8 = D/d)$ tells us that $D = 8d$. Therefore, with proper substitutions in Equation 12.2, D and d can be found. From Figure 12.9, when $C = 8$, $K = 1.18$:

$$\tau_{max} = \frac{8KP(8d)}{\pi d^3}$$

$$d = \sqrt{\left(\frac{64 \times 1.18 \times 50}{3.14 \times 80,000}\right)}$$

$$d = 0.1226 \text{ in.} \quad D = 8d = 0.9808 \text{ in.}$$

We can now substitute data into Equation 12.3:

$$n = \frac{\delta G d^4}{8PD^3} = \frac{5 \times 12(10)^6 \times (0.1226)^4}{8 \times 50 \times (0.9808)^3}$$

$$n = 35.9 \text{ turns}$$

Compression springs can buckle when the free height is greater than four times the coil mean diameter. If the spring is properly guided, such as inside a tube or over a bar, the amount of buckling can be greatly reduced. However, the use of a guide may be undesirable because of the friction between the spring and guide.

So far, we have been analyzing the equations used in spring design. The actual design of a spring from scratch is somewhat of an educated trial-and-error approach. When in industry you will usually be given the load (or loads at various deflections), deflection, available space, types of ends, tolerances and environmental conditions or type of wire desired. One of the first calculations is for wire size. Once this is determined the designer must adjust his calculations for the nearest standard size available.

For our design purposes, we will deal only with round wire springs. When asked to design a spring you will be given load and deflection information, type of wire, and the limit for the coil outside diameter. You are to find the mean coil diameter, the wire diameter, spring rate, number of active coils, the actual shear stress, and the allowable shear stress.

PROCEDURE

Designing a Helical Coil Spring

Step 1   Go to Figure 12.10 and obtain the tensile stress for your material. Since the stress varies inversely with the wire size, select a stress value near the right end of the curve. Turn to the column relating to allowable shear stress in Table 12.1 and multiply the proper percentage value by the tensile stress to obtain the allowable shear stress.

Step 2 Rearrange Equation 12.1 to solve for the wire diameter. For this step, use the allowable shear stress and assume the coil mean diameter is the same as the maximum outside diameter allowed.

Step 3 With this wire diameter and a recalculated mean diameter, find the Wahl Factor and solve Equation 12.2 for the actual stress on this configuration. Also recalculate the allowable stress for this wire diameter and compare.

    (a) If the wire is stressed above the allowable you must adjust the wire diameter and/or the coil diameter and go through this step again.

    (b) If the actual wire stress is below the allowable and within 5% we will consider the problem solved. If not, repeat step three with an adjusted wire and/or coil size.

Step 4 Use Equation 12.3 or 12.4 to solve for the number of active coils.

Step 5 For your answer, state **(a)** mean coil diameter, **(b)** wire diameter, **(c)** actual maximum stress in the wire, **(d)** allowable stress, **(e)** spring rate, and **(f)** number of coils.

SAMPLE PROBLEM 12.3

Spring Design

PROBLEM: Design a spring to take a maximum load of 50 lb with a deflection of 0.435 in. The maximum outside diameter of the spring cannot exceed 0.925. Use oil tempered steel type ASTM A229.

Solution

Step 1 From Fig. 12.10 the min tensile stress = approx. 160,000 psi. From Table 12.1 the multiplying factor is 0.45.

$$\tau_{allow} = 0.45 \times 160,000 = 72,000 \text{ psi}$$

Step 2 Eq 12.1:

$$d = \sqrt[3]{\left(\frac{8PD}{\pi \times \tau}\right)} = \sqrt[3]{\left(\frac{8 \times 50 \times .925}{\pi \times 72,000}\right)}$$

$$d = 0.118 \text{ in.}$$

Step 3 *New D* = 0.925 − 0.118 = 0.807 in.

$$C = \frac{.807}{.118} = 6.84$$

From Fig. 12.9,

$$K = 1.20$$

Using Equation 12.2:

$$\tau_{actual} = \frac{8KPD}{\pi d^3} = \frac{8 \times 1.2 \times 50 \times .807}{\pi \times (.118)^3}$$

$$\tau_{act} = 75,000 \text{ psi}$$

The allowable shear stress for a diameter of 0.118 is

$$0.45 \times 220,000 = 99,000 \text{ psi}$$

Our actual stress is about 25% too low. If we decrease the wire diameter about 8% (3 × 8% = 24%) we can get closer to the allowable. However, in this case, a reduction in wire diameter increases the coil diameter if we maintain the outside diameter. Therefore we will drop about 7% to a wire diameter of 0.110. Then,

$$d = .110 \text{ in.}; D = .925 - .110 = .815 \text{ in.};$$

$$C = .815/.110 = 7.4; K = 1.23.$$

and

$$\tau_{act} = \frac{8 \times 1.23 \times 50 \times .815}{\pi \times (.110)^3} = 95,900 \text{ psi}$$

This is within 5% of the allowable stress of 99,000 psi.

Step 4 $k = \dfrac{P}{\delta} = \dfrac{50}{.435} = 115 \text{ lb/in.}$

From Eq. 12.4,

$$n = \frac{Gd}{8C^3k} = \frac{11.5(10)^6 \times .110}{8(7.4)^3 \times 115}$$

$$n = 3.4 \text{ coils}$$

Step 5   **(a)** $D = 0.815$ in.; **(b)** $d = 0.110$ in.; **(c)** $\tau_{act} = 95,900$ psi;

(d) $\tau_{allow} = 99,000$ *psi*, **(e)** spring rate $k = 115$ lb/in.; **(f)** 3.4 active coils.

12.6 SERIES AND PARALLEL ARRANGEMENTS ▬▬▬

Springs may be connected in series or parallel arrangements similar to that done with electrical components. A *series spring system* is one that has the following two characteristics:

1. Each spring absorbs an equal force.
2. The total system deflection equals the sum of the deflections of the individual springs.

Figure 12.14 shows a series spring system for which an applied force P_T produces a total system deflection δ_T.

The following derivation is presented to find the total equivalent spring rate k_T of a series system. From mechanics, we know that the applied force equals the force in each spring:

$$P_T = P_1 = P_2 = P_3$$

Also, the total deflection equals the sum of the individual deflections:

$$\delta_T = \delta_1 + \delta_2 + \delta_3$$

or

$$\delta_T = \frac{P_1}{k_1} + \frac{P_2}{k_2} + \frac{P_3}{k_3} = P_T \left(\frac{1}{k_1} + \frac{1}{k_2} + \frac{1}{k_3} \right)$$

The equivalent stiffness k_T of the system is then found:

$$k_T = \frac{P_T}{\delta_T} = \frac{1}{(1/k_1) + (1/k_2) + (1/k_3)}$$

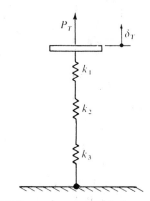

FIGURE 12.14 Series spring system

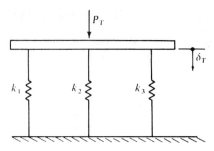

FIGURE 12.15   Parallel spring system

or, in a more common form:

$$\frac{1}{k_T} = \frac{1}{k_1} + \frac{1}{k_2} + \frac{1}{k_3}$$

(12.5)

Careful examination of Equation 12.5 reveals the following law: **The total equivalent spring rate of a series spring system is less than the smallest of the individual rates comprising the system.**

A *parallel spring system* is one that has the following two characteristics:

1. Each spring experiences an equal amount of deflection.
2. Each spring absorbs a portion of the applied load.

Figure 12.15 shows a parallel spring system which will be analyzed as follows.

The deflection is the same for each spring:

$$\delta_T = \delta_1 = \delta_2 = \delta_3$$

From mechanics, we know that the sum of each of the spring forces equals the applied load:

$$P_T = P_1 + P_2 + P_3$$

or

$$P_T = k_1\delta_1 + k_2\delta_2 + k_3\delta_3$$
$$P_T = \delta_T(k_1 + k_2 + k_3)$$

The total equivalent spring rate is found next:

$$k_T = \frac{P_T}{\delta_T} = k_1 + k_2 + k_3$$

(12.6)

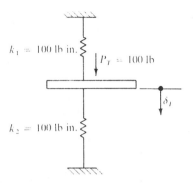

FIGURE 12.16   Sketch for Sample Problem 12.4

Therefore, we have a second law: **The total spring rate of a parallel system equals the sum of the rates of the individual springs of the parallel system.**

SAMPLE PROBLEM 12.4

Parallel Connected Springs

PROBLEM: Two springs are connected to a weightless central plate as shown in Figure 12.16. If a force P_T of 100 lb is applied as shown, how much will the plate move?

Solution

This problem can be confusing. On first glance, the springs appear to be in series. However, because of the way the load is applied, they are in parallel. The way to visualize this is to mentally remove the lower spring. The upper spring will now deflect 1 in. Adding the lower spring gives additional support to the load and decreases the deflection of the upper spring.

In a parallel system, each spring experiences an equal amount of deflection and therefore:

$$k_T = k_1 + k_2 = 100 + 100 = 200 \text{ lb/in.}$$

$$\delta_T = \frac{P_T}{k_T} = \frac{100 \text{ lb}}{200 \text{ lb/in.}} = 0.5 \text{ in.}$$

SAMPLE PROBLEM 12.5

Series and Parallel Systems

PROBLEM: Four identical springs are joined together as shown in Figure 12.17a. If P_T equals 50 lb, what is the system total deflection?

Solution

Since springs 1 and 2 are connected in parallel, they can be replaced by a spring whose rate is $k_1 + k_2$, or 100 lb/in. Thus an equivalent system

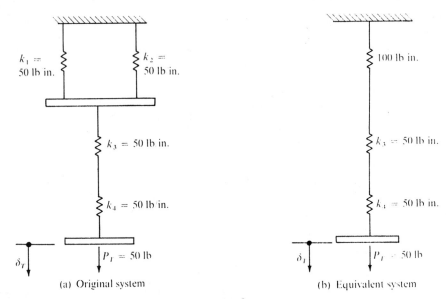

FIGURE 12.17   Sketch for Sample Problem 12.5 (a) Original system. (b) Equivalent system

(Figure 12.17b) is one having a 100 lb/in. spring in series with k_3 and k_4. Therefore, from Figure 12.17b, we have

$$\frac{1}{k_T} = \frac{1}{100} + \frac{1}{50} + \frac{1}{50} = \frac{5}{100}$$

$$k_T = 20 \text{ lb/in.}$$

$$\delta_T = \frac{P_T}{k_T} = \frac{50 \text{ lb}}{20 \text{ lb/in.}} = 2.5 \text{ in.}$$

12.7 DYNAMIC LOADING

Sometimes springs are used to absorb energy from shock or impact loads. Applications include automotive suspension system springs and bumper springs. In addition, elevator chutes contain springs to protect passengers in case the cable breaks and the elevator falls to the bottom of the chute.

Figure 12.18 shows a spring (having a stiffness k) in three possible conditions as follows:

1. Figure 12.18a. The spring is in its free-length condition.

2. Figure 12.18b. A weight W is gradually lowered on the spring until it comes to rest. This produces a static load W, and static deflection and Equation 12.4 is rewritten as

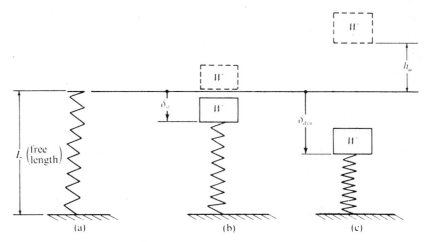

FIGURE 12.18 Dynamic versus static loading on a spring

$$\delta_{st} = \frac{W}{k}$$

where δ_{st} = the static deflection of the spring in inches

 k = the spring rate (or gradient) lb/in.

 W = the weight of the object in pounds

3. Figure 12.18c. The same weight W is dropped from a free-falling distance h_o and strikes the spring with impact. The maximum deflection of the spring is called the *dynamic deflection*, δ_{dyn}. Likewise, an equivalent static load that will produce the same deflection as the dropped weight is called the *dynamic load* P_{dyn}.

Let's review what happens. Weight W is dropped on the spring. Momentarily, the spring is deflected to its maximum, δ_{dyn} (as shown in Figure 12.18c). The spring then repositions itself to its static position under the load as shown in Figure 12.18b. As we proceed, you will see that P_{dyn} can be many times greater than the static load W.

Our analysis of dynamic loads must make use of the energy (or work) relationship to force and distance. So, let's have a short review. If a force of 10 lb pushes a large box 10 ft across the floor, the work done or energy expended equals force times distance, or 100 ft-lb. If force and distance are plotted on a graph, as in Figure 12.19a, we can see that the work done is equal to the *area* enclosed by the 10 lb and 10 ft lines. The concept of using the area of a graph to represent the energy expended will aid us in developing our dynamic equations. Let us now consider a helical spring with $k = 1$ lb/in. How much energy is required to extend the spring a full 10 inches? The answer is not 100 in.-lb because a force of 10 lb was not applied through a distance of 10 in. The graph of force versus deflection for this spring is plotted in Figure 12.19b (compare this to the previous Figure 12.13). The area under the curve is the energy

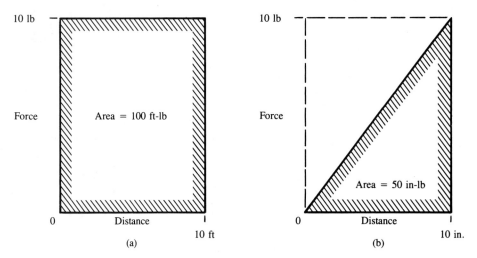

FIGURE 12.19   Graphical illustration of energy expended to move a box and to stretch a spring (a) Box (b) Spring

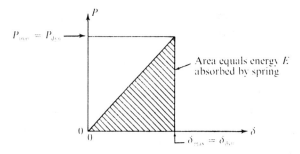

FIGURE 12.20   Energy absorbed by a spring

expended and also represents the area of a triangle ($A = \frac{1}{2} \times$ height $\times$ base). The energy expended is 50 in.-lb.

The energy absorbed E can be represented by

$$E = \frac{1}{2} (P_{dyn} \times \delta_{dyn}) \tag{12.7}$$

where δ_{dyn} = the deflection of the spring resulting from the dynamic load

This equation is represented by the graph in Figure 12.20.

Dynamic loads are caused by moving objects. The kinetic energy of the object can be obtained by

$$KE = \left(\frac{1}{2}\right) mv^2$$

where
KE = the kinetic energy in foot-pounds
m = the mass of the object in slugs
v = the velocity in feet per second

This expression is set equal to Equation 12.7 because the spring must absorb the energy of the object. Note that in most cases the kinetic energy must be changed from ft-lb to in.-lb because spring deflections are usually measured in inches.

We will now develop an equation for the particular case where the dynamic load is caused by falling weight. The kinetic energy of the falling weight need not be calculated, since the change in potential energy offers a more direct solution.

$$\begin{bmatrix} \text{The change in potential} \\ \text{energy of the weight } W \end{bmatrix} = \begin{bmatrix} \text{The potential engery} \\ \text{absorbed by the spring} \end{bmatrix}$$

$$W(h_o + \delta_{dyn}) = \frac{1}{2}(P_{dyn} \times \delta_{dyn})$$

where h_o = height of an object above the spring

Solving for P_{dyn} we have

$$P_{dyn} = \frac{2W(h_o + \delta_{dyn})}{\delta_{dyn}}$$

In most spring problems, δ_{dyn} will not known. Instead, h_o, W, and k will be the given parameters. Therefore, a direct substitution of $\delta_{dyn} = P_{dyn}/k$ can be made as follows:

$$P_{dyn} = \frac{2W[h_o + (P_{dyn}/k)]}{P_{dyn}/k}$$

Upon rearranging, we obtain a quadratic equation in terms of P_{dyn}:

$$(P_{dyn})^2 - 2WP_{dyn} - 2Wh_ok = 0$$

Using the quadratic formula,

$$P_{dyn} = \frac{2W \pm \sqrt{(4W^2 - (4)(1)(-2Wh_ok))}}{2}$$

Since P_{dyn} is a positive quantity, the minus sign is eliminated and the final result becomes

$$P_{dyn} = W + \sqrt{(W^2 + 2Wh_ok)} \tag{12.8}$$

Notice that the dynamic load consists of a static component W and a dynamic component $\sqrt{(W^2 + 2Wh_o k)}$. For the specific case of a suddenly applied load where $h_o = 0$, the magnitude of the suddenly applied load equals twice the static load:

$$P_{dyn} = 2W.$$

There are several points to consider when evaluating Equation 12.8.

1. For $h_o = 0$, the magnitude of a suddenly applied load does not depend on the spring stiffness (spring rate).

2. As h_o increases in value, the dynamic load increases. If k is very large (a very stiff spring), the dynamic load can become very great. Thus, for example, we would like to have soft springs (low k value) under an elevator, but the springs should not be so soft that the solid height is reached under load. The value k in effect becomes infinitely large when each coil has metal-to-metal contact with its neighboring coils.

3. The dynamic load increases as h_o increases.

SAMPLE PROBLEM 12.6

Dynamic Load

PROBLEM: A weight of 500 lb falls freely for a distance of 10 inches and then strikes a spring having a spring rate of 100 lb/in. Find the maximum load (the dynamic load).

Solution

Substitute directly into Equation 12.8:

$$P_{dyn} = 500 + \sqrt{500^2 + 2 \times 500 \times 10 \times 100}$$
$$= 500 + 1120 = 1620 \text{ lb}$$

SAMPLE PROBLEM 12.7

Dynamic Load and Deflection

PROBLEM: A 3000 lb automobile is supported by four coil springs, each having a spring rate of 500 lb/in. During a rapid run over a hill, the automobile leaves the road by a 6 in. elevation.

  (a) What will be the dynamic load in each spring when the automobile lands back on the road? Assume that each spring supports an equal share of the load.

  (b) How much will the springs compress if they do not bottom or reach solid height?

Solution

  (a) The static load per spring equals 3000/4 = 750 lb. Substituting directly into Equation 12.8,

$$P_{dyn} = 750 + \sqrt{(750^2 + 2 \times 750 \times 6 \times 500)}$$
$$= 750 + 2250 = 3000 \text{ lb}$$

(b) $\delta_{dyn} = \dfrac{P_{dyn}}{k} = \dfrac{3000 \text{ lb}}{500 \text{ lb/in.}} = 6 \text{ in.}$

12.8 LEAF SPRINGS

The design of single leaf springs for size and deflection is simply a matter of using the beam deflection and stress equations that you are already familiar with. These equations can be found readily in handbooks. Considerations for typical multiple leaf spring require adjustments to these basic equations. The automotive suspension system of Figure 12.21 uses leaf springs in its rear end. The design of such a leaf spring is based on the premise that a triangular-shaped cantilever beam is a spring of constant stress. Figure 12.22 shows such a beam undergoing an applied load P at the free end. The width of the beam at the fixed end is W, the length of the beam is l, and the beam thickness is h.

As can be seen from Figure 12.22d, the bending moment increases linearly from zero at the free end to a maximum value at the fixed end. Moreover, the width of the beam also increases linearly from zero at the free end to a maximum value at the fixed end. We conclude that the bending stress is a constant throughout the beam length.

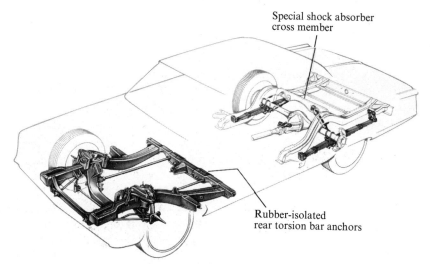

Special shock absorber
cross member

Rubber-isolated
rear torsion bar anchors

FIGURE 12.21 Automotive suspension system using front-end torsion bars and rear end leaf springs on the rear axle *(Courtesy of Chrysler Corp.)*

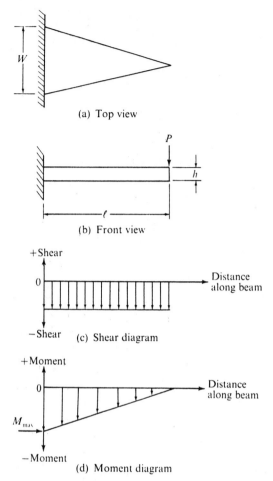

FIGURE 12.22   Triangular-shaped cantilever beam (a)
Top view (b) Front view (c) Shear dia-
gram (d) Moment diagram

A leaf spring is essentially equivalent to a triangular cantilever spring where strips are cut out and stacked on top of each other. This is illustrated in Figure 12.23 where the cantilever beam is cut into the total equivalent of four strips. The maximum width of the cantilever beam is $4b$ where b is the width of each leaf. The result is a leaf spring having four leaves each with width b. The same technique can be used to develop a simply supported leaf spring of constant stress, as illustrated in Figure 12.24, where once again four leaves have been generated for the sake of simplicity.

The actual construction of the automotive leaf spring is shown in Figure 12.21. The master leaf has an eye at each end for attachment to the automobile frame. The center of the spring has U-bolts for attachment to the rear axle. Notice that the leaves are banded together at two locations in addition to the U-bolt center position.

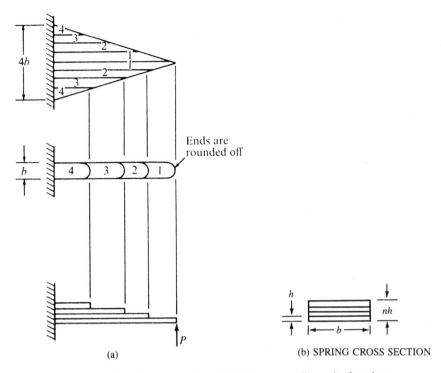

(a)

(b) SPRING CROSS SECTION

FIGURE 12.23   Development of constant-stress cantilever leaf spring

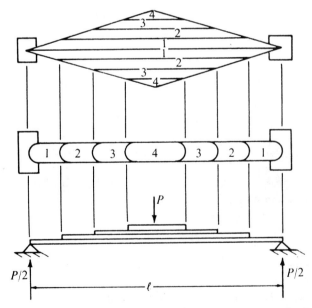

FIGURE 12.24   Development of constant-stress simply supported leaf spring

One obvious advantage of a leaf spring over a triangular plate spring is the savings in space requirements. The analysis of leaf springs is presented below.

For a simple beam with a concentrated load at its midspan, the maximum bending moment occurs at the midpoint. The maximum bending moment is

$$M = \frac{Pl}{4}$$

where M = the maximum bending moment in pound-inches

P = the concentrated load in pounds

l = the span length in inches

Since we have bending stress, the flexure formula applies:

$$S = \frac{Mc}{I}$$

where S = the bending stress in pounds per inch^2

c = $h/2$ for rectangular cross sections

h = the leaf thickness in inches

I = the moment of inertia in inches^4

The moment of inertia for leaf springs is taken at the location with the largest number of leaves. The formula is

$$I = \frac{(nb)h^3}{12}$$

where n = number of leaves (see Fig. 12.23)

b = width of each leaf in inches

Using the leaf spring development technique shown in Figure 12.24 and substituting into the flexure formula, we obtain

$$S = \frac{(Pl/4)\,(h/2)}{(nb)h^3/12} = \frac{3Pl}{2nbh^2} \qquad\qquad (12.9)$$

When solving for any one of the variables in Equation 12.9, simply rewrite the equation to solve for the unknown quantity.

12.9 OTHER COMMON SPRING CONFIGURATIONS ▰▰▰▰▰

Torsion Bars.   In Figure 12.12, we showed how a coil spring can be unwrapped and thereby converted into a torsion bar. It should therefore not be surprising that *torsion bars* have become popular for use in the suspension systems of automobiles. Figure 12.21 illustrates an automotive suspension system where torsion bars are used in the front end. Notice the rubber-isolated rear torsion bar anchors.

Helical Torsion Springs.   The *helical torsion spring* (see Figures 12.25 and 12.26) is widely used in door and cover hinges, automotive starters and in various electrical mechanisms. Contrary to its name, a torsion spring is subjected to bending stresses rather than torsional stresses.

Belleville Springs.   Figure 12.27 shows the configuration of a *Belleville spring,* which is a conical spring washer. Belleville springs have applications in buffers and cushions, where high loads and small deflections are required. When a load is applied, the conical washer tends to flatten out.

A Belleville spring should be designed so that it can be compressed to the flat position (h = 0) without exceeding the elastic limit. Because the deflection in a single Belleville spring is quite small, it is sometimes necessary to stack Belleville springs in series. This increases the deflection in direct proportion to the number of springs used. If it becomes necessary to increase the load, the washers can be stacked in parallel. Although the load carrying capacity should theoretically increase in direct proportion to the number of parallel connected washers, friction between the washers changes the load-deflection characteristics. A series-parallel arrangement can be used to produce a combination of increased load and increased deflection. Figure 12.28 illustrates the series, parallel, and series-parallel arrangements. Figure 12.29 illustrates various sizes of Belleville washers, as well as the series and parallel arrangements mentioned.

Conical and Volute Springs.   *Conical springs* are special versions of the conventional compression coil spring. The conical spring can be designed so that each coil can fit into the adjacent larger coil as shown in Figure 12.30a. Therefore, the solid height of a conical spring can be as small as one wire diameter. As a result, conical springs have an increasing stiffness as each turn becomes successively inactive during

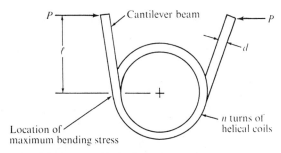

FIGURE 12.25   Helical torsion spring

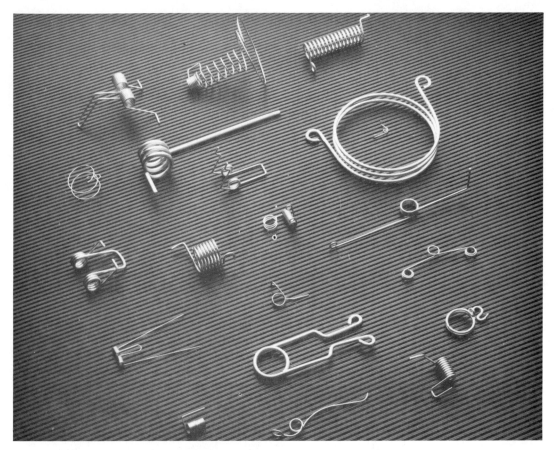

FIGURE 12.26   Various types of helical torsion springs *(Courtesy of Hardware Products Co., Inc., Chelsea, Massachusetts)*

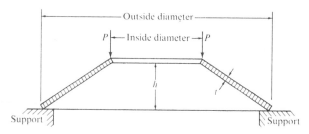

FIGURE 12.27   Belleville spring

compression. If the turns are variably spaced, the conical spring can be designed to have a constant spring rate. Variable-pitch conical springs are commonly used in dynamic applications such as valve springs, since the natural frequency of the spring

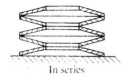
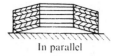
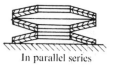

In series

In parallel

In parallel series

FIGURE 12.28 Various arrangements for Belleville washers *(Courtesy of Associated Spring Corp., Bristol, Connecticut)*

FIGURE 12.29 Various sizes and arrangements of Belleville washers *(Courtesy of Associated Spring Corp., Bristol, Connecticut)*

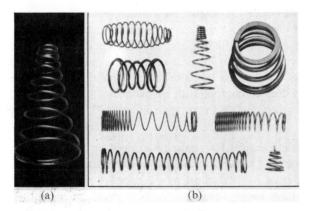

(a)   (b)

FIGURE 12.30 Conical, variable-pitch, barrel, and hourglass compression springs *(Courtesy of Associated Spring Corp., Bristol, Connecticut)*

changes as the it is cycled. This characteristic reduces surging (violent and sudden action), especially when the load frequency is near the natural frequency of the spring. Figure 12.30*b* shows a variable-pitch conical spring. Also illustrated are the barrel and hourglass configurations, which similarly minimize dynamic surging problems.

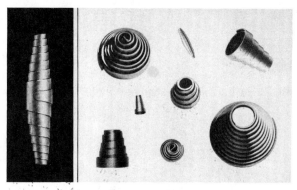

FIGURE 12.31   Miscellaneous types of volute springs
*(Courtesy of Associated Spring Corp.,
Bristol, Connecticut)*

Volute springs (see Figure 12.31) operate like conical compression springs. As shown, the volute spring consists of tapered metallic strips wound in such a way that each turn telescopes inside the preceding one. Since each coil contacts the preceding one, sliding friction occurs, which provides vibration dampening. The solid height of a volute spring equals essentially the width of the metallic strip. As in the case of the conical compression spring, volute springs provide an increasing stiffness as compression progresses.

Constant-Force Springs.   A *constant-force spring* is one whose force does not change as it is extended. Such a spring has a spring rate equal to zero. Figure 12.32 shows several sizes of a Neg'ator constant-force spring. The spring metal is prestressed in a special way to maintain a constant output force throughout its entire deflection range. As a result, the Neg'ator constant-force spring provides a much greater deflection than does an conventional extension coil spring.

The following is a list of applications for the Neg'ator extension spring:

—   Feeding devices for vending machines
—   Control of force of follower roller on a cam
—   Powering of mechanical movie cameras Neg'ator springs provide smooth operation and longer running times per winding. Figure 12.33 shows this application, where a Neg'ator spring drives a mechanical movie camera at constant torque.
—   Retracting of hose reels and electrical cords
—   Driving of toys as a long-running constant torque power source
—   Electric motor brush springs
—   Window or door counterbalances

Power Springs.   *Power springs,* commonly called *clock springs,* store energy after winding and then release the energy to provide torque through an output shaft or drum. Figure 12.34 shows a power spring which is called Spir'ator by its manufacturer. The

FIGURE 12.32   Various sizes of Neg'ator constant-force springs. *(Courtesy of Hunter Spring Div., Ametek, Inc., Hatfield, Pennsylvania)*

inner end of the spring is wound around a shaft, and the outer end is attached to the inside of a rotating drum. The spring can be wound by rotating either the shaft or the

FIGURE 12.33   Mechanical movie camera powered by Neg'ator spring *(Courtesy of Hunter Spring Div., Ametek, Inc., Hatfield, Pennsylvania)*

FIGURE 12.34   Spir'ator power spring *(Courtesy of Hunter Spring Div., Ametek, Inc., Hatfield, Pennsylvania)*

drum. The Spir'ator spring is an efficient, long-running, low spring-rate, compact torque device. The following is a list of common applications of power springs:

— Seat belt retracting mechanisms
— Ordinary clock springs
— Power lawn mowers having rewind starters
— Timing mechanisms in parking meters
— Energy drive source for toys
— Hose retrieval reels on gasoline pumps

Garter Springs.   Garter springs are coil extension or compression springs whose ends are connected so that the spring forms a complete circle, or garter. In operation, garter springs exert radial forces similar to a rubber band. Figure 12.35 shows three different sizes of garter springs. One very common application is in oil seals where the inside diameter of the spring is expanded to fit over a larger-diameter shaft. In this way, the spring is actually stretched, and therefore it behaves as an extension coil

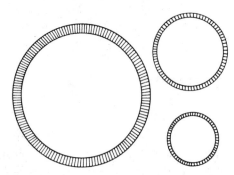

FIGURE 12.35   Garter springs *(Courtesy of Associated Spring Corp., Bristol, Connecticut)*

spring. As a result, a compressive radial force is exerted on the shaft. Other applications include small motor belts, electrical connectors and piston-ring expanders.

12.10 SI UNIT SPRINGS

Springs may be obtained in metric dimensions. Table 12.2 lists the preferred metric spring wire dimensions. In order to maintain consistency when using the spring equations, loads must be in newtons; dimensions in meters; and stress, the modulus of elasticity, and the shear modulus of elasticity in pascals. For some equations it is possible to use other units, but when starting out it is best to stay with the SI standard units.

SAMPLE PROBLEM 12.8

Spring Deflection With SI Units

PROBLEM: A tension spring is to elongate 100 mm under a 50 kg load. Choose a suitable steel wire diameter and specify the number of active coils. The following parameters are known: Spring index $C = 8$; $G = 83$ GPa; allowable shear stress $\tau_{max} = 500$ MPa.

Solution

From Figure 12.9, $C = 8$, $K = 1.18$
From Equation 12.2

$$\tau_{max} = \frac{8KPD}{\pi d^3}$$

$$500 \times 10^6 = \frac{8 \times 1.18 \times (50 \times 9.81) \times (8d)}{3.14 \times d^3}$$

$$d = \sqrt{\frac{37{,}043}{1.57 \times 10^9}} = 0.004\ 857\ m = 4.86\ mm$$

Choose the standard size of 5.0 mm diameter. (from Table 12.2)
(Note: $k = P / \delta$.) Therefore,

$$k = 50 \times \frac{9.81}{0.1} = 4905$$

Using Equation 12.4

$$k = \frac{Gd}{8C^3 n}$$

$$n = \frac{83 \times 10^9 \times 0.005}{8 \times 8^3 \times 4905} = 20.7\ \text{coils}$$

TABLE 12.2 Preferred Metric Sizes (From Handbook of Spring Design by the Spring Manufacturers Institute, Inc.)

Preferred Thickness Or Diameter	Second Preference	Inch Equivalent
.025		.0010
.050		.0020
.060		.0024
.080		.0032
.10		.0039
.12		.0047
	.14	.0055
.16		.0063
	.18	.0071
.20		.0079
	.22	.0087
.25		.0098
	.28	.0110
.30		.0118
	.35	.0138
.40		.0158
	.45	.0177
.50		.0197
	.55	.0217
.60		.0236
	.65	.0256
	.70	.0276
.80		.0315
	.90	.0354
1.0		.0394
	1.1	.0433
1.2		.0472
	1.4	.0551
1.6		.0630
	1.8	.0709

Preferred Thickness Or Diameter	Second Preference	Inch Equivalent
2.0		.0787
	2.2	.0866
2.5		.0984
	2.8	.1102
3.0		.1181
	3.2	.1260
3.5		.1378
	3.8	.1496
4.0		.1575
4.5		.1772
	4.8	.1890
5.0		.1969
	5.5	.2165
6.0		.2362
7.0		.2756
8.0		.3150
	9.0	.3543
10.0		.3937
	11.0	.4331
12.0		.4724
	14.0	.5512
16.0		.6299

Rectangular Wire Cross Section

Preferred Width	Second Preference	Inch Equivalent
1.5		.059
3.0		.118
	4.5	.177
6.0		.236
	8.0	.315
10.0		.394
12.0		.472
16.0		.630
20.0		.787
25.0		.984
30.0		1.181
35.0		1.378
40.0		1.575
	45.0	1.772
50.0		1.969
60.0		2.362
80.0		3.150
	90.0	3.543
100.0		3.937
	110.0	4.331
120.0		4.724
140.0		5.512
	150.0	5.906
160.0		6.299
180.0		7.087
200.0		7.874
	250.0	9.843
	300.0	11.811

12.11 SUMMARY

Springs are mechanical devices designed to produce relatively large deflections when subjected to externally applied loads. The four main purposes for using springs are

— Control of machine motion
— Reduction of forces from impact loading
— Energy storage
— Force measurement

Most springs are made of steel. Helical coil steel springs may be formed cold or formed hot and heat treated to increase strength and hardness. Compression-type helical coil springs have four types of ends. The best type of end for resisting bowing of the spring under load is the *squared-and-ground end*. However, it is also the most expensive to manufacture.

The stress and deflection equations for helical coil springs are derived from basic strength of materials equations. The limiting stress is most commonly torsional shear stress, which is unaffected by the number of coils. The deflection of a coil spring (for a given load) is directly proportional to the number of coils. The ratio of load to deflection is referred to as spring rate or spring gradient and the symbol is k. A similar term, but with an entirely different meaning, is spring index, which is the ratio of the coil mean diameter to the coil wire diameter (its symbol is C).

Springs may be connected in series or parallel. When placed in series, each spring supports the total load. When placed in parallel, the springs share the load. Springs are used to absorb shock or impact loads. Such loads are grouped under the term dynamic loading to differentiate them from static loading. When a weight is dropped on a spring, the dynamic load can be many times greater than if the same weight were placed gently on it.

The types of springs discussed in this chapter, other than helical coil springs are

— Torsion bars
— Leaf springs
— Helical torsion springs
— Belleville springs
— Conical and volute springs
— Constant-force springs
— Power springs
— Garter springs

12.12 QUESTIONS AND PROBLEMS

Questions

1. Name four general application areas for springs.
2. What are the four types of ends used for helical compression springs?

3. What is the significance of the phrase *number of active turns of a coil spring?*
4. Differentiate between the free length and the solid height of a helical coil spring.
5. Name one advantage and one disadvantage of rectangular wire as compared to round wire for compression coil springs.
6. How does a compression coil spring differ physically from an extension coil spring?
7. What kind of stress does a helical coil spring experience?
8. What is the significance of the Wahl Factor when designing coil springs?
9. Why are springs frequently heat-treated to produce stress-relieving?
10. What is meant by the term cold-drawn when used to describe wire?
11. Name two factors that determine the allowable stresses for springs.
12. Relative to springs, what is the purpose of shot peening?
13. Differentiate between the terms spring index and spring rate.
14. Differentiate between a series spring system and a parallel spring system.
15. What is a dynamic spring load?
16. Differentiate between a torsion bar and a torsion spring.
17. Discuss the construction of an automotive leaf spring.
18. Sketch a Belleville spring. When would Belleville springs normally be used?
19. Differentiate between conical and volute springs. Name one advantage of each type.
20. Show the force-deflection curve of a constant-force spring. What is the value of the spring rate for such a spring?
21. What is a power spring? Give two applications.
22. What are garter springs? Give two applications.

Problems

Stress, Deflection, and Design

1. Refer to Sample Problem 12.1. Decrease the wire diameter 4%. All other data remain the same. By what percentage does the shear stress change?
2. Repeat Problem 1 except decrease the coil diameter by 4% instead of the wire diameter.
3. A load of 100 lb deflects a spring 5 in. What load will deflect the spring 2 in.?
4. Solve Sample Problem 12.2, except change the maximum allowable shear stress from 80,000 lb/in.^{2} to 50,000 lb/in.^{2}
5. An extension coil spring is to elongate 4 in. under an 80 lb load. The following data are given: spring index = 8, shear modulus of elasticity = 11.5×10^6 lb/in.^{2}, and the allowable shear stress = 80,000 lb/in.^{2}. Find the number of active coils required.
6. Solve Sample Problem 12.3 if phosphor bronze ASTM B159 is required for the spring material.
7. The exhaust and intake valves on an automobile engine are to open ½ in. and the valve-spring forces are to be 75 lb when the valve is closed and 150 lb when the valve is open. Design the required helical coil compression springs. The follow-

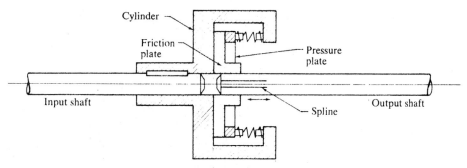

Cylinder

Friction plate

Pressure plate

Input shaft

Output shaft

Spline

FIGURE 12.36   Sketch for Problem 8

ing data are given: use ASTM A230 steel; the maximum outside diameter for the coil is 1.00 in.

Design and Parallel Applications

8. An automotive-type clutch uses friction in transmitting a torque as shown in Figure 12.36. The normal force on the friction surfaces is supplied by coil compression springs. Thus the clutch is always engaged until an external force is applied to compress the spring the additional amount necessary to release the clutch. The following data are given: use ASTM A231 material; the coil outside diameter is to be 1.3 in.; the maximum normal force = 2000 lb when the clutch is new; the minimum normal force = 1600 lb when the clutch facing is worn $\frac{1}{8}$ in.; the total number of springs to be used = 12.
    Design the springs for this application.

Series and Parallel Applications

9. Four extension coil springs are hooked in series and support a weight of 100 lb. Two of the springs have a spring rate of 30 lb/in., and the other two have spring rates of 15 lb/in. **(a)** What will be the total deflection? **(b)** What is the equivalent stiffness k_T?

10. Four compression coil springs are hooked in parallel and support a 400 lb load. If each spring has a gradient of 40 lb/in., how much deflection will occur?

11. A helical coil spring has a gradient of 600 lb/in. It is cut into quarters and the four pieces are then combined in parallel. What is the total equivalent spring rate of the parallel system?

12. Show how the four quarter pieces from Problem 11 can be connected to produce a total equivalent spring rate of 960 lb/in.

13. Figure 12.37 shows four identical springs (k = 50 lb/in.) connected together in a special arrangement. How much deflection will occur at points A, B, and C if P equals 100 lb?

14. Figure 12.38 shows a simulated automotive suspension system. Locate the center of gravity as measured by x if the 2000 lb automobile is to remain horizontal.

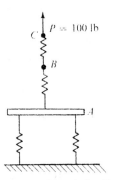

FIGURE 12.37   Sketch for Problem 13

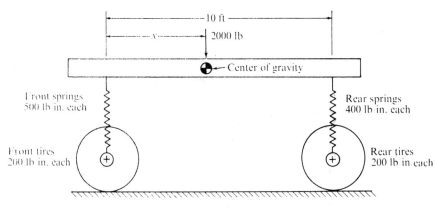

FIGURE 12.38   Sketch for Problem 14

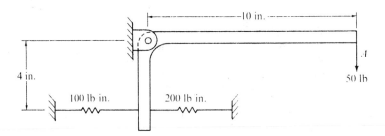

FIGURE 12.39   Sketch for Problem 15

15.   Two springs are connected to a lever with no preload, as shown in Figure 12.39.
      Find the deflection at point A.
16.   Figure 12.40 shows a concentric spring design in the free-length position. If the
      allowable stress for each steel spring is 50,000 lb/in.^{2} and G = 11.6 × 10^6 lb/in.^{2}
      (a) what is rated load and spring rate for each spring?

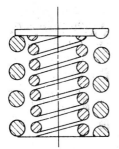

Spring Data—Dimensions in inches

	Inner Spring	Outer Spring
Inside diameter	4	6
Outside diameter	5	8
Wire diameter	$\frac{1}{2}$	1
Number of coils	12	6
Free length	7	7

FIGURE 12.40   Sketch for Problem 16

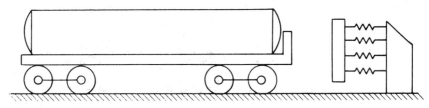

FIGURE 12.41   Sketch for Problem 19

(b) what total load would cause each spring to reach solid height?

Dynamic Loads

17. A 200 lb weight falls freely for 10 in., strikes a compression coil spring and deflects it 4 in. The following data are given: spring index = 8; shear modulus of elasticity = 11.5×10^6 lb/in.^{2}; allowable shear stress = 80,000 lb/in.^{2}
Find the wire diameter, the mean coil diameter and the number of active coils.

18. Figure 12.41 shows four bumper springs used to stop the motion of a 100,000 lb railroad car. By what amount will each spring length change if the springs are to stop the car traveling at 7.5 mph? The spring gradient of each spring is 2000 lb/in.

Leaf Springs

19. An automotive leaf spring (similar to that in Figure 12.24) must sustain a load P of 1000 lb at its midpoint. The following data are given: distance between supports $l = 30$ in.; design stress = 40,000 lb/in.^{2}; number of leaves $n = 6$; width of each leaf $b = 2.5$ in.
What must be the thickness h of each leaf?

SI Unit Problems

20. Solve Sample Problem 12.8 if the load = 75 kg and the spring index $C = 10$.

21. A helical coil spring has a 10 mm diameter wire and a mean coil dia. of 75 mm.
     Find the maximum shear stress when the spring supports a 500 kg load.

22. Is the shear stress in Problem 21 within the allowable limits for ASTM A227
     steel? Use Figure 12.10 and Table 12.1.

23. Find the deflection of a high carbon steel spring supporting a 25 kg load. Wire
     diameter = 3 mm; coil diameter = 25 mm; there are 36 active coils.

13

Dynamic Loads on Machine Members and Dynamic Balancing of Shafts

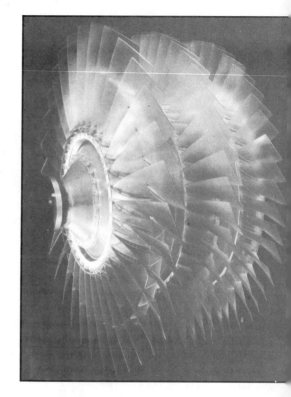

Objectives

(A) RELATING TO DYNAMIC LOADS ON MACHINE MEMBERS:

The objectives of this chapter on dynamic loads include impressing on you the large forces resulting when relatively small weights are dropped from a height of a few feet or inches. Upon completion of this material, you will be able to determine stresses and loads in certain machine members resulting from impact loads, and you will be able to specify the minimum size or length of a member necessary to withstand a given impact.

(B) RELATING TO DYNAMIC BALANCE:

This chapter is intended to explain the meaning of the term *dynamic balance* and how to achieve it. On completing this part of the chapter you will be able to explain the difference between dynamic balance and static balance and you will be able to apply the proper equations to dynamically balance rotating shafts.

13.1   INTRODUCTION

You are now familiar with the behavior of springs subjected to an impact load. Now our interest is in regular structural shapes, not springs. The machine members act like springs except that the deflections (deformations) are much smaller. Also, we are interested in relating the energy absorbed by the member to tensile and compressive stress and strain and to the member's volume. Equations will be developed for impact from a falling weight.

13.2   IMPACT EQUATIONS

When solving for stress in a member subject to impact, the basic assumption is that the strain produced by the falling object can be compared to the same strain produced by an equivalent static load. The object of the comparison is to keep the stress developed below the allowable limit. An accepted method of doing this is to say that the energy of the falling object is equal (or transferred) to the energy absorbed by the machine member. Our assumption in this text will be that all of the energy of the falling weight is absorbed as strain energy in the resisting part, and no energy is dissipated as heat or absorbed in the supports of the part. As in Section 12.7 of Chapter 12, we have this relationship:

$$\begin{bmatrix} \text{The change in potential} \\ \text{energy of weight } W \end{bmatrix} = \begin{bmatrix} \text{The potential energy} \\ \text{absorbed by the part} \\ \text{(strain energy)} \end{bmatrix}$$

$$W(h_o + \delta_{\text{dyn}}) = \left(\frac{1}{2}\right)(P_{\text{dyn}} \times \delta_{\text{dyn}})$$

This equation is not satisfactory, as other data must be used. Figure 12.20 showing a graph of the energy absorbed by a spring will be redrawn on a stress-strain diagram (see Figure 13.1). The area under the straight line part of the curve is called the mod-

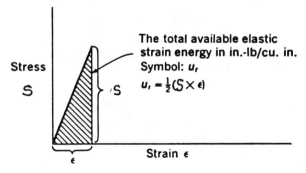

FIGURE 13.1   Modulus of resilience u_r

ulus of resilience and its units are inch-pounds per inch.[3] The symbol for the modulus of resilience is u_r. Since this is the strain energy that can be absorbed (for a given stress value) for one cubic inch, u_r must be multiplied by the volume of the part to obtain the total amount of energy that can be absorbed. Therefore the next step in obtaining our desired equation is

$$W(h_o + \delta_{dyn}) = u_r \times V \qquad\qquad\qquad (13.1)$$

where V = the volume in cubic inches.

h_o = the height above the part in inches

δ_{dyn} = the maximum deformation of the part from the impact (in inches)

Since the strain energy area on the Stress-Strain diagram is a triangle, a substitution can be made for u_r

$$u_r = \frac{1}{2} S\epsilon$$

and

$$W(h_0 + \delta_{dyn}) = \frac{1}{2} S\epsilon \times V \qquad\qquad\qquad (13.2)$$

From the basic strength of materials relationship:

strain = stress/modulus of electricity

$$\epsilon = \frac{S}{E}$$

A substitution is made for the strain ϵ:

$$W(h_0 + \delta_{dyn}) = \left(\frac{1}{2}\right) \frac{S^2 V}{E} \qquad\qquad\qquad (13.3)$$

where S = the tensile or compressive stress in lb/in.^{2}

E = the modulus of elasticity in lb/in.^{2}

Frequently δ_{dyn} is so small compared to h_0 in Equation 13.3 that it is neglected. In Sample Problem 13.1 below δ_{dyn} could be neglected with no significant error.

Impact Equation 13.3 indicates that a machine member can absorb more energy with an increase in volume. For example, Figure 13.2 shows two prisms with equal cross sectional areas, but B is twice as long as A. We know that both can support the same static load but the impact equation tells us that prism B can absorb a greater amount of energy because its volume is greater. Therefore it can support a greater dynamic load.

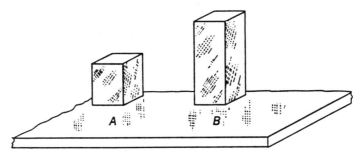

FIGURE 13.2   Two prisms with equal cross-sectional areas

Figure 13.3 shows how the advantage of increased volume may be utilized. The figure illustrates two pressure vessels subject to impact pressures. Assuming the same size bolts in both cases, the bolts in sketch (b) will withstand higher impact loads.

13.3   VOLUME OF LEAST AREA

An important consideration when designing a machine member for impact loads concerns the volume of the *least area* of the member. Note Figure 13.4. Part B can support the same static load as A but B can support a higher impact load. The reason is that every cross section of part B can be stressed to the allowable value and thus the strain for every inch of length will reach a maximum. However, the lower portion of part A is stressed to only one-half the value of the upper portion. Since part B has a larger volume stressed to the maximum allowable, it can absorb a larger amount of energy. Sample Problem 13.1 makes this comparison.

The effort to increase the volume of least area of a part leads to distinctive shapes for members subjected to impact loads. For instance, head studs on automobile engines are generally shaped as in Figure 13.5a. The shank is machined to the same diameter as the thread minor diameter so that the volume of the least cross-sectional area is increased. Figure 13.5b shows that part *B* is designed to give the greatest volume to the least cross section.

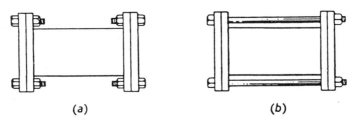

(a)   (b)

FIGURE 13.3   Pressure vessels subject to impact loads

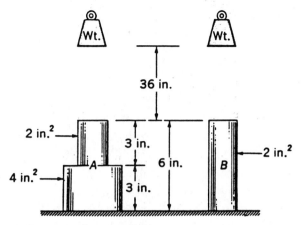

FIGURE 13.4   The importance of Least Area Volume in impact loads

PROCEDURE

For Solving Impact Problems

Step 1   If deflection must be determined, use the deflection equation: $\delta = SL/E$.

Step 2   Apply Equation 13.3 to solve for the unknown quantity.

SAMPLE PROBLEM 13.1

Volume of Least Area

PROBLEM: Please refer to Figure 13.4. If the members A and B are aluminum, and the allowable stress is 15,000 lb/in.^{2}:

(a) Determine the allowable weight to be dropped on A.

(b) Determine the allowable weight to be dropped on B.

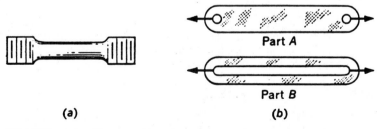

(a)   (b)

FIGURE 13.5   The consideration of Least Area in design for impact loads

Solution

(a) In this situation, the total height of the fall is 36 in. plus the deformation of *A*. For all practical purposes the deformation of *A* is insignificant compared to the 36 in. and could be omitted. The calculations will include the deformation for illustration only.

Observe that the top section of *A* will be stressed at 15,000 psi, while the lower section (which has twice the cross-sectional area of the top) will be stressed at 1/2 that value.

$$(h_o + \delta_{dyn}) = 36 \text{ in.} + \delta_{top} + \delta_{bottom}$$

Step 1 From stress-strain relationships:

$$\delta = \frac{S}{E} L$$

where   L = the length of the part.

$$(h_o + \delta) = 36 + \frac{15{,}000}{10.6(10)^6} \times 3 + \frac{7\,500}{10.6(10)^6} \times 3 = 36.0064 \text{ in.}$$

Step 2 The maximum weight is determined by

$$W = \frac{1}{(h_o + \delta_{dyn})}\left[\left(\frac{1}{2} \times \frac{S_{top}^2}{E} \times V_{top}\right) + \left(\frac{1}{2} \times \frac{S_{bot}^2}{E} \times V_{bot}\right)\right]$$

$$W = \frac{1 \times [15{,}000^2 \times (3 \times 2) + 7\,500^2 \times (3 \times 4)]}{36.0064 \times 2 \times 10.6(10)^6}$$

$$W = 2.65 \text{ lb}$$

(b) **Step 1.** $(h_o + \delta_{dyn}) = 36 \text{ in.} + \left(\frac{S}{E} \times L\right) = 36 + \left(\frac{15{,}000}{10.6(10)^6} \times 6\right)$

$$(h_o + \delta_{dyn}) = 36.009 \text{ in.}$$

Step 2. $W = \dfrac{1 \times 15{,}000^2 \times (6 \times 2)}{36.009 \times 2 \times 10.6(10)^6}$

$$= 3.54 \text{ lb}$$

Note that the reason the thinner part can take a larger impact load is because a longer length was stressed to the allowable limit of 15,000 psi. Also, note that the 3.54 lb weight dropped 3 ft is equivalent to a 30,000 lb static load!

SAMPLE PROBLEM 13.2

Impact Load

PROBLEM: Refer to Figure 13.6. A weight is to be held in contact with its supporting rod such a manner that the rod is not loaded until the weight is suddenly released. If the stress in the rod is not to exceed 30,000 psi

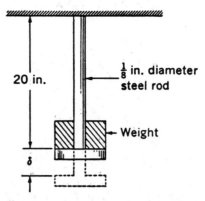

FIGURE 13.6   Sketch for Sample Problem 13.2

(a) What weight should be used?

(b) What is the equivalent static weight?

Solution

(a) The volume of the 1/8 in. diameter steel rod is

$$V = A \times L = \left[\frac{\pi d^2}{4}\right] \times L = 3.14 \times \left(\frac{1}{8}\right)^2 \times \frac{20}{4}$$

$$= 0.245 \text{ in.}^3$$

$$h_o = 0$$

$$\delta = \frac{S \times L}{E} = \frac{30{,}000 \times 20}{30 \times 10^6} = 0.02 \text{ in.}$$

Rearranging Equation 13.3, we have

$$W = \frac{1 \times S^2 \times V}{\delta \times 2 \times E} = \frac{(30{,}000)^2 \times 0.245}{0.02 \times 2 \times 30 \times 10^6}$$

$$W = 184 \text{ lb}$$

(b) $W = S \times A = 30{,}000 \times \pi \dfrac{(1/8)^2}{4}$

$$W = 368 \text{ lb}$$

Note that the dynamic weight is 1/2 of the static weight.

13.4 DYNAMIC BALANCING OF SHAFTS

High speed rotating parts in modern machinery must be carefully balanced to avoid destructive vibrations. The rotating elements of motors, generators, centrifugal pumps,

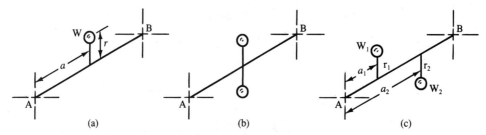

FIGURE 13.7   Static and dynamic balance. (a) Static & dynamic unbalance. (b) Static & dynamic balance. (c) Static balance, dynamic unbalance

and fans must be balanced because manufacturing tolerances may add up to an unbalanced situation. On the other hand, rotating parts such as cams are inherently unbalanced because of their design.

Your previous courses have given you a solid understanding of static balance, or static equilibrium. So in looking at Figure 13.7a, you know that this weight and shaft assembly is out of balance and that the weight W will tend to rotate to a rest position below the shaft. You also know that the simplest way to balance this system is to attach an equal weight an equal distance in the opposite direction, as in Figure 13.7b. Thus, if the assembly is rotated to any position through 360°, there is no tendency to move to some other position. The same static balance is obtained if the weights are offset along the shaft as in Figure 13.7c.

However, when a shaft is rotated at some given speed, static balance may not be enough to provide smooth, vibration free operation. We say that the assembly is dynamically unbalanced, and that the unbalance is caused by centrifugal force. As an aid to understanding dynamic balance (or unbalance) let us take the shaft assembly in Figure 13.7c and place it in open journal bearings at the shaft ends. By open journal bearings, we mean that the bearings are open at the top; we can readily remove the shaft simply by picking it up. The shaft and weights are now rotated at some given speed. Think of Figure 13.8 as a photograph taken at the instant the weights are in a vertical position. The centrifugal force F_c of W_1 is directed up, and that of W_2 is directed down. This creates a couple tending to rotate the whole assembly about bearing B in the plane of the paper. If the couple is great enough the shaft can be thrown out of bearing A. We must balance the system dynamically by placing weights on the shaft to provide an equal and opposite couple.

Let us now develop the procedure for dynamically balancing a rotating system of weights. The basic equation is

$$F = ma$$

where F = a force acting on an object

m = the mass of the object

a = the acceleration of the object

For circular motion the equation for centrifugal force becomes

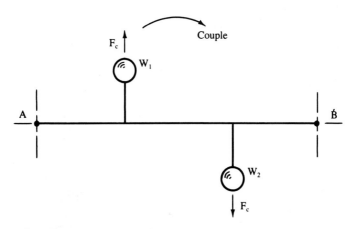

FIGURE 13.8   A couple is created in the longitudinal plane

$$F_c = \frac{mV^2}{r}$$

where V = the linear velocity of the rotating object

             r = the radius of the object's center of gravity about the point of rotation

and

$$V = r\omega$$

where ω = the rotational speed

Therefore,

$$F_c = \frac{m(r\omega)^2}{r} = mr\omega^2$$

Since

$$m = \frac{W}{g}$$

where   g = 386 inches per second^2

the equation for centrifugal force can be written as

$$F_c = \frac{Wr\omega^2}{g} \qquad\qquad\qquad\qquad (13.4)$$

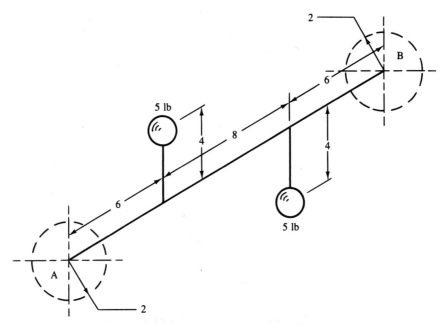

FIGURE 13.9   Balance weight locations for Sample Problem 13.3

Since the centrifugal force of an added weight must be set equal to those already in the system the rotational speed ω and g, can be factored out. So for dynamic balance

Added system = Original system

$$(Wr)_{add} = (Wr)_{orig} \qquad\qquad \textbf{(13.5)}$$

where W = the weight in pounds

r = the radius in inches

and Wr therefore represents the centrifugal force acting on the shaft.

Figure 13.8 indicates that an unbalanced couple in the vertical longitudinal plane of the shaft must be balanced. This can be done by taking moments. It is convenient to take moments about the positions where the added weights will be placed.

For example, let points A and B on the shaft in Figure 13.9 be the positions where weights will be added. In this case, the moment about A for the added weight at B must equal the sum of the moments of the original weights.

$$(Wra)_{add} = (Wra)_1 + (Wra)_2 \qquad\qquad \textbf{(13.6)}$$

where a = distance from moment center A to the position on the shaft in inches

and Wra therefore represents the moment of the centrifugal force

Actually, the above equations are simply rearrangements of the standard equilibrium equations:

$$\sum F = 0 = \sum (Wr)_{orig} - (Wr)_{add} \qquad (13.7)$$

and

$$\sum M = 0 = \sum (Wra)_{orig} - (Wra)_{add} \qquad (13.8)$$

The minus signs in Equations 13.7 and 13.8 simply indicate that the added values must be of opposite sign to the sum of the original values.

PROCEDURE

Solving Shaft Balancing Problems

The procedure is the same as the one you have used for solving reaction forces on loaded beams.

Step 1 Use the moment equation (Equation 13.8) to find the unknown weight values at the two specified locations on the shaft.

Step 2 Check you results with Equation 13.7, the force equation.

Note: Placing the data in tabular form, as in Sample Problem 13.3, may help to keep it organized.

SAMPLE PROBLEM 13.3

Dynamic Balance of Weights, All in the Same Longitudinal Plane With the Shaft

PROBLEM: Figure 13.9 shows two equal weights having their centers of gravity the same distance from the shaft and in the same vertical plane. Counterweights are to be placed at A and B and at a radius of 2 inches from the shaft. Determine the magnitude of each weight and its position with respect to the positions of the other weights.

Solution

Cross plane at	W lb	r in	a Dist. from A in.	Wr Equivalent F_c lb-in.	Wra lb-in.^{2}
A	W_A	reference	location	—	—
1	5	4	6	20	-120
2	5	4	14	-20	$+280$
B	W_B	2	20	$2W_B$	$40W_B$

Moments will be taken about A to find W_B.

Step 1 $\sum M_A = 0 = -120 + 280 - 40W_B.$

$W_B = 4$ lb up

Moments will now be taken about B to find W_A. $(Wra)_2 = 120$; $(Wra)_1 = -280$; and $(Wra)_A = 40W_A$.

$\sum M_B = 0 = 120 - 280 + 40W_A.$

$W_A = 4$ lb down

Step 2 $\sum F = 0 = \sum (Wr)_{orig} - \sum (Wr)_{add}$

$0 = [+(5 \times 4) + (-5 \times 4)] - [-(4 \times 2) + (4 \times 2)]$

One note of caution: If you are working on a bearing design that must take into account an unbalanced dynamic force, you must use Equation 13.4 to find the true centrifugal force.

Let us now investigate a situation where the unbalanced weights are not in the same longitudinal plane, such as in Figure 13.10. The procedure in this situation is to use the vertical and horizontal components of the centrifugal forces (represented by Wr) and solve for the vertical and horizontal components of the unknown balancing force. Once we have both components we simply find the resultant. A student problem will refer to Figure 13.10.

13.5 SI UNITS

SAMPLE PROBLEM 13.4 _____

Impact on a Steel Rod

PROBLEM: A steel rod with the configuration in Figure 13.6 in 800 mm long and has a diameter of 4 mm. The allowable tensile stress is 210 MPa. $E = 207$ GPa (from Appendix).

(a) What static load in kilograms can be placed on the rod?

(b) What mass can be dropped from a height of 10 mm to produce the same stress?

Solution:

(a) $S = \dfrac{P}{A}, \quad P = SA$

$A = \dfrac{\pi \times (0.004)^2}{4} = 12.566 \times 10^{-6}\, m^2$

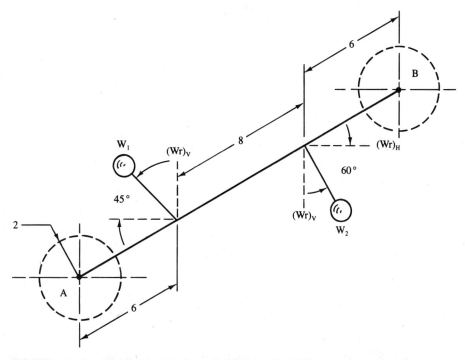

FIGURE 13.10   Weights not in the same longitudinal plane

$$P = (210 \times 10^6) \times (12.566 \times 10^{-6}) = 2638.9 \text{ N}$$

$$\frac{2638.9 \text{ N}}{9.81 \text{ N/kg}} = 269 \text{ kg}$$

(b) $\delta = \dfrac{SL}{E} = (210 \times 10^6) \times \dfrac{0.8 \text{ m}}{207 \times 10^9}$

$\delta = 0.81159 \times 10^{-3}$

$V = L \times A = 0.8 \times (12.566 \times 10^{-6}) = 10.0528 \times 10^{-6} \text{ m}^3$

$$W = \frac{(1/2)\,(S^2)V}{(h_o + \delta)E} = \frac{.5 \times (210 \times 10^6)^2 \times (10.0528 \times 10^{-6})}{(0.01 + 0.81159 \times 10^{-3})(207 \times 20^9)}$$

$= 99\text{N} = 10 \text{ kg}$

13.6 SUMMARY

A falling weight is used to demonstrate the effect of impact because height and weight can be used in the equations and kinetic energy is not needed. The impact equation is developed from energy relationships. Remember that the stress involved occurs at the instant the member is at its greatest deformation (dynamic deformation), not when the weight is resting on the member. A machine member can absorb the maximum energy when the volume of its least cross-sectional area is at a maximum.

Rotating machinery may be dynamically unbalanced even though statically balanced. When solving dynamic balance problems, centrifugal forces must be used in the statics equations:

$$\sum F = 0$$

$$\sum M = 0$$

Since all parts of the rotating machine have the same rotational speed and the same gravity affecting them, the centrifugal force is proportional to the weight times the radius:

$$F_c \propto W \times r$$

where W = weight of the object

r = radius from the center of the shaft to the center of gravity of the object

13.7 QUESTIONS AND PROBLEMS

Questions

1. What is meant by the term *equivalent static load?*
2. What happens to the potential energy of some object when it is dropped on a structural or machine member?
3. What is meant by the term *volume of least area?*
4. Explain how a rotating shaft can be statically balanced but not dynamically balanced.
5. When a shaft is not rotating, does dynamic balancing apply? Explain.
6. Why is it unnecessary to use the complete centrifugal force equation (Equation 13.4) when solving for the required shaft balance weights?

Problems

Dynamic loads on machine members

1. A steel prism 2 × 3 × 5 in. high has an allowable compressive stress of 30,000 psi. How much total strain energy can be absorbed when an object is dropped on top of it?

2. How long should the length of a 1/2 in. diameter steel rod be if it must absorb 300 in·lb of energy and the tensile stress must not exceed 28,000 psi?

3. Find the diameter of a 30 inch long steel circular rod sufficient to withstand an impact energy of 400 in.-lb and an allowable tensile stress of 25,000 psi.

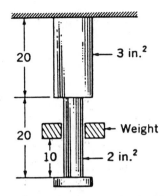

FIGURE 13.11   Sketch for Problem 6

4. Consider a steel rod similar to that in Sample Problem 13.2, but the diameter is 3/4 in. and it is 60 in. long. What weight can be dropped 10 in. to the support so that a tensile stress of 35,000 psi is not exceeded? Neglect the deformation on the member when considering the height of the drop.

5. Solve Problem 4 above except use a brass rod. The Modulus of Elasticity for brass is 16,000,000 psi.

6. Refer to Figure 13.11. A weight W is dropped 10 in. onto the end of the steel support. The allowable stress is 15,000 psi. Determine the weight W.

Dynamic balancing of shafts

7. Refer to Figure 13.12 for the given data. Place counterweights at A and B, each on a 2 in. radius from the shaft. Find the magnitude of the counterweights and their positions.

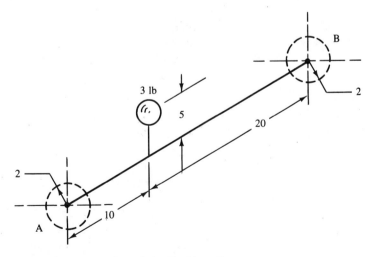

FIGURE 13.12   Sketch for Problem 7

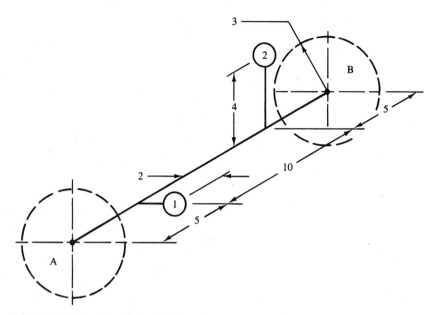

FIGURE 13.13   Sketch for Problem 9

8. Solve Sample Problem 13.3 except make the *a* distance of Weight 1 (the distance from *A*) just 3 in. instead of 6 in. (W_2 is still 14 in. from A and B is 20 in. from A.)

9. Figure 13.13 provides data for this problem. Weights 1 and 2 are 20 lb each. Counterweights are to be placed in the cross planes at *A* and *B*. Each counterweight is to be placed on a radius arm of 3 in. Find **(a)** the magnitude of each counterweight and **(b)** its angle to the horizontal.

10. Solve Problem 8 above, except add a third weight to the midpoint of the shaft. The weight is 20 lb, its radius is 5 in., and its angle to the horizontal is 45°.

11. Refer to Figure 13.10 for dimensional data. Counterweights are to be placed at locations *A* and *B* on a 2 in. radius. $W_1 = 5$ lb and $r_1 = 4$ in. $W_2 = 5$ lb and $r_2 = 4$ in. Solve for the counterweight at *B* and its angle to the horizontal.

SI Unit Problems

12. How much strain energy can an A36 steel cube 10 mm on a side absorb if the yield stress is not exceeded? The yield stress for A36 steel is 248 MPa.

13. How much strain energy can a brass block, 40 mm on a side, absorb if the yield stress of 103 MPa is not exceeded? $E = 110$ GPa.

14. Solve Sample Problem 13.4, except change the rod diameter to 8 mm and the allowable stress to 124 MPa.

15. Use the data in Sample Problem 13.4 to determine how high a 4 kg mass can be dropped.

APPENDICES

APPENDIX 1 ▬▬▬

PROPERTIES OF MATERIALS

This information is from Baumeister & Marks *Standard Handbook for Mechanical Engineers*, Eighth Edition, © 1978 by McGraw-Hill Book Company and from ASME's *Handbook of Metal Properties*. The values are typical, unless otherwise noted.

Material	Ultimate Tensile Strength (Stress) psi	MPa	Yield Strength (Tensile or Compressive Stress) psi	MPa	Percent Elongation	Modulus of Elasticity E psi	GPa	Modulus of Rigidity G psi	GPa
Class 30 cast iron	30 000 (minimum) (115 000 comp) (44 000 shear)	207 793 303	—	—	—	15 000 000	103	5 000 000	34
ASTM A36 low-carbon structural steel	58 000 to 80 000	400 to 552	36 000	248	20	30 000 000	207	12 000 000	83
AISI 1020 cold drawn (0.2% carbon steel)	75 000	517 379	64 000	441	20	30 000 000	207	12 000 000	83
AISI 1020 hot worked	65 000	448	48 000	331	36	30 000 000	207	12 000 000	83
AISI 1040 cold drawn (0.4% carbon steel)	97 000	669	82 400	568	16	30 000 000	207	12 000 000	83
AISI 4340 oil quenched steel	212 000	1460	200 000	1378	12.5	30 000 000	207	12 000 000	83
AISI 302 stainless steel	90 000	621	40 000	276	50	28 000 000	193	10 000 000	69
Copper (soft)	32 000	221	10 000	69	45	17 000 000	117	6 000 000	41
Brass, 30% zinc (hard)	76 000	524	63 000*	434		16 000 000	110	6 000 000	41

(continued)

APPENDIX 1 (Continued)

PROPERTIES OF MATERIALS (continued)

Material	Ultimate Tensile Strength (Stress)		Yield Strength (Tensile or Compressive Stress)		Percent Elongation	Modulus of Elasticity E		Modulus of Rigidity G	
	psi	MPa	psi	MPa		psi	GPa	psi	GPa
Brass, 30% zinc (soft)	47 000	324	15 000*	103	62	10 600 000	73.0	4 000 000	28
6061–T6 aluminum alloy for structural shapes	45 000	310	40 000	276	17				
Douglas fir (subjected to bending forces)	12 400	85.4	—	—	—	1 760 000	12.1	—	—

*0.5 percent extension under load

Average weight of steel = 0.284 lb/in.^{3}

Average weight of wood = 0.0231 lb/in.^{3}

APPENDIX 2 ▬▬▬▬
BASIC NUMBERING SYSTEM FOR SAE AND SISI STEELS

Material	SAE or AISI Number*
Carbon steels	1xxx
plain carbon	10xx
free-cutting, screw stock	11xx
Chromium steels	5xxx
low chromium	51xx
medium chromium	52xxx
corrosion and heat resisting	51xxx
Chromium-nickel-molybdenum steels	86xx
Chromium-nickel-molybdenum steels	87xx
Chromium-vanadium steels	6xxx
1.00% Cr	61xx
Manganese steels	13xx
Molybdenum steels	4xxx
carbon-molybdenum	40xx
chromium-molybdenum	41xx
chromium-nickel-molybdenum	43xx
nickel-molybdenum: 1.75% Ni	46xx
nickel-molybdenum; 3.50% Ni	48xx
Nickel-chromium steels	3xxx
1.25% Ni, 0.60% Cr	31xx
1.75% Ni, 1.00% Cr	32xx
3.50% Ni, 1.50% Cr	33xx
Silicon-manganese steels	9xxx
2.00% Si	92xx

* This numbering system uses a number composed of four or five digits. The first two digits indicate the type or alloy classification. The last two or three digits give the carbon content. For example, plain carbon steel has 1 and 0 as its first two digits. A steel designated as 1045 is therefore a carbon steel containing 0.45 percent carbon.

APPENDIX 3 ▬▬▬▬▬▬▬▬▬▬▬▬▬▬▬▬▬▬

UNIFIED SCREW THREADS

Size Designation	Nominal Major Diameter (in.)	Threads per Inch— Course Series (UNC)	Threads per Inch— Fine Series (UNF)
0	0.0600	—	80
1	0.0730	64	72
2	0.0860	56	64
3	0.0990	48	56
4	0.1120	40	48
5	0.1250	40	44
6	0.1380	32	40
8	0.1640	32	36
10	0.1900	24	32
12	0.2160	24	28
$\frac{1}{4}$	0.2500	20	28
$\frac{5}{16}$	0.3125	18	24
$\frac{3}{8}$	0.3750	16	24
$\frac{7}{16}$	0.4375	14	20
$\frac{1}{2}$	0.5000	13	20
$\frac{9}{16}$	0.5625	12	18
$\frac{5}{8}$	0.6250	11	18
$\frac{3}{4}$	0.7500	10	16
$\frac{7}{8}$	0.8750	9	14
1	1.0000	8	12
$1\frac{1}{4}$	1.2500	7	12
$1\frac{1}{2}$	1.5000	6	12
$1\frac{3}{4}$	1.7500	6	12
2	2.0000	5	12
$2\frac{1}{4}$	2.2500	$4\frac{1}{2}$	12
$2\frac{1}{2}$	2.5000	$4\frac{1}{2}$	12
$2\frac{3}{4}$	2.7500	4	12
3	3.000	4	12
$3\frac{1}{4}$	3.2500	4	12
$3\frac{1}{2}$	3.5000	4	12
$3\frac{3}{4}$	3.7500	4	12
4	4.0000	4	12

APPENDIX 4

TABLE OF EXPONENTIALS

x	e^x	x	e^x	x	e^x	x	e^x	x	e^x
0.00	1.000	1.00	2.718	2.00	7.389	3.00	20.086	4.00	54.598
0.05	1.051	1.05	2.858	2.05	7.768	3.05	21.115	4.05	57.397
0.10	1.105	1.10	3.004	2.10	8.166	3.10	22.198	4.10	60.340
0.15	1.162	1.15	3.158	2.15	8.585	3.15	23.336	4.15	63.434
0.20	1.221	1.20	3.320	2.20	9.025	3.20	24.533	4.20	66.686
0.25	1.284	1.25	3.490	2.25	9.488	3.25	25.790	4.25	70.105
0.30	1.350	1.30	3.669	2.30	9.974	3.30	27.113	4.30	73.700
0.35	1.419	1.35	3.857	2.35	10.486	3.35	28.503	4.35	77.487
0.40	1.492	1.40	4.005	2.40	11.023	3.40	29.964	4.40	81.451
0.45	1.568	1.45	4.263	2.45	11.588	3.45	31.500	4.45	85.627
0.50	1.649	1.50	4.482	2.50	12.182	3.50	33.115	4.50	90.017
0.55	1.733	1.55	4.712	2.55	12.807	3.55	34.813	4.55	94.632
0.60	1.822	1.60	4.953	2.60	13.464	3.60	36.598	4.60	99.484
0.65	1.916	1.65	5.207	2.65	14.154	3.65	38.475	4.65	104.58
0.70	2.014	1.70	5.474	2.70	14.880	3.70	40.447	4.70	109.95
0.75	2.117	1.75	5.755	2.75	15.643	3.75	42.521	4.75	115.58
0.80	2.226	1.80	6.050	2.80	16.445	3.80	44.701	4.80	121.51
0.85	2.34	1.85	6.360	2.85	17.288	3.85	46.993	4.85	127.74
0.90	2.460	1.90	6.686	2.90	18.174	3.90	49.902	4.90	134.29
0.95	2.586	1.95	7.029	2.95	19.106	3.95	51.935	4.95	141.17
1.00	2.718	2.00	7.389	3.00	20.086	4.00	54.598	5.00	148.41

APPENDIX 5 ▬▬▬▬▬▬▬

SI CONVERSION FACTORS

To Convert from	to	Multiply by
STRESS/PRESSURE		
pounds per square inch	pascal	6895
pascal	pounds per square inch	0.000 145
ENERGY		
foot·pound	joule	1.36
joule	foot·pound	0.738
Btu	joule	1055.
Btu	calorie	252.
calorie	joule	4.19
calorie	Btu	0.003 97
kilowatthour	kilocalorie	860.
kilowatthour	Btu	3413.
POWER		
horsepower	kilowatt	0.746
kilowatt	horsepower	1.34
heat rate (Btu/hr)	horsepower	0.000 393
MOMENT (TORQUE)		
pound·foot	newton·meter	1.36
pound·inch	newton·meter	0.113
VELOCITY		
feet per second	miles per hour	0.682
miles per hour	feet per second	1.47
feet per second	meters per second	0.305
meters per second	feet per second	3.28
VOLUME FLOW		
cubic feet per minute	gallons per minute	7.48
cubic feet per minute	pounds per minute	62.4 (water only)
gallons per minute	cubic feet per minute	0.1337

Note: 1 joule = 1 newton·meter

APPENDIX 5 (Continued) ▬▬▬▬▬▬
SI CONVERSION FACTORS

To Convert from	to	Multiply by
FORCE		
pound	newton	4.45
newton	pound	0.225
MASS TO FORCE/FORCE TO MASS		
newton (force unit)	kilogram	0.102
kilogram	newton (force unit)	9.81
kilogram	pound (force unit)	2.20
pound (force unit)	kilogram	0.454
pound (force unit)	slug	0.0311
slug	pound (force unit)	32.2
LENGTH		
foot	meter	0.305
inch	meter	0.0254
inch	millimeter	25.4
kilometer	mile	0.621
millimeter	inch	0.0394
meter	foot	3.28
mile	kilometer	1.61
VOLUME		
cubic centimeter	cubic inch	0.061 0
cubic centimeter	liter	0.001 00
cubic inch	cubic centimeter	16.39
cubic inch	liter	0.016 4
cubic inch	quart	0.017 3
gallon	liter	3.78
liter	cubic centimeter	1000.0
liter	cubic inch	61.0
liter	gallon	0.264
liter	quart	1.06
quart	cubic inch	57.8
quart	liter	0.946

INDEX

NOTE: Italicized references refer to material in figures or tables, not to text material.